AIRCRAFT TECHNICAL

JEPPESEN
Sanderson Training Products

JS312625C

ISBN 0-89100-410-6

A&B system — A form of emergency brake design used in some of the larger multiple-disc brake systems. These brakes have a number of cylinders, with each alternate cylinder connected to the aircraft's "A" hydraulic system, and the others to the aircraft's "B" hydraulic system. There would still be normal brake actuation if one system should fail.

A&P technician — An aircraft technician certified by the FAA after meeting the experience and knowledge requirements for certification. The A&P technician is qualified to return licensed United States airplanes to service after performing certain inspections and maintenance procedures.

A-battery — Usually has a voltage in the range of 1.5 to 6.0 volts and are capable of supplying a reasonable amount of current.

A, B, C, D, E chambers — Pressure carburetor. A diaphragm controlled unit of a pressure type carburetor which is divided into five chambers and contains two regulating diaphragms and a poppet valve assembly. Chamber A is regulated air-inlet pressure. Chamber B is boost venturi pressure. Chamber C contains metered fuel pressure. Chamber D contains unmetered fuel pressure. Chamber E is fuel pump pressure.

abeam — At right angles relative to the longitudinal axis of an aircraft.

abort — A condition that fails to develop. An operation that prematurely terminates, such as an airplane that aborts a takeoff.

aborted start — Termination of engine starting if no combustion "light off" occurs in a gas turbine engine within the prescribed time.

abradable seal — A general description of knife-edge seals which can wear away slightly and still function. Abradable seals in a gas turbine engine abrade slightly to produce a close fit.

abradable shroud — Generally a honeycomb type turbine shroud ring set into the outer turbine case. It can abrade without damage if the turbine blades creep and contact the shroud.

abrade — To scrape or wear away a surface or a part by mechanical or chemical action.

abrasion –
[1]An area of roughened scratches or marks usually caused by foreign matter between moving parts or surfaces.
[2]The wearing or rubbing away of a surface by any substance used for grinding, or polishing, etc.
[3]A roughening or wearing away of a surface by scratched or marks usually caused by foreign matter between moving parts or surfaces.

abrasive – A substance used to wear away other substances or surfaces by friction. In grinding wheels, the abrasives most commonly used are silicon carbide or aluminum oxide.

abrasive blasting – The removal of carbon and other deposits from machine parts using a high velocity blast of air that contains fine particles of abrasive sand, glass bead, or walnut shell.

abscissa – The horizontal reference line of a graph or curve by which a point is located with reference to a system of coordinates such as the X-axis (abscissa) and the Y-axis (ordinate).

absolute altimeter – A radar altimeter used to indicate the exact height of an aircraft over the terrain. *See radio also altimeter.*

absolute altitude – The actual distance the aircraft is above the ground. It is measured with an electronic altimeter.

absolute ceiling – Maximum height above sea level at which an aircraft can maintain level flight under standard atmospheric conditions.

absolute humidity – The actual water vapor that is present in a given volume of air. If one cubic meter of air contains 200 grams of water, the absolute humidity of the air is 200 grams per cubic meter.

absolute pressure – Pressure measured relative to zero pressure or a vacuum. Absolute pressure is often measured in inches of mercury, as, for example, manifold pressure.

absolute pressure controller – Abbrev.: APC. In a reciprocating engine turbocharger system, controls the maximum turbocharger compressor discharge pressure (34 + or –.5 in. Hg to critical altitude, approximately 16,000 ft).

absolute temperature – Temperature referenced from absolute zero, (–273.18°C or –459.6°F) or the temperature at which molecular motion stops. There are two absolute scales in use, the Rankine scale using Fahrenheit degrees and the Kelvin scale using Celsius degrees. *See also absolute zero.*

absolute value – The value of a number without considering its sign, whether it is plus or minus.

absolute zero – The hypothetical temperature (–273.18°C) at which molecular motion ceases.

AC 43.13-1a – An Advisory Circular in book form issued by the FAA which covers acceptable methods. techniques, and practices for aircraft inspection and repair. The procedures that are described in this advisory circular are considered by the FAA to be approved data for aircraft inspection, maintenance, and alteration.

AC fittings – Air Corps. Replaced by the AN (Army/Navy) standard and MS (Military Standard fittings). The AN fittings have a slight shoulder between the cone and the first thread. The AC does not have this shoulder. Other differences include sleeve design and pitch of threads in most cases.

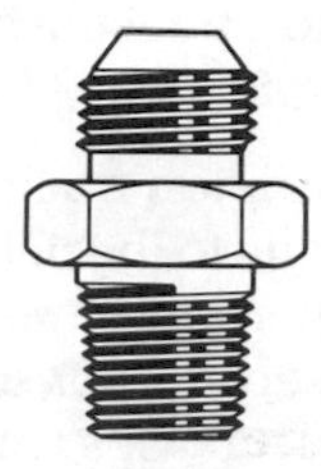

AC plate resistance – The ratio of the change in the voltage of a plate, to the small change in the plate current that it causes. The AC plate resistance is actually the internal resistance of the tube to the flow of alternating current. AC plate resistance is measured in ohms.

accelerate – To increase the speed of an object, or to make it move faster.

accelerated life test – An operational test of a system or a component, accelerated under severe conditions, to find any weak points and to help predict the service life that the system or component will likely have under normal operating conditions.

accelerate-stop distance – The overall length of runway that is needed for an aircraft to accelerate and be able to stop in case of engine failure.

accelerating agent – A substance that is used to hasten a chemical action or change.

accelerating pump – A small pump in a carburetor used to supply a momentarily rich mixture to the engine when the throttle is suddenly opened. This is to prevent hesitation during the time of transition between operating on the idle system and the main metering system.

accelerating system – A common type of accelerating system used in float carburetors that supplies extra fuel during increases in engine power. This is usually accomplished by a small fuel pump called an accelerating pump. *See also accelerating pump.*

accelerating well – Secondary tank built into the main oil tank. The well, or hopper, retains only that portion being circulated through the engine. Thus, oil warm-up is hastened during engine warm-up. The well also makes oil dilution practical.

acceleration – The rate of change in velocity of a body. If the velocity of an object is increased from 20 MPH to 30 MPH, the object has been accelerated. Acceleration is usually expressed in such terms as feet per second, per second.

acceleration check – A maintenance check of the time an engine takes to spool-up from idle to rated power without hesitation and no evidence of backfire. *See also reciprocating engine.*

acceleration due to gravity – The acceleration of a freely falling body due to the pull of gravity, expressed as the rate of increase of velocity per unit time. In a vacuum, the rate is 32.2 feet (9.8 meters) per second per second near sea level. Acceleration caused by gravity decreases with an increase in altitude until it becomes zero outside of the earth's gravitational field.

acceleration error – An error inherent in magnetic compasses, caused by the force of acceleration acting on the dipcompensating weight when the aircraft accelerates or decelerates on an easterly or westerly heading. When the aircraft accelerates on an easterly or westerly heading, the compass gives the indication that the aircraft is turning to the north. When the aircraft decelerates on either of these headings, the compass gives the indication that the aircraft is turning to the south.

acceleration of gravity – The acceleration of a freely falling body due to the attraction of gravity expressed as the rate of increase of velocity per unit of time (32.17 feet per second at seal level at 45° latitude). The acceleration decreases with an increase in altitude until it becomes zero on leaving the earth's gravitational field.

acceleration thermostat – A bimetallic probe positioned in the exhaust stream of an auxiliary power unit. When overheated it expands to dump a Pb fuel control signal and reduce fuel flow.

acceleration well – An enlarged space around the discharge nozzle of some float-type carburetors When the throttle is suddenly opened, this fuel is rapidly discharged from the main discharge nozzle.

accelerator – A substance that is added to a catalyzed resin to shorten the time that is needed for the resin to cure.

accelerator system — A system in an aircraft carburetor that supplies additional fuel to the engine when the throttle is suddenly opened.

accelerator winding — A series winding used in vibrating-type voltage regulators, which, when the points open, decreases the magnetic field immediately, allowing the points to close more rapidly.

accelerometer — A sensitive instrument calibrated in G-units, which measures the amount of force exerted by acceleration on a body. One G-unit force is equal to the weight of the object.

acceptance test — A test performed on an airplane or piece of equipment to be sure it is in the condition specified in the purchase contract. All large and expensive aircraft are given extensive acceptance tests before a customer accepts them.

acceptor atom — An impurity atom in a semiconductor material which will receive or accept electrons. Germanium with an acceptor impurity is called p-type germanium because it has a positive nature.

access cover — *See access panel.*

access door — A door which provides normal or emergency entrance into or exit from the aircraft. It also provides access to servicing points and manually operated drains.

access panel — A panel that is easily removed to facilitate inspection and maintenance.

accessories — Components that are used with an engine, but are not a part of the engine itself. Units such as magnetos, carburetors, generators, and fuel pumps are commonly installed engine accessories.

accessory drive gearbox — A section of an aircraft engine that contains the drive to operate such accessories as fuel pumps, air pumps, and generators.

accessory drive shaft — A shaft which drives the accessory gearbox from a bevel gear connected to the compressor shaft in some gas turbine engines.

accessory gear trains — Containing both spur and bevel-type gears, are used in the different types of engines for driving engine components and accessories. *See also accessory section, accessory drive gearbox, accessory section turbine.*

accessory section — That part of an engine which provides the necessary mounting pads for such accessory units as magnetos, fuel pumps, oil pumps, generators, etc. are mounted.

accident – An event that happens by chance or from some known or unknown cause.

accumulated error – The sum of all of the errors that occur in the operation of a system or in the manufacture of a part.

accumulator – A hydraulic component consisting of two chambers separated by a piston, diaphragm or bladder. Compressed air in one chamber holds pressure on hydraulic fluid in the other chamber. Hydraulic fluid under pressure can thus be stored in a system. Accumulator types include bladder, diaphragm, and piston type.

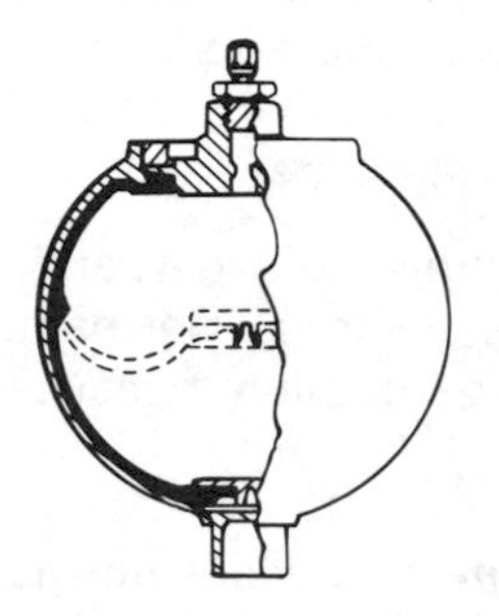

accumulator precharge – Compressed air that is stored in the air chamber of a hydraulic accumulator without producing an increase in hydraulic system pressure.

accuracy – The state of being exact, and free from mistakes. Conformance to a standard without error.

accurate – Free from error.

AC/DC – Electrical components that can operate equally well on alternating current or on direct current electricity

ace – A term which originated in World War I to acknowledge pilots who shot down five or more enemy aircraft.

acetone – Liquid ketone (C_3H_6O). A flammable, fast-evaporating solvent used as a constituent in many types of aircraft finishes.

acetylene cylinder – A seamless steel shell with welded ends, approximately 12 inches in diameter and 36 inches long. A fully charged acetylene cylinder of this size contains approximately 225 cu. ft. of gas at pressures up to 250 PSI.

acetylene gas – A flammable, colorless non-toxic gas which has a distinctive, disagreeable odor. Calcium carbide is made to react with water to produce acetylene. Mixed with oxygen in the proper proportions and ignited, will produce temperatures which range from 5,700 to 6,300°F for welding purposes.

A Chamber – One of the chambers of a pressure type carburetor regulator unit. Chamber A is regulated air-inlet pressure from the air intake.

acid – A chemical substance that contains hydrogen, has a characteristically sour taste, and is prone to react with a base or an alkali to form a salt and to accept electrons from the alkali.

acid diluent – A constituent of a wash primer that is used to mildly etch the surface of the metal that is being primed to provide a good bond between the finishing system and the metal.

acid-resistant paint – A paint used on battery boxes and surrounding areas.

acorn nut – A dome-shaped nut whose threaded hole does not go completely through the nut to produce a finished and a smooth appearance by the dome. Acorn nuts and cap nuts are terms used interchangeably.

acrobatic category airplane – One that is certificated for flight without restrictions, except those found necessary as a result of flight test.

acrobatics – Flight maneuvers which are not considered to be necessary for normal utility category flight such as loops and rolls.

acrylic – A glossy, transparent thermoplastic material used for cast or molded parts such as aircraft windshields and side windows.

acrylic lacquer – An aircraft finish that consists of an acrylic resin vehicle and certain volatile solvents.

acrylic nitro lacquer finish – An aircraft finish which is applied in a specified sequence. It includes a wash primer coat, modified zinc chromate primer coat, and an acrylic nitrocellulose lacquer topcoat.

acrylic nitrocellulose lacquer – A common topcoat for aircraft, available either as a nonspecular material or glossy finish.

acrylic resin – A clear thermoplastic produced by polymerizing acrylic acid and used for cast and molded aircraft windshields, windows and parts or as coatings and adhesives.

activated charcoal – Used as a filter for liquids and as a medium to absorb gases.

active electrical component – An electrical part that controls current or voltage for switching or for amplification.

actuating cylinder – A cylinder and piston arrangement used to convert hydraulic or pneumatic pressure into work by the fluid under pressure moving the piston. The force applied is equal to the piston area times the pressure on the fluid. Actuating cylinders can be double-action or single-action actuating types.

actuating horns – The levers to which control cables are attached to move the control surfaces.

actuator – A device which transforms fluid pressure into mechanical force. The action may be linear, rotary, or oscillating, may be actuated by either hydraulic or pneumatic pressure.

actuator piston – The movable part of a hydraulic or pneumatic linear actuator.

acute angle – An angle of less than 90°, also called a closed angle.

adapter – A device which fits or adapts one component to another.

adcock radio antenna – A directional radio transmitting antenna made up of two vertical conductors from which electromagnetic energy radiates. These conductors are connected in a way that radiates signals in opposite phases.

adhesive – A substance used to bond two materials together by chemical means.

adiabatic – Refers to airflow being compressed without loss or gain of heat.

adiabatic change – A physical change in state or condition which occurs within a material with no loss or gain of heat.

adiabatic lapse rate – The decrease in temperature with altitude when no heat is added to or taken from the air. It is nominally 5.4°F per 1,000 ft.

adjacent – Something that is near or close to the adjacent side in mathematics are the two sides of a triangle that have a common angle.

adjust – To change a condition to make it more satisfactory or to make it operate better.

adjustable stabilizer – A kind of horizontal stabilizer that can be adjusted in flight to trim the airplane thereby allowing the airplane to fly hands-off at any given airspeed

adjustable-pitch propeller – A propeller which has means of adjusting the pitch of the blades on the ground with the engine not running, but is not adjustable in flight.

adjustable-split die – A tool used for cutting external threads on round stock. The die is split on one side and an adjusting screw is used to spread the die to adjust the fit of the threads.

adjusting idle mixture 144 (AC 65-12) – Adjusting carburetor idle mixture tailored for the particular engine installation for best operation.

Administrator – Any officer, inspector, etc., of the FAA designated to have appropriate authority in the activity concerned.

admittance – Measured in siemens, is a measure of the ease with which alternating current can flow in an electrical circuit. Admittance is the current divided by the voltage, and is the reciprocal of impedance.

advance – To move forward.

advanced firing – *See advanced timing.*

advanced timing – A term used in reference to ignition timing that describes the ignition event taking place before the piston reaches top dead center.

advancing blade – Any blade located in a semicircular part of the rotor disc where the blade direction is the same as the direction of flight.

advection – A method of heat transfer by horizontal movement of air. Advection is different from convection in that convection transfers heat by vertical movement, rather than horizontal movement of the air.

advection currents – Air currents that move horizontally over a surface.

advection fog – Fog that forms when moist air is moved horizontally across a surface that is cold enough to cool the air to a temperature that is below its dew point. Moisture condenses from the air and remains suspended in the air to form fog.

adverse yaw – A condition of flight in which the nose of an airplane starts to move in the direction opposite that in which a turn is to be made. It is caused by the downward deflected aileron producing induced drag. This often called aileron drag.

aerated — Any fluid mixed or exposed to air. When lubricating oil is used in an engine, it picks up a good deal of air, and it is said to be aerated.

aeration — The process of mixing air in a liquid.

aerial — A term having to do with aircraft in flight. It is used in terms such as aerial photograph.

aerial photograph — Any photograph that is made from an aircraft in flight.

aerial refueling; air refueling — A method of refueling aircraft while in flight in order to extend the airplanes range. Military aircraft use flying tankers that meet with the aircraft that is to be refueled, and large amounts of fuel are transferred in flight.

aerodrome — An early British term used for airfield and airport.

aerodynamic balance — The portion of a control surface on an airplane that extends out ahead of the hinge line. This utilizes the airflow about the aircraft to aid in moving the surface.

aerodynamic blockage thrust reverser — A configuration of thrust reverser used in turbojet engines in which thin airfoils or obstructions are placed in the engine's exhaust stream to duct the high-velocity exhaust gases forward. This creates a great amount of aerodynamic drag and decreases the airplane's landing roll.

aerodynamic braking — The generation of a great deal of aerodynamic drag by reversing the pitch of the propeller blades or ducting forward some of the jet engine exhaust to reduce the roll after landing or to allow the aircraft to descend at a steep angle without building up excessive airspeed.

aerodynamic center — That point within the airfoil section located at a point approximately one-fourth of the way back from the leading edge. It is the point at which the (pitching) moment coefficient is relatively constant for all angles of attack.

aerodynamic center of horizontal tail — Under certain conditions of speed. load, and angle of attack, the flow of air over the horizontal stabilizer creates a force which pushes the tail up or down. This point is the aerodynamic center of the horizontal tail.

aerodynamic coefficient — Nondimensional coefficients for aerodynamic forces and moments.

aerodynamic design point — (of the compressor) The most efficient compression ratio which occurs at altitude.

aerodynamic drag – Drag caused by turbulent airflow on an airfoil, such as compressor blade or vane.

aerodynamic factors –
[1]Those factors which affect the amount of lift or drag produced by an airfoil.
[2]The forces acting on a propeller while rotating through the air as it transforms the rotary power of the engine into thrust.

aerodynamic heating – The temperature rise caused by high-speed air flowing over an aerodynamic surface.

aerodynamic lift – A force produced by air moving over an airfoil. Aerodynamic lift acts in a direction that is perpendicular to the direction of the moving air.

aerodynamic shape – The shape of an object with reference to the airflow over it. Certain shapes cause air pressure differentials which produce lift; others are designed for minimum resistance to the flow of the air.

aerodynamic twisting force – One of the five forces acting on a rotating propeller. The aerodynamic twisting force tends to twist the blade angle towards the feather position.

aerodynamics – The science of the action of air on an object, and with the motion of air on other gases. Aerodynamics deals with the production of lift by the movement of the aircraft, the relative wind, and the atmosphere.

Aerofiche – Registered trade name for a form of microfiche used in the aircraft industry. 288 frames of information may be placed on a single 4″ × 8″ card of film.

aeronaut – A person who operates or travels in airships or balloons.

aeronautical beacon – *See Airport Rotating Beacon.*

aeronautical chart – A detailed navigational map exhibiting geographical aspects of the earth's topographical features, airports and designated airspace.

aeronautics – The science of making and flying airplanes. A term that applies to anything that is in any way associated with the design, construction, or operation of an aircraft. Aerodynamics and aerostatics are both branches of aeronautics.

aerosol – A liquid that is broken up into tiny drops divided into extremely fine particles and dispersed or sprayed into the air by the use of a propellant such as carbon dioxide, nitrogen, or Freon.

aerospace — Space from the earth's surface extending outward beyond the earth into space.

aerospace industry — That portion of our economy associated with such devices as aircraft, space ships, missiles, and their associated parts.

aerospace vehicle — Any controllable device capable of flight in the aerospace.

aerostat — A device which is supported in the air by displacing more than its weight, such as a balloon or dirigible. *See also aerostatics.*

aerostatics — The branch of science that deals with the generation of lift by the displacement of air by a body lighter than the air it displaces. Balloons and dirigibles that are filled with hot air or gas fall under the science of aerostatics.

aft — To the rear, back, dorsal, or tail of the aircraft.

aft flap — The aft section of a triple-slotted, segmented wing flap.

after bottom center — Abbrev.: ABC. In a reciprocating engine, the number of degrees of crankshaft rotation after the piston has stopped at the bottom of its stroke.

after top center — Abbrev.: ATC. In a reciprocating engine, the number of degrees of engine crankshaft rotation after the piston has passed the top of its stroke.

afterburner — A portion of a jet engine into which additional fuel is sprayed into the hot, oxygen-rich exhaust in the afterburner, where it burns and produces additional thrust. Afterburners provide a great amount of additional thrust with a minimum of additional weight.

after-firing — A condition in a reciprocating engine often resulting from either too rich fuel/air mixture or unburned fuel being pumped into the exhaust system of a reciprocating engine and ignited when it comes in contact with some hot component. This condition is sometimes called afterburning or torching.

afterglow — The glow, that remains on the phosphorescent screen of a cathode ray tube after the electron beam passes.

aft-fan engine — A turbofan engine that has the fan constructed as an extension of the turbine blades.

age hardening – The process of increasing the hardness of aluminum after the solution-heat-treat process. Age hardening occurs at room temperature and continues for a period of several days until the metal reaches its full hard state. *See also precipitation heat treating.*

aging – *See age hardening.*

agitate – To stir something, or to shake it up in order to mix its ingredients.

AGL altitude – The vertical elevation above ground expressed in feet.

agonic line – An irregular imaginary line across the surface of the earth along which the magnetic and geographic poles are in line, and where there is "agonic", "no-angle", or no variation error.

agricultural aircraft – Aircraft that are especially designed and built for use in agricultural application of chemicals to crops for insect and weed control.

agronautics – Branch of the aviation industry that deals with agriculture.

aileron – A primary control surface located near the wing tip, which makes up part of the total wing area. Ailerons are operated by the lateral motion of the control wheel which used causes rotation of the aircraft about the longitudinal axis.

aileron angle – The angle of displacement of an aileron from its neutral, or trailing, position. The aileron angle is positive when the trailing edge of the aileron is below its neutral position.

aileron spar – Also "false spar". A spar that extends only part of the spanwise length of the wing and provides a hinge attachment point for the aileron.

aileron station – Distances measured outboard from the root end of an aileron, parallel to the aileron spar.

air – A mixture of gases that comprises the earth's atmosphere. Pure dry air contains approximately 78% nitrogen by volume, 21% oxygen, and 1% argon. The remainder consists of 0.03% carbon dioxide, 0.03% hydrogen, traces of neon and helium. Dry air weighs 0.07651 pound per cubic foot at sea level with a temperature of 59°F at 40° latitude and a barometric pressure of 14.69 PSI.

air adapters – A component in a centrifugal compressor gas turbine engine. Its purpose is to deliver air from the diffuser to the individual can-type combustion chambers at the proper angle.

air brake — A plate or series of plates that may be projected into the airplane's slipstream to provide turbulence and drag to slow the airplane during descent, glide, or when maneuvering. Air brakes differ from flaps in that they produce no useful lift.

air bus — A multi-engine airplane in the jumbo jet category having a range of approximately 1,400 miles with a gross takeoff weight of 275,600 lbs. carrying 261 passengers.

air capacitor — A capacitor that uses air as the dielectric.

air carrier — Any company which provides commercial air transportation of passengers and cargo.

Air Carrier District Office — Abbrev.: ACDO. A local FAA office whose purpose is to serve the public and aviation industry on matters related to aviation safety, certification and enforcement of procedures.

air commerce —
[1]Interstate, overseas, or foreign air commerce or the transportation of mail by aircraft or any operation or navigation of aircraft within the limits of any Federal airway or any operation or navigation of aircraft when directly affects, or which may endanger safety in, interstate, overseas, or foreign air commerce.
[2]Transportation by aircraft of persons or property for hire or compensation.

air conditioning system — A system consisting of cabin air conditioning and pressurization which supplies conditioned air for heating and cooling the cockpit and cabin spaces. This air also provides pressurization to maintain a safe, comfortable cabin environment.

air conditioning — The process of treating air so as to control simultaneously its temperature, humidity, cleanliness and distribution to meet the requirements of a conditioned space. In combination with pressurization, a complete environmental control is possible.

air cycle cooling system — One of several cooling systems consisting of an expansion turbine, an air-to-air heat exchanger, and various valves which control airflow through the system. Used to provide a comfortable atmosphere within the aircraft cabin. *See also air conditioning.*

Air Defense Identification Zone — Abbrev.: ADIZ. The land and airspace within the United Statesa boundaries.

air density — The weight of air in a given volume.

air filter — Any filtering device that filters dust and dirt from entering the intake or induction system. *See also air filter carburetor; air filter system.*

air filter system – An air filter system normally consists of a filter element, a door, and an electrically operated actuator. When the filter system is operating, air is drawn through a louvered access panel that does not face directly into the airstream. With this entrance location, considerable dust is removed as the air is forced to turn and enter the duct. *See also air filter (carbuertor).*

air impingement – A fault that resembles haze in an enamel or lacquer paint finish. It is caused by microscopic sized bubbles in the paint that form when the paint is applied with too high an atomizing air pressure.

air impingement starter – A type of starter used on small gas turbine engines in which a stream of high-pressure compressed air is directed onto the blades of the compressor or the turbine in order to rotate the engine for starting.

air inlet – A portion of the engine designed to conduct incoming air to the compressor with a minimum energy loss resulting from drag or ram pressure loss.

air lock – A trapped pocket of air that is trapped in a line, which blocks the flow of fluid.

air mass – A body of air with nearly uniform conditions of temperature and humidity at any given level.

AIRMET – An amended weather advisory issued to warn of potentially hazardous flight conditions.

air metering force – The force used in Bendix pressure carburetors and fuel injection systems in which venturi and ram air pressures control the amount of fuel metered.

air scoop –

[1]A hooded opening to an engine carburetor, or other device, the purpose of which is to receive the ram air during flight which increases the amount of air that is taken into the structure.

[2]A specially designed scoop or ducting that conducts air to the carburetor and intake manifold of a reciprocating engine induction system.

air seal – A seal around a rotating shaft that is used to keep air from passing out of the housing that holds the shaft. Usually air seals are thin rotating or stationary rims designed to act as air dams that reduce airflow leakage between the gas path and the internal engine or over blade tips.

air start – The process of starting an aircraft engine in flight. In an air start, aerodynamic forces cause the propeller or the compressor to turn the engine. A starter is not generally used during an air start.

air starter – Same as air turbine starter.

air strip – *See airfield.*

air taxi – The movement of a helicopter in flight within 100 ft. above the ground under the direction of a ground signalman using standard helicopter taxi operating procedures.

air temperature control – An air control door or valve, near the entrance of the carburetor, which admits alternate heated air to the carburetor to prevent carburetor ice. *See also air temperature gage.*

air temperature gage – The carburetor air temperature gage indicates the temperature of the air before it enters the carburetor. The temperature reading is sensed by a bulb located in the air intake passage to the engine.

air traffic – Aircraft operating in the air or on an airport surface, exclusive of loading ramps and parking areas.

air traffic clearance – Authorization by air traffic control for the purpose of preventing collision between known aircraft, and for an aircraft to proceed under specified traffic conditions within controlled airspace.

air traffic control – A service operated by appropriate authority to promote the safe, orderly, and expeditious flow of air traffic.

air transportation – Interstate, overseas, or foreign air transportation or the transportation of mail by aircraft.

airbleed –

[1]A small hole in the fuel passage between the float bowl and the discharge nozzle of a float carburetor, through which air is introduced into the liquid fuel. It serves as an aid to atomization.

[2]Used in gas turbine engines for a variety of purposes and is taken from the engine's compressor section. *See also bleed air.*

airborne – The condition of an airplane, glider, or balloon when off of the ground and borne aloft by the air.

airborne navigation equipment – A phrase embracing many systems and instruments. These systems include VHF omnirange (VOR), instrument landing systems, distance-measuring equipment, automatic direction finders, Doppler systems, and inertial navigation systems.

airborne weather radar — A device used to see certain objects in darkness, fog, or storms, as well as in clear weather. In addition to the appearance of these objects on the radar scope, their range and relative position are also indicated.

air-breathing engine — An engine that requires an intake of air to supply the oxygen it needs in order to operate. Reciprocating and turbine engines are both air-breathing engines.

air-cool — To remove excess heat from a body by transferring it directly into the airstream passing over or around the body.

air-cooled blades and vanes — Hollow airfoils in the hot section which receive air from the cold section so that they can operate in a much higher temperature environment. Names such as gill holes, film holes and tip holes direct air back to the gas path. *See also convection, transpiration.*

air-cooled engine — A reciprocating aircraft engine whose waste heat is transferred directly into the airstream by means of fins on the cylinders.

air-cooled oil cooler — A type of heat exchanger in the lubrication system of an aircraft engine that removes heat from the oil and transfers it into the air that flows through the cooler.

air-cooled turbine blades — Hollow turbine wheel blades of certain high-powered gas turbine engines that are cooled by passing compressor bleed air through them.

air-core electrical transformer — A transformer made up of two or more coils wound on a core of non-magnetic material. Air-core transformers are normally used for radio-frequency alternating current.

aircraft — Any weight-carrying device designed to be supported by the air or intended to be used for flight in the air.

aircraft accident — Any damage or injury that occurs when an aircraft is moving with the intention of flight.

aircraft alteration — The modification of an aircraft, its structure or components, which changes its physical or flight characteristics. Alterations are classified as major or minor, in accordance with Federal Aviation Regulations, Part 43.

aircraft basic operating weight — The established basic weight of an aircraft available for flight without its fuel and payload.

aircraft battery — A source of electrical energy for an aircraft that may be used as an auxiliary source of power when the engine generator is not operating.

aircraft cable — Basic component of a cable is a wire. Diameter of the wire determines the total diameter of the cable. Most common aircraft cables are the 7×7 and 7×19.

aircraft dope — A colloidal solution of cellulose acetate or nitrate, combined with sufficient plasticizers to produce a smooth, flexible, homogeneous film. The dope imparts to the fabric cover additional qualities of increased tensile strength, airtightness, weather-proofing, and tautness of the fabric cover.

aircraft engine — An engine that is used or intended to be used for propelling aircraft. It includes turbosuperchargers, appurtenances, and accessories necessary for its functioning, but does not include propellers.

aircraft inspection — A systematic check of an aircraft and its components. Its purpose is to detect any defects or malfunctions before they become serious. Annual inspections, 100 hour inspections, progressive inspections, and preflight inspections are the most commonly used type of aircraft inspections.

aircraft lighting system — Provide illumination for both exterior and interior use. Some of these include instruments, cockpits, cabins and other sections occupied by crewmembers and passengers.

aircraft listings — Information sheets that are published by the FAA. They contain essential information on particular models of aircraft.

aircraft log — A record which contains the operational or maintenance history of the aircraft.

aircraft operating weight — The basic weight of an aircraft plus the weight of the crew members, equipment, fuel, oil, and passengers.

aircraft plumbing — The hoses, tubing, fittings, and connections that are used for carrying fluids through an aircraft.

aircraft quality — Indicates that aircraft equipment or materials are to be produced under closely controlled, special, and restricted methods of manufacture and inspection.

aircraft records — Documentation of the flight time and the maintenance that has been performed on an aircraft, its engines, or its components.

aircraft repair – Restoration of an aircraft and/or its components to a condition of airworthiness after a failure, damage, or wear has occurred.

aircraft rigging – The final adjustment and alignment of the various components of an aircraft to give it the proper aerodynamic characteristics.

aircraft steel structure – A truss-type fuselage frame usually constructed of steel tubing welded together in such a manner that all members of the truss can carry both tension and compression loads.

aircraft tires – Tubeless or tube-type, provide a cushion of air that helps absorb the shocks and roughness of landings and takeoffs; they support the weight of the aircraft while on the ground and provide the necessary traction for braking and stopping aircraft on landing.

aircraft welding – The process of joining metal by fusing the materials while they are in a plastic or molten state. There are three general types of welding: gas, electric arc, and electric resistance.

aircraft wooden structures – An early aircraft structure in which wood was used as the structural material.

air-dry – The process of removing moisture from a material by exposing it to the air.

airfield – Any area in which aircraft may land, take off, and park. An airfield may also be called an airstrip. The term airfield includes the buildings, equipment, and maintenance facilities that are used to store or service aircraft.

air filter – A device or system that removes dust particles before they enter the carburetor air intake.

airflow over wing section – Air flowing over the top surface of the wing must reach the trailing edge of the wing in the same amount of time as the air flowing under the wing. The increased velocity of air over a surface designed to obtain a desirable reaction from the air produces lift.

airfoil – Any surface designed to obtain a desirable reaction from the air through which it moves. The airfoil converts air resistance into a force useful for flight. Wings, control surfaces, propeller blades, and helicopter rotors are examples of airfoils.

airfoil profile – The outline of an airfoil section such as a wing.

airfoil section — The cross-sectional shape of an airfoil, viewed as if it were cut through in a fore-and-aft plane.

airframe — The structure of an aircraft without the powerplant. It is generally considered to consist of five principle units, the fuselage, wings, stabilizers, flight control surfaces, and landing gear. Helicopter airframes consist of the fuselage, main rotor and related gearbox, tail rotor, and the landing gear.

airframe technician — Any person who holds a certificate from the FAA authorizing him to perform maintenance or inspections on the airframe of certificated aircraft.

airline — A company or organization that operates aircraft for the transportation of persons or cargo.

airliner — One of the large transport type airplanes used in air commerce for the transportation of passengers or cargo.

Airloc fastener — A patented form of cowling fastener in which the actual locking is done by turning a steel cross-pin in a spring steel receptacle.

airman — A person involved in flying, maintaining, or operating aircraft.

airman certificate — Issued by the FAA authorizing a person to perform certain aviation related duties. Certificates are issued to pilots, technicians, and parachute riggers.

air-oil separator — A device in the vent portion of the lubrication system of a gas turbine engine that separates any oil from the air before the air is vented overboard.

airplane — An engine-driven, heavier-than-air, fixed wing aircraft, that is supported in flight by the dynamic reaction of the air against its wings.

airport — An area of land or water that is used or intended to be used for the landing and takeoff of aircraft, and includes its buildings and facilities, if any. *See also airfield.*

airport traffic area — Unless otherwise specifically designated in Part 93, that airspace within a horizontal radius of 5 statute miles from the geographic center of any airport at which a control tower is operating, extending from the surface up to, but not including, an altitude of 3,000 ft above the elevation of the airport.

airscrew — The name given to the aircraft propeller by the British.

airship — An engine-driven, lighter-than-air aircraft that can be steered.

airspace — The space lying above a certain geographical area.

airspeed — The rate at which an aircraft is moving through the air.

airspeed indicator — A differential air pressure gage which measures the difference between ram, or impact, air pressure and the static pressure of the air to indicate the speed of the aircraft through the air.

airspeed/mach indicators — Also called machmeters. They indicate the ratio of aircraft to the speed of sound at the particular altitude and temperature existing at any time during flight.

airstream direction detection — A unit of an angle-of-attack indicating system. The airstream direction detector contains the sensing element which measures local airflow direction relative to the angle of attack by determining the angular difference between local airflow and the fuselage reference plane.

air-turbine starter — A large volume of compressed air from an auxiliary power unit or bleed air from an operating engine is directed into the air-turbine starter. This air spins the turbine inside the starter, and the starter, which is geared to the main engine compressor, spins the engine fast enough for it to start. *See also air impingement starter.*

Airworthiness Certificate — Issued by the FAA to all aircraft that have proven to meet the minimum standards for services or by an aircraft manufacturer for use in an aircraft.

Airworthiness Directive — A regulatory notice that is sent out by the FAA to the registered owner of an aircraft informing him of the discovery of a condition that keeps his aircraft from continuing to meet its conditions for airworthiness. Airworthiness Directives (AD notes) must be complied with within the required time limit, and the fact of compliance, the date of compliance, and the method of compliance must be recorded in the aircraft maintenance records.

airworthy — The condition of an FAA certificated aircraft, engine, or component that meets all of the requirements for its original certification.

Alclad — A clad structural aluminum alloy. Alclad is a corrosion protection coating of pure aluminum, making up approximately 5% of the thickness on each side, which is rolled onto the alloy sheet in the rolling mill.

alcohol — A colorless, volatile, flammable liquid that is produced by the fermentation of certain types of grain, fruit, or wood pulp. Alcohol is used as a cleaning fluid, as a solvent in many of aircraft finishes, and as a fuel for certain types of specialized engines. *See also isopropyl alcohol.*

alcohol deicing – The preventing or controlling of ice formation by the use of alcohol spayed to the surface, in the case of windshields, or to the inlet airstream of a carburetor.

algebra – The branch of mathematics that uses letters or symbols to represent numbers in formulas and equations.

algebraic expression – A quantity that is made up of letters, numbers, and symbols. The parts of the expression that are separated by a plus or a minus sign are called the terms of the expression. An algebraic expression that has only one term is called a monomial, and one that has two or more terms is called a polynomial.

algorithm – A system or procedure to use in solving a problem.

alignment – The arrangement, or position, of parts of an object or a system in the correct relationship to each other.

alignment pin – Installed in a helicopter rotor blade to serve as an index when aligning the blades of a semi-rigid rotor system.

alkali – A chemical substance, usually the hydroxide of a metal. An alkali has a characteristically bitter taste, and it is prone to react with an acid to form salt, supplying electrons to the acid.

alkaline – Having the property of reacting with an acid to form a salt and of giving up electrons to the acidic material.

alkaline cell – An electrochemical cell that uses powdered zinc as the anode, powdered graphite and manganese dioxide as the cathode, and potassium hydroxide as the electrolyte. An alkaline cell has an open-circuit voltage of 1.5 volts, and it has from 50 to 100 percent more capacity than a carbon-zinc cell of comparable size.

alkyd resin – A type of synthetic resin that is used as the base for certain enamels and primers.

Allen head bolt – A bolt with a hexagonal hollow or hole in its head to accommodate an Allen wrench for turning.

Allen wrench – A hexagonal shaft, bar or rod used to turn an Allen head bolt.

alligator clip – A spring-loaded clip that has long, narrow, spring-loaded jaws with meshing teeth. It is used on the end of an electrical wire to make temporary connections in an electrical circuit.

allowable – Permissible.

allowance – The amount of difference that is permissible between the dimensions of two mating parts.

alloy – A substance of different elements, one of which is metal.

alloy steel – Steel into which certain chemical elements have been mixed. Alloy steel has different characteristics from those of simple carbon steel.

alloying agent – A chemical element used to change the characteristics of the base metal to form an alloy.

all-weather spark plug – A shielded spark plug for use in an aircraft reciprocating engine, in which the ceramic insulator is recessed into the shell so that a resilient cigarette on the ignition harness lead can provide a watertight seal. All-weather spark plugs are identified by their 3/4", 20-thread per inch shielding.

Alnico – An alloy of iron, aluminum, nickel, and cobalt. Alnico has an extremely high permeability and excellent retentivity for use in magnets.

Alodine – A registered trade-mark of Amchem Products, Inc., for a conversion coating chemical which forms a hard, unbroken aluminum oxide film, chemically deposited on a piece of aluminum alloy. Alodining serves the same function as anodizing, but does not require an electrolytic bath. It conforms to specification MIL-C-5541B.

alpha – Symbol: α. Something that is first.

alpha hinge – The hinge at the root of a helicopter rotor blade that allows the tip of the blade to move back and forth in its plane of rotation. The axis of the alpha hinge is perpendicular to the plane of rotor rotation. *See also lead-lag hinge.*

alpha mode of operation – The operation of a turboprop engine that includes all of the flight operations, from takeoff to landing. Alpha operation is normally between about 95% to 100% of the engine operating speed. *See also Alpha range.*

alpha particle – A positively charged nuclear particle that has the same mass as the nucleus of a helium atom. Alpha particles consist of two protons and two neutrons.

alpha range – The pitch of a turbopropellor system which maintains a constant RPM of the engine in flight idle conditions.

alpha transistor operation – A measure of emitter-to-collector current gain in a transistor that is connected in a common-base amplifier circuit. The alpha of a junction transistor is never greater than one. Its output is always less than its input.

alphanumeric – Consisting of numbers and letters.

alteration — Any changes in configuration and/or design of an aircraft, engine, or component.

alternate air door — *See alternate air valve.*

alternate air valve — An automatic or cockpit operated valve in the induction system of a reciprocating engine used for selecting an alternate warm air source to prevent carburetor icing. The alternate air valve on an engine that is equipped with a carburetor is called a carburetor heat valve.

alternate airport — Means an airport at which an aircraft may land if a landing at the intended airport becomes inadvisable.

alternate source — A secondary source that achieves approximately the same level of operation.

alternate static air valve — A valve in the static air system that supplies reference air pressure to the altimeter, the airspeed indicator, and the vertical speed indicator. Normally these instruments pick up their reference air pressure from flush-mounted static vents on the outside of the aircraft.

alternating current — Abbrev.: AC. The flow of electrons in a circuit whose direction periodically reverses and whose amplitude continually changes.

alternation — One half of the cycle of alternating current or voltage. There are two alternations in one cycle: the positive alternation, and the negative alternation.

alternator — A type of generator in which alternating current electricity is generated in the fixed windings and is converted to direct current by solid-state rectifying diodes.

altimeter —

[1]An aneroid barometer calibrated in feet or meters above a reference pressure level. It measures the weight of the column of air above it, thus indicating the height at which the aircraft is flying above a given reference level.

[2]A radio altimeter used to measure the distance from the aircraft to the ground. This is accomplished by transmitting radio frequency energy to the ground and receiving the reflected energy at the aircraft.

altimeter setting — Station pressure corrected to sea level. When this number is placed in the barometric window of an altimeter, the instrument will read indicated altitude.

altitude – The actual height above sea level at which an aircraft is flying.

altitude engine – A reciprocating aircraft engine having a rated takeoff power that is producible from sea level to an established higher altitude.

altitude indicator – A type of attitude indicator system that is motor driven by 115 v, 400 Hz alternating current. The motor drives a gyro at approximately 21,000 RPM, and is supported by a yoke and pivot assembly which is linked with a horizon bar and stabilizer kidney-shaped sphere having pitch attitude markings. The pictorial readings on the indicator give a continuous presentation of the aircraft attitude in pitch and roll with respect to the earth's surface.

alumina – An oxide of aluminum (Al_2O_3). Alumina occurs in nature in the form of corundum, emery, sapphires, or in the form of bauxite, a clay.

aluminium – The British name for aluminum.

aluminizing –
[1]A form of corrosion protection for steel parts.
[2]A metal coating process which bonds either a corrosion resistant or a wear resistant surface to a base metal. Older use being aluminum coating for hot section parts.

aluminum – A metallic chemical element. Aluminum is a bluish, silvery-white metal that is lightweight, malleable, and ductile. It is the chief metal used in aircraft construction. It is produced from the clay bauxite, which is a form of aluminum oxide. In its natural form, aluminum is soft and weak, but it can be alloyed with copper, magnesium, manganese, and zinc to give it strength. Aluminum is a good conductor of both electricity and heat and is a good reflector of heat and light. Pure aluminum is highly resistant to corrosion.

aluminum alloy – Pure aluminum to which one or more alloying elements has been added to increase its hardness, toughness, durability, and resistance to fatigue.

aluminum oxide – A compound of aluminum and oxygen (Al_2O_3). It is extremely hard and is used as an abrasive.

aluminum paste – Extremely small flakes of aluminum metal suspended in a vehicle in the form of a paste. Aluminum paste is mixed with clear dope to make aluminum pigmented dope, which is applied over clear dope that is used on aircraft fabric to exclude the ultraviolet rays of the sun from damaging the clear dope and fabric underneath.

aluminum welding — The welding of weldable aluminum and aluminum alloys used in aircraft construction using equipment and techniques acceptable to the FAA.

aluminum wool — Shavings of aluminum metal that are formed into a pad. Aluminum wool can be used to remove corrosion products from aluminum alloy parts, and also to smooth out minor scratches from the surface of aluminum sheets or tubing.

aluminum pigmented dope — Clear aircraft dope in which extremely tiny flakes of aluminum metal are suspended. When sprayed on aircraft fabric, the flakes leaf out and form an opaque covering. protecting the fabric and clear dope from the harmful effect of the ultraviolet rays of the sun.

amalgam — Any mixture of mercury with another metal.

amalgamate — To combine or join or mix ingredients.

amber — A hard, yellowish, translucent, fossilized tree resin that is used in jewelry.

ambient — Refers to the condition of the atmosphere as it exists at the time of observation.

ambient air — *See ambient.*

ambient pressure — The pressure of the air that immediately surrounds an object.

ambient temperature — The temperature of the air that surrounds an object.

ambiguity — Something that does not have a clear meaning.

American Society of Testing Materials — Abbrev.: ASTM. An organization that sets up various standards that are used in the aircraft industry.

American Standards — Dimensional standards for fasteners that are issued by the American Standard Association.

American Wire Gage — Abbrev.: AWG. The standard that is used for measuring the diameter of round wires and the thickness of non-ferrous metal sheets. The American Wire Gage is also known as the Brown and Sharpe gage.

ammeter — An electrical measuring instrument used to measure electron flow in amperes. Ammeters that measure very small rates of flow are called milliammeters (thousandths of an ampere) and microammeters (millionths of an ampere).

ammeter shunt – A low-resistance resistor installed in parallel with an ammeter to allow the meter to read a flow of current that exceeds the current limit of the instrument. The ammeter, acting as a millivoltmeter, measures the voltage drop across the shunt, and indicates, on a scale, the amount of current that is flowing through the circuit.

ammonia – A invisible gas that is made up of one atom of nitrogen and three atoms of hydrogen (NH_3). Ammonia becomes a liquid at –28°F, and it freezes at –107°F. Ammonia is used to case-harden steel by a process called nitriding.

amorphous – Without shape.

ampere – Abbrev.: Amp or A. A measure of electron flow. One ampere is equal to a flow of one coulomb (6.28 billion billion electrons) past a point in one second. One ampere is the amount of current that can be forced through one ohm of resistance by a pressure of one volt.

ampere-hour capacity – A measure or rating of a battery that tells the capacity of electrical energy the battery can supply. One ampere-hour is the product of the current flow in amperes, multiplied by the length of time, in hours, that the battery can supply this current.

ampere-hour meter – An electrical measurement instrument that measures the rate of current flow per unit of time.

ampere turn – A measure of magnetomotive force (mmf) of an electromagnet. It is the force produced when one ampere of current flows through one turn of wire in a coil. One ampere turn is equal to 1.26 gilberts.

amphibian aircraft – An aircraft with the capability of landing on and taking off from either land or water.

amphibious floats – Floats which may be attached to an aircraft to allow it to operate from either land or water. Retractable wheels which are mounted in the floats can be extended for operation on the land as well as water.

amplification – The increase in either voltage or current that takes place in a device or in an electrical circuit. *See also amplification factor.*

amplification factor – The ratio of the amplitude of the output of an electrical or electronic circuit, to the amplitude of its input.

amplifier – An electronic device that increases the amplitude of the signal that is put into its input.

amplitude – The magnitude or amount a value changes from its at-rest condition, or its normal condition, to its maximum condition.

amplitude modulation — In a radio frequency carrier wave, a system of changing the voltage of a radio-frequency carrier wave so that it can carry information.

AMS specifications — Materials and process specifications for aircraft components that conform to established engineering and metallurgical practices in the aircraft industries. AMS specifications are developed by the SAE Aeronautics Committee.

AN aeronautical standard drawings — Dimensional standards for aircraft fasteners that were developed by the Aeronautical Standards Group. AN is the part number prefix for all fasteners that are described in these drawings.

AN fittings — A series of fittings for flared tubing, using a 37° flare angle and having a small shoulder between the ends of the threads and the beginning of the flare cone.

AN hardware — Standard hardware items such as bolts, nuts, washers, etc., whose design and material have been approved by both the air Force and Navy and are acceptable for use in civilian and military airplanes.

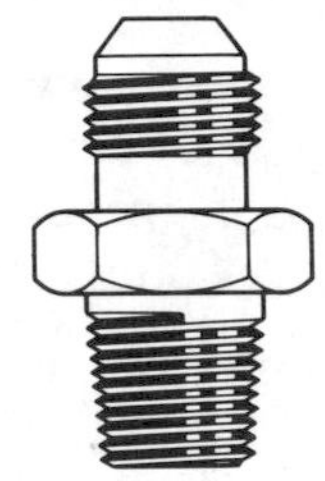

anaerobic resin — A single-component polyester resin that hardens when all air is restricted from it.

analog — A physical variable that keeps a fixed relationship with another variable as it changes. For example, the position of the hands of a clock keeps a fixed relationship with time. It is because of this relationship that we can tell the time of day by knowing the positions of the hands of the clock. The position of the clock hands is an analog of time.

analog computer — An electronic computer that operates by converting different levels of voltage or current into numerical values.

analog data — Data that is represented by a continuously varying voltage or current.

analog-to-digital conversion — A conversion that changes analog information into a digital form.

analyzer, engine — A portable or permanently installed instrument, whose function is to detect, locate, and identify engine operating abnormalities such as those caused by a faulty ignition system, detonation, sticking valves, poor fuel injection, or the like.

anchor nut – A nut that is riveted or welded to a structure in such a way that a screw or bolt can be screwed into it. An anchor nut does not have to be held with a wrench to keep it from turning.

anemometer – An instrument that measures the velocity of moving air. An anemometer uses a series of hemispherical metal cups mounted on arms on a shaft. The air blows the cups and rotates the shaft. A counter measures and converts this into wind speed which may be marked in feet per second, meters per second, kilometers per hour, miles per hour, or knots.

aneroid –
[1]A sealed flexible container which expands or contracts in relation to the surrounding air pressure. It is used in an altimeter or a barometer to measure the pressure of the air.
[2]A thin disc shaped box or capsule, usually metallic, that is partially evacuated of air and sealed. It expands or contracts with changes of the surrounding air or gas.

aneroid barometer – An instrument that measures the absolute pressure of the atmosphere by balancing the weight of the column of air above it against the spring action of a specially shaped evacuated metal bellows.

angle – A figure formed by two lines or two plane surfaces extending from the same point.

angle drill – A form of sheet metal drilling tool in which the twist drill is held at an angle to the spindle of the drill motor.

angle of attack –
[1]The acute angle formed between the relative wind striking an airfoil and the zero lift line of the airfoil. The chord line of the airfoil is often substituted for the zero-lift line.
[2](Absolute) The angle of attack of an airfoil, measured from the attitude of zero lift.
[3](Critical) The angle of attack at which the flow about an airfoil changes abruptly as shown by corresponding abrupt changes in the lift and drag.
[4](For infinite aspect ratio) The angle of attack at which an airfoil produces a given lift coefficient in a two-dimensional flow. Also called "effective angle of attack".
[5](Turbine compressor) The acute angle formed between the relative wind striking the chord line of the compressor blades and the direction of the air that strikes the blades.

angle of attack indicator system – Detects the local angle of attack of the aircraft from a point on the side of the fuselage and furnishes reference information to an angle-of-attack indicator.

angle of head – In countersunk heads, the included angles of the conical under portion or bearing surface, usually 82° or 100°.

angle of incidence –
[1]The acute angle which the wing chord makes with the longitudinal axis of the aircraft.
[2]Angle at which blades are set into the compressor disk. A fixed angle in all cases except the variable pitch fan. Angles set for optimum airflow at altitude cruise and RPM.

angle of roll – Angle of bank. The angle through which an aircraft must be rotated about its longitudinal axis in order to bring its lateral axis into the horizontal plane. The angle is positive when the left side is higher than the right.

angle of stabilizer setting – The acute angle between the longitudinal axis of an airplane and the chord of the stabilizer. The angle is positive when the leading edge is higher than the trailing edge.

angle of wing setting – The acute angle between the plane of the wing chord and the longitudinal axis of the airplane. The angle is positive when the leading edge is higher than the trailing edge.

angle of yaw – The acute angle between the direction of the relative wind and the plane of symmetry of an aircraft. The angle is positive when the aircraft turns to the right.

angled gearbox – Same function as transfer gearbox. Receives its name because the driveshaft is angled, usually 90° up to the main rotor bevel gear.

angular acceleration – The rate at which a rotating object increases its rotational speed.

angular measurement – The measured displacement between two lines that project from the same point.

angular momentum – The product of an object's mass directed along a rotating axis.

angular type piston pump – A type of pump whose housing is angular which causes a corresponding angle to exist between the cylinder block and the drive shaft plate to which the pistons are attached. It is this angular configuration of the pump that causes the pistons to stroke as the shaft is turned.

angular velocity —
[1]The rate of change of an angle as a shaft rotates. It is expressed in revolutions per minute or in radians per second.
[2]The velocity of an object moving around a center at a given distance from the center, expressed in radians.

anhydrous — Any material that does not contain water.

anion — A negative ion that moves toward an anode in an electrolysis process.

anneal — To soften by means of heat treatment.

annealed wire — Wire softened by heat after its diameter has been decreased by drawing it through dies.

annealing — A method of heat treatment in which a metal is softened, losing some of its hardness. *See also annealing process.*

annealing process — Heating of an alloy to a temperature called solid solution temperature. This is followed by allowing it to cool slowly at a controlled rate through its critical range for the purpose of inducing softness. This results in the removal of former heat-treatment strain hardening, and internal stresses.

annual inspection — A complete inspection of an aircraft and engine, required by the Federal Aviation Regulations to be accomplished every twelve calendar months on all certificated aircraft. Only an A&P technician holding an Inspection Authorization can conduct an annual inspection.

annual rings — The rings that appear in the end of a log that has been cut from a tree. The more rings there are, and the closer they are together, the stronger the wood.

annular, basket type — One of several basic types of combustion chambers used in turbine engines. It consists of a housing and a liner. The liner is a one piece shroud (combustion chamber) extending all the way around the outside of the turbine shaft housing. Fuel is sprayed from nozzles mounted around a full manifold into the inner liner of the combustor. Here it is mixed with air from the compressor and is burned.

annular, can type — One of several basic types of combustion chambers used in turbine engines. Each of the can-type combustion chambers consist of an outer case or housing, within which there is a perforated stainless steel combustion chamber liner or inner liner. Interconnector tubes connect each can for flame propagation which spread combustion during the initial starting operation.

annular combustor –
[1]Annular refers to ring shaped. Therefore, a ring shaped or cylindrical one-piece combustion liner inside a combustion outer case as seen on the jumbo jet engines CFs, JTs, RB-211 and many corporate jet engines.
[2]A cylindrical one-piece combustion chamber, sometimes referred to as a single basket type combustor.

annulus – Opening between two concentric rings, e.g. the space between the compressor disk and outer case could be referred to as the compressor annulus.

annunciator panel – Sometimes called a master warning system. A panel of warning lights that are in plain sight of the pilot of an airplane. All the lights are identified by the name of the system they represent, and they are usually covered with a colored lens to show the meaning of the condition they announce. Red lights are used to indicate a dangerous condition, amber lights show that some system is armed, or active, and green lights show that a condition is safe.

annunciator system – *See Annunciator panel.*

anode –
[1]The positive plate of an electrochemical combination, such as a battery or electroplating tank. Electrons leaving the anode cause it to be less negative, or positively charged. When electrons leave an anodic material, the chemical composition of the anode changes from a metal to a salt caused by the reaction with the electrolyte. In the process, the anode is corroded or eaten away.
[2]The electrode in a vacuum tube or a semiconductor diode to which the electrons travel after they leave the cathode. The anode in a vacuum tube (an electron tube) is called the plate. The anode of a semiconductor diode is the end that is made of P-type material, and it is not marked (the cathode has the mark). In the diode symbol, the anode is shown by the arrowhead.

anodic – The component of an electrolytic cell which is positive is said to be anodic, because it has supplied electrons for the reaction and is the material that is corroded or eaten away.

anodizing – The formation of a hard, unbroken film of aluminum oxide on the surface of an aluminum alloy. This film is electrolytically deposited by using the aluminum as the anode and chromic acid as the electrolyte.

anoxia – A severe case of hypoxia which can cause permanent damage on the human body that is caused by a lack of oxygen.

antenna — A special type of electrical circuit designed to radiate and receive electromagnetic energy. Antennas vary in shape and design depending upon the frequency to be transmitted, and specific purposes they serve.

antenna matching — The process of matching the impedance of a radio antenna with the impedance of the transmission line that carries the signal from the radio transmitter to the antenna.

antenna wire — A type of wire that has a low electrical resistance and a high tensile strength. Copperweld, which is a form of wire that has copper plated over a core of strong steel wire, is often used as antenna wire.

antiblush thinner —
1 A slow-drying thinner that is used with certain types of lacquer or dope to slow the evaporation of the solvents thereby preventing blush.
2 A slow-drying thinner used in conditions of high humidity to prevent blushing of the aircraft dope.

anticollision light — A flashing light on the vertical fin or underneath the aircraft. Used to increase the visibility of the aircraft.

anticyclone — In the northern hemisphere, air flows out of an anticyclone with a clockwise direction. In the southern hemisphere, the direction is counterclockwise.

antidetonant fluid — A fluid, such as a water/alcohol mix, which when injected into the fuel/air stream enables more power to be obtained from the engine, especially at takeoff, than is otherwise possible. The fluid itself, does not increase the engine power, it merely replaces formerly excess fuel, allowing for a cooler operating engine as the water/alcohol dissipates heat more rapidly than does fuel.

antidrag wire — A diagonal, load carrying member of a Pratt truss wing. It runs from the rear spar inboard to the front spar outboard, and opposes tensile loads tending to pull forward on the wing.

antifreeze — Ethylene glycol. A chemical that is added to a liquid to lower its freezing point.

antifriction bearings — Ball or roller bearings that have a special low drag quality.

antiglare paint — A black or dark paint that dries to a dull or nonspecular finish. It is applied to a surface to prevent glare interfering with the function of the aircraft crew.

anti-icing — The prevention of the formation of ice on a surface. Ice may be prevented by using heat or by covering the surface with a chemical that prevents water reaching the surface. Anti-icing should not be confused with deicing, which is the removal of ice after it has formed on the surface.

anti-icing fluid — A fluid comprised of alcohol and glycerine, used to prevent the formation of ice on the leading edge of propellers, in the throat of a carburetor, or on the windshield.

anti-icing system —

[1]Any system or method used to provide heat to critical external surfaces in order to prevent ice formation.

[2]A system in a gas turbine engine in which some of the hot compressor bleed air is routed through the engine air inlet system to warm it and to keep ice from forming.

antiknock rating — The rating of fuel that refers to the ability of the fuel to resist detonation.

antileak check valve —

[1]A check valve with a crack pressure setting just above gravity pressure. Used to prevent oil tank seepage to lower portions of the lube system during periods on engine inactivity.

[2]An oil tank check valve that is used in some aircraft reciprocating engines that have a dry-sump lubrication system. Anti-leak check valves hold the oil in the tank against the pull of gravity. When the oil pump puts a low pressure on the check valve, it opens and allows oil to flow from the tank into the engine.

antipropeller end — The end of an engine away from the propeller.

anti-servo tab — An adjustable tab attached to the trailing edge of a stabilator which moves in the same direction as the primary control. It is used to make the stabilator less sensitive.

antiskid system — A system of controls for aircraft brakes that releases the hydraulic pressure to the brake in the event the wheel begins to lock up or to skid.

antitear strips — Strips of aircraft fabric of the same material as the airplane is covered with, laid over the wing rib under the reinforcing tape before the fabric is stitched.

antitorque pedals – The foot pedals used to control the pitch of the anti-torque rotor on the tail of a single-rotor helicopter. Controlling the pitch of the anti-torque rotor allows the helicopter to rotate about its vertical axis.

antitorque rotor – *See tail rotor.*

antiwindmilling brake – A friction brake fitted as an accessory to the main gearbox of some older engines. Seldom seen today. Note that the turboprops include a brake in their propeller mechanisms which is not the friction brake described here.

anvil – A hard-faced block on which a material is hammered or shaped.

aperiodic damping – Damping that prevents an object over swinging, or moving past its at-rest position. Aperiodic damping is also called dead-bead damping.

aperiodic-type compass – A magnetic compass in which the floating magnet assembly is fitted with damping vanes to increase the period of its oscillations.

API scale – A scale that has been developed by the American Petroleum Institute to measure the specific gravity of a liquid.

apogee – The point at which an orbiting vehicle is the greatest distance from the center of the object it is circling.

apparent power – In an AC circuit, the product of rms current and rms voltage, expressed in volt-amperes.

apparent weight – The apparent weight of an object that is immersed in a liquid is the weight of the object, less the weight of the fluid that the object has displaced.

appliance – Any instrument, equipment, mechanism, part, apparatus, or accessory, including communications equipment, that is used or intended to be used in operating or controlling an aircraft in flight, and is installed or attached to the aircraft, but is not part of an engine, airframe, or propeller.

applicability – Something that applies to and/or affects another.

approach – The flight of an airplane just preceding the landing.

approach lights – High intensity lights along the approach end of an instrument runway to aid the pilot in the transition from instruments to visual flight at the end of an instrument approach.

approved – Unless used with reference to another person, means approved by the administrator.

approved data — Data which may be used as an authorization for the techniques or procedures for making a repair or an alteration to a certificated aircraft. Approved data may consist of such documents as Advisory Circular 43.13-1a and -2a, Manufacturer's Service Bulletins, manufacturer's kit, instructions, Airworthiness Directives, or specific details of a repair issued by the engineering department of the manufacturer.

approved inspection system — A maintenance program consisting of the inspection and maintenance necessary to maintain an aircraft in airworthy condition in accordance with approved FAA practices.

approved repair station — A facility approved by the FAA for certain types of repair to certificated aircraft.

approved type certificate — An approval issued by the FAA for the design of an airplane, engine, or propeller. This certifies that the product has met at least the minimum design standards.

apron — The surface area in front of a hanger building where aircraft can be parked and tied down. Aprons may also be called ramps or tarmacs.

Aqua-dag — A lubricant used for components in an oxygen system. Oil or other petroleum products cannot be used with oxygen system components because of its high probability to ignite when they are in contact with each other.

arbor press — A press with either a mechanically or hydraulically operated ram.

arc —
[1]A portion of the circumference of a circle.
[2]A sustained luminous discharge of electricity across a gap.

arc lamp — A source of light that is produced by an electric arc. The arc is produced when electrons flow through ionized gases between two electrodes.

arc welding — A form of welding in which the heat required to melt the metal is produced by an electric arc.

Archimedes' principle — The principle of buoyancy which states that a body emersed in a fluid is buoyed up with a force equal to the weight of the fluid it displaces.

area — The number of square units in a surface.

area navigation high route — An area navigation route within the airspace extending upward from, and including, 18,000 ft. MSL to flight level 450.

area navigation low route — An area navigation route within the airspace extending upward from 1,200 ft. above the surface of the earth to, but not including 18,000 ft MSL.

area navigation — Abbrev.: RNAV. A form of electronic navigation system in which waypoints are established by electronically moving the signal from the VORTAC station to any longitude and latitude to which the pilot wishes to fly. The new location of the signal from the VORTAC is called a waypoint.

arm — The horizontal distance in inches between the reference datum line and the center of gravity of an object. If the object is behind the datum, the arm is positive, and if in front, the arm is negative.

armature — The rotating element of an aircraft motor or generator. Load current is generated in the armature.

armature reaction — The distortion of the generator field flux by the current flowing in the windings of the armature. Armature reaction is the effect of the magnetic field that surrounds the windings in the armature that distorts the magnetic field produced by the field coils. Armature reaction causes the brushes to pick up current from the armature at a point on the commutator where there is a potential difference. This causes the brushes to spark.

armed — A condition in which a device is made ready for actuation.

Armed Force — The Army, Navy, Air Force, Marine Corps, and Coast Guard, including their regular and reserve components and members serving without component status.

aromatic gasoline — Gasoline that has had its anti-detonation characteristics improved by blending in aromatic additives such as benzene, toluene, or xylene.

aromatics — Chemical compounds such as toluene, xylene, and benzene. These aromatics may be blended with gasoline to improve its anti-detonation qualities. Aromatics have the bad feature of softening rubber hoses and diaphragms and seals that are used in fuel metering system components.

arrival time — The time when an aircraft in flight touches the ground upon landing.

arsenic — A chemical element that has five valence electrons. An extremely small amount of arsenic is alloyed with silicon or germanium to make N-type semiconductor material.

articulated connecting rod — A link rod which connects the pistons in a radial engine to the master rod. There is one less articulated rod than there are cylinders in each row of cylinders in a radial engine.

articulated rod assembly — *See articulated connecting rod.*

articulated rotor — A helicopter rotor in which each of the blades is connected to the rotor hub in such a way that it is free to move up and down, move back and forth in its plane of rotation, and change its pitch angle. It is sometimes called a fully articulated rotor.

artificial aging — A process of increasing the strength of a heat-treated aluminum alloy by holding it at an elevated temperature after it has been solution heat treated. Artificial aging is also called precipitation heat treatment. *See also precipitation heat treatment.*

artificial feel — A type of force feedback or cushioning effect that is used in the automatic flight control systems of some aircraft. Artificial feel produces an opposition to the movement of the controls that are proportional to the aerodynamic loads that are acting on the control surfaces.

artificial horizon — A flight instrument in which a bar or a display is held in a constant relationship with the earth's horizon. An artificial horizon provides the pilot with a visual reference when the natural horizon is not visible.

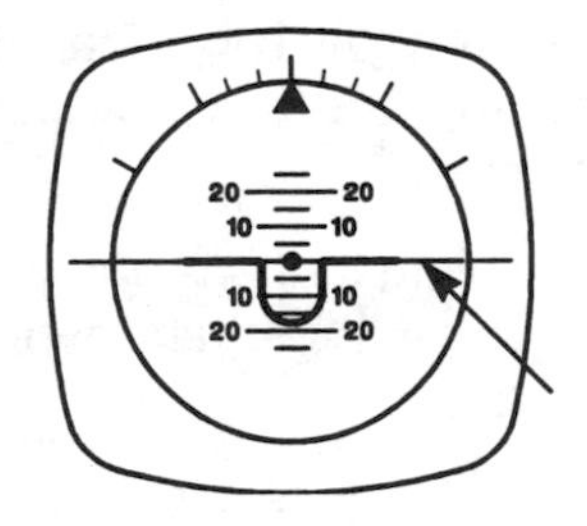

artificial radio antenna — A device that is attached to the output of a radio transmitter when adjusting the transmitter. The artificial antenna has the same impedance as the antenna, but the radio signal put into the artificial antenna is not radiated.

asbestos — A fiber of magnesium silicate that has a high resistance to fire. It has good insulating qualities.

ashless dispersant oil – Abbrev.: AD oil. A popular mineral oil that is used as a lubricant for aircraft reciprocating engines. Ashless dispersant oil does not contain any metallic ash-forming ingredients. It has additives in it that keep the contaminants that form in the oil dispersed throughout the oil so they will not join together and clog the oil filters.

aspect ratio – The ratio of the span of the wing to its chord. The aspect ratio of a tapered wing is found by dividing the square of the wing span by its area.

asphalt – A heavy, brownish-black mineral that is found in crude oil. Asphalt is used as a base for some acid-resistant paints.

assembly – The fitting together of parts to form a complete structure or unit.

assembly break – A joint in the structure of an aircraft that is formed when two sub-assemblies are removed from their jigs and joined together to form a single unit.

assembly drawing – An aircraft drawing which shows a group of parts laid out in the relationship they will have when they are assembled.

assembly line – Any arrangement of work stations in an aircraft factory that allows certain functions to be performed on aircraft that are being manufactured.

assigned radio frequency – The frequency of a carrier signal that is produced by a radio transmitter. The frequency is assigned by the Federal Communications Commission for a particular transmitter or particular type of transmission.

astable – Not stable. An astable electronic device has two conditions of temporary stability, but no condition of permanent stability.

astable multivibrator – A free-running multivibrator. They contain two devices such as transistors to control the flow of electrons. When one transistor is conducting, the other is not conducting. The two transistors alternate between a condition of conducting and not conducting.

astern – A location or direction to the rear or rear part of an aircraft.

ASTM specifications – Standards developed by the American Society for Testing Materials.

astronaut – A person who travels beyond the earth's atmosphere.

astronautics – The field of science that deals with space flight and includes the design, construction, and operation of space vehicles and their related support activities.

astronomical unit – One astronomical unit is the average distance between the earth and the sun. This is approximately 93 million miles.

astrophysics – The field of science that deals with the physical characteristics of the heavenly bodies. These include their mass, density, temperature, size, luminosity, and origin.

asymmetrical – A condition in which the shape of a body is NOT the same on both sides of its center line.

asymmetrical lift – A condition of unequal lift produced by the rotor disc of a helicopter in motion. The advancing blade travels at its peripheral speed plus the forward speed while the airspeed of the retreating blade is the difference between the two speeds.

ATA system – Airline Transport Association's standardized format for maintenance manuals.

athodyd – Aero Thermo Dynamic Duct. An open tube shaped to produce thrust when fuel is ignited inside it. Fuel is added to incoming air as the athodyd moves through the air at a high speed. This burning causes air expansion that speeds up the air and produces thrust.

athwartships – At right angles to the longitudinal axis of an aircraft.

atmosphere – The entire mass of air surrounding the earth. *See also air.*

atmospheric pressure – The pressure that is exerted on the surface of the earth by the air that surrounds the earth. Within standard conditions, at sea level, the atmospheric pressure is 14.69 PSI.

atom – The smallest particle of an element, consisting of a positively charged nucleus orbited by one or more negatively charged electrons.

atomic number – The number of protons in the nucleus of an atom. The atomic number of an element determines the position of the element in the periodic table of elements.

atomize – To reduce a liquid to a fine spray.

atomizing nozzle – *See fuel nozzle.*

ATR racking system – A widely accepted size and mounting standard for airborne electronic equipment.

attenuation — A reduction in the strength of a signal, the flow of current, flux, or other energy in an electronic system.

attitude — The position of an aircraft as determined by the relationship of its axes and a reference, usually the earth's horizon.

attitude gyro — A gyro-actuated flight instrument that shows the pilot the attitude of the aircraft relative to the earth's horizon.

attitude indicator — An instrument that gives a pilot an artificial reference of the airplane's attitude in pitch and roll with respect to the earth's surface.

attraction — A force acting mutually between particles of matter, tending to draw them together.

audio frequency — Abbrev.: af. Frequency in a range that can normally be heard by the human ear, ranging from about 16 Hz to 16,000,000 Hz.

audio-frequency amplifier — An electronic amplifier that is capable of amplifying alternating current with a frequency in the range of human hearing.

augmentation — Any designed method of increasing the basic thrust of an engine for a short period such as for takeoff or combat; usually accomplished by coolant injection or afterburning.

augmenter — A device used to draw cooling air through an engine cowling by the use of a low pressure created by the rapid moving exhaust gases as they leave the exhaust pipes.

augmenter tube — A specially shaped tube mounted around the exhaust tail pipe of an aircraft reciprocating engine. When exhaust gases flow through the augmenter tube, they produce a low pressure in the engine compartment that draws cooling air into the compartment through the fins of the engine.

aural warning system — A bell or horn type warning sounding system that alerts the pilot during an abnormal takeoff condition, landing condition, pressurization condition, mach-speed condition, an engine or wheel well fire, calls from the crew call system, and calls from the SECAL system.

austenite — A supersaturated solution of carbon is dissolved in the iron. Austenite exists in the iron only when the iron is at a high temperature.

authorized — Having a legal right to act or perform certain functions by the FAA.

auto lean – A lean fuel-air mixture whose ratio is kept constant by an automatic mixture control in the carburetor.

auto rich – A rich fuel-air mixture whose ratio is kept constant by an automatic mixture control in the carburetor.

autoclave – A pressure vessel in which the air inside can be heated to a high temperature and the pressure raised to a high value in order to decrease the amount of time needed to cure plastic resins. They also improve the quality of the curing process.

autoclave molding – A method of molding laminated reinforced plastic components. *See also autoclave.*

autofeather – A portion of a propeller control system that causes the propeller to feather automatically if the engine on which it is mounted is shut down in flight.

autogiro – A heavier than air rotorcraft whose rotor is spun by aerodynamic forces which act on the blades. No engine power is used to drive the rotor in flight.

automatic – An operation that has the ability to perform by itself or to self regulate. An automatic operation has all the necessary signals built into it so that it will perform its function without any external decisions having to be made.

automatic adjusters – A portion of the return spring system of disk brakes that maintain a constant clearance between the disk and the brake linings as the lining wears. The automatic adjusters allow the brake piston to move back a specific amount each time the brake is released.

automatic direction finders – Abbrev.: ADF. A radio receiver equipped with directional antennas which are used to determine the direction from which signals are received. This direction is shown on an instrument that looks much like the dial of a compass, with zero degrees representing the nose of the aircraft, rather than North. The needle on the ADF indicator shows the pilot the number of degrees between the nose of the aircraft and a line to the radio station that is being monitored.

automatic flight control system – Abbrev.: AFCS. A complete instrument system that includes the automatic pilot which is coupled with radio navigation and approach equipment. An aircraft with an AFCS can be flown in a completely automatic mode.

automatic frequency control – Abbrev.: AFC. A circuit in a radio receiver that keeps the receiver tuned to a desired frequency within specific limits.

automatic gain control – An electronic circuit within a radio receiver that keeps the output volume relatively constant.

automatic mixture control: – Abbrev.: AMC. A device in a fuel metering system (carburetor or a fuel injection system) that keeps the fuel-air mixture ratio constant as the density of the air changes with altitude.

automatic pilot – An automatic flight control system which keeps an aircraft in level flight or on a set course. Automatic pilots can be directed by the human pilot, or they may be coupled to a radio navigation signal.

automatic volume control – Abbrev.: AVC. Circuit which regulates a relatively constant volume as the input signal strength varies.

automatic-reset circuit breaker – An electrical circuit protection device that opens a circuit when a current overload occurs, and will automatically reset itself and restore the circuit when the overload is no longer present. Automatic reset circuit breakers are installed in some electric motors, but they are not approved for use in aircraft.

autorotation –
[1]A condition of rotorwing flight in which the rotors are turned by aerodynamic action, rather than by engine power supplied to the rotors.
[2]The property of a rotor system for maintaining its angular velocity without engine power, the relative force being provided by the forward component of the lift forces acting on the rotor blades.

autorotation region – The portion of the rotor disk of a helicopter which produces an autorotative force. *See also autorotative force.*

autorotative force – An aerodynamic force that causes an autogiro or helicopter rotor to turn when no power is supplied to it.

Autosyn – A form of an alternating current remote-indicating system using an electromagnet excited by 400 Hz AC as its rotor and having a three-phase stator.

autotransformer – A single winding transformer having a carbon brush which can tap off any number of turns for the secondary. It produces variable voltage AC output.

auxiliary – A supplement, or an addition, to a main unit.

auxiliary flight surfaces — Lift-modifying devices on an airfoil, such as flaps, slots, or slats.

auxiliary fuel pump — An electrically operated fuel pump that is used to supply fuel to the engine for starting, takeoff, or in the case of engine-driven fuel pump failure. Auxiliary fuel pumps are also used to pressurize the fuel in the line to the engine-driven fuel pump in order to prevent vapor lock at altitude when the fuel is warm.

auxiliary hydraulic pump — A hydraulic pump that is used as an alternate source of hydraulic pressure for emergencies, or to produce hydraulic pressure when the aircraft engines are not in operation.

auxiliary ignition units — An auxiliary ignition system that facilitates engine starting. The auxiliary device is connected to the magneto to provide a high ignition voltage. Reciprocating engine starting systems normally include one of the following types of auxiliary starting system: booster coil, induction vibrator, impulse coupling, or other specialized retard breaker and vibrator starting systems.

auxiliary power unit — Abbrev.: APU. A self-contained power unit which is driven by a small reciprocating or turbine engine, used to provide auxiliary air, hydraulic, and electrical power.

auxiliary pump — Any pump that is used as an alternate source, or for an emergency. An auxiliary pump provides assistance or support to the main pump.

auxiliary rotor — A rotor that serves either to counteract the effect of the main rotor torque on a rotorcraft or to maneuver the rotorcraft about one or more of its three principle axes.

auxiliary view — A view used in an aircraft drawing that is made at some angle to one of the three views of an orthographic drawing. It is used to show details that would not otherwise be visible.

avalanche diode — Another name for a zener diode.

avalanche voltage — The reverse voltage required to cause a zener diode to break down.

average value — Sine wave alternating current. 0.637 times the peak value of alternating current or voltage, measured from the zero reference line.

aviation — The branch of science, business, or technology that deals with any part of the operation of machines that fly through the air.

aviation medicine – A special field of medicine which establishes standards of physical fitness for airmen.

aviation shears – *See aviation snips.*

aviation snips – Compound-action hand shears used for cutting sheet metal. They normally come in sets of three: one which cuts to the left, one which cuts to the right, and one that cuts straight.

aviation weather – Specific characteristics of weather that pertain to flight or to the operation of aircraft.

avionics – A name coined to indicate airborne electronic equipment.

Avogadro's principle – A principle of physics which states that under equal pressure and temperature, equal volumes of all gases will contain equal numbers of molecules.

avoirdupois weight – The system of weight for measuring the weight of most substances. In avoirdupois weight, one pound is equal to 453.6 gms. One ounce is equal to 1/16 lb., and 1 g. is equal to 1/16 oz.

awl – A sharp-pointed tool that is used to make holes in soft materials such as leather, plastic, or wood.

axes of an aircraft – The axes of an aircraft can be considered as imaginary axles around which the aircraft turns like a wheel. At the center, where all three axes intersect, each is perpendicular to the other two. *See also axes of an airplane.*

axes of an airplane – Three mutually perpendicular imaginary lines about which an airplane is free to rotate. The longitudinal axis passes through the center of gravity of the aircraft from front to rear. The lateral axis passes through the center of gravity from wingtip to wingtip, and the vertical axis passes through the center of gravity from top to bottom. *See also axes of an aircraft.*

axial –

[1]Motion along a real or imaginary straight line on which an object supposedly or actually rotates.

[2]The engine center line.

axial-centrifugal compressor – A combination axial and centrifugal compressor usually fitted together with the axial portion as the front stages and the centrifugal portion as the rear stage. This is a popular design for smaller engines in corporate-size jets.

axial flow – The straight-through flow of a fluid. In an axial-flow compressor, the air flows through the compressor parallel to the engine and the stages of compression do not essentially change the direction of the flow.

axial flow compressor – A type of compressor used in a turbine engine in which the airflow through the compressor is essentially linear. An axial-flow compressor is made up of several stages of alternate rotors and stators. The compressor ratio is determined by the decrease in area of the succeeding stages.

axial flow turbine engine – A turbine engine in which the air is compressed by a series of rotating airfoils. The airflow through the engine is essentially in a straight line.

axial load – A load on a bearing that acts parallel to the shaft supported in the bearings. The thrust load produced by a propeller is an axial load. Axial loads are usually carried by ball bearings or by tapered roller bearings, and are carried into the engine crankcase through the thrust bearing.

axial loading – An aerodynamic force that tries to move the compressor forward. Axial loading is supported in a gas turbine engine in ball bearings.

axial-lead resistor – A disparate electronic component that provides a given amount of resistance to a circuit. The wire leads of an axial-lead resistor extend from the ends in a direction that is parallel to the axis of the resistor.

axis – A straight line about which a body can rotate.

axis of symmetry – An imaginary center line about which a body or object is symmetrical in either weight or area.

axle – The shaft on which one or more wheels are mounted and about which the wheels are free to rotate.

axonometric projection – A projection used in mechanical drawing that shows a solid rectangular object inclined in a way that three of its faces are visible. An isometric projection is a form of axonometric projection.

azimuth – The angular measurement in a horizontal plane and in a clockwise direction.

B

babbitt metal – An alloy of tin, lead, copper, and antimony used for lining engine bearings. It is used because of its exceptional anti-friction qualities.

back – The curved portion of a propeller blade that corresponds to the curved upper surface of an airfoil.

back plate – A floating plate on which the wheel cylinder and brake shoes attach on an energizing-type brake.

back pressure – The pressure caused by the exhaust system of a reciprocating engine that opposes the evacuation of the burned gases from the cylinders of a reciprocating engine.

back voltage – Counter-electromotive force generated in a conductor by the action of changing lines of flux cutting across the conductor. As the magnetic field produced by the changing current builds up and decays, it cuts across the conductor and induces a voltage in it. The polarity of this induced voltage is opposite to that of the voltage that caused the original current to flow.

backfire –
[1]In welding a momentary backward flow of the gases at the torch tip which causes the flame to go out. A backfire may be caused by touching the tip against the work, by overheating the tip, or by operating the torch at other than recommended pressures.
[2]A burning or explosion within the cylinder of a reciprocating engine when the fuel-air mixture in the induction system is ignited by gases that are still burning inside the cylinder when the intake valve opens.

backfiring – *See backfire.*

backhand welding – The technique of pointing the torch flame toward the finished weld and moving away in the direction of the unwelded area, welding the edges of the joint as it is moved. The welding rod is added to the puddle between the flame and the finished weld.

backing plate – A reinforcing plate used when making a sheet metal repair. It is often a doubler.

backlash – The clearance measured between the meshing teeth of two gears. Usually units such as selector valves, pumps, etc. are required to have a predetermined amount of backlash. However, too much movement due to backlash can be undesirable.

backsaw – A fine-toothed wood cutting saw that has a stiff lip along its upper edge. Backsaws are used to make straight or angled cuts across a board.

backup ring – An anti-extrusion ring used on the side of an O-ring packing away from the pressure. This stiff ring, usually made of some material such as Teflon®, prevents the resilient O-ring from being extruded into the space between the cylinder wall and the piston by hydraulic pressure.

bacteria – Microscopic plant life that lives in the water entrapped in fuel tanks. The growth of bacteria in jet aircraft fuel tanks causes a film which holds water against the aluminum alloy surface.

bactericide – An agent that is used to destroy bacteria.

baffle –

[1]A series of partitions inside an aircraft fuel tank. These baffles have holes in them that allow the fuel to feed to the tank outlet, but they keep the fuel from surging enough to uncover the fuel outlet.

[2]A partition separating the upper portion of the tank which prevents oil pump cavitation as the oil tends to rush to the top of the tank during periods of deceleration. The baffle is fitted with a weighted swivel outlet control valve which is free to swing below the baffle. The valve is normally open; but closes only when the oil in the bottom of the tank rushes to the top of the tank during deceleration.

[3]Sheet metal shields used to direct the flow of air between and around the cylinders of an air-cooled reciprocating engine. Also called cylinder deflectors.

[4]A structure used to impede, regulate, or alter the flow direction of a gas, fluid, or sound.

bag molding – A method of applying pressure to a piece of laminated plastic material so that all of the layers are held in tight contact with each other. The reinforcing material is injected with plastic resin which is laid up over a rigid mold in as many layers as are needed. A sheet of flexible, air-tight plastic material is placed over the mold and the edges are sealed to form a bag over the part. The entire assembly is then placed in an autoclave for curing. In the absence of an autoclave, a vacuum pump can be attached to the inside of the bag, and the air pumped out. The atmospheric pressure pressing on the outside of the bag will supply the needed force.

Bakelite – A phenol resin, often used as electrical insulation, made by the Bakelite Corporation..

baking soda – The common term for bicarbonate of soda ($NaHCO_3$).

balance —
[1]A state of equilibrium.
[2]Any object that is in a state of equilibrium.

balance — *See dynamic balance.*

balance cable — The cable that links the up-side of both ailerons together. When the control wheel or stick pulls on the control cable to move one aileron these forces are equal in amount, but they act on the opposite sides of the balance point.

balance chamber — An internal air chamber in a turbine engine used to absorb some of the compressor axial loading "thrust".

balance checks — A check or series of inspections performed on rotating components after overhaul to statically and dynamically check for correct balancing.

balance point —
[1]A point within an object at which the sum of all moments which tend to cause rotation is zero.
[2]The point about which an object will balance.

balance pressure torch — A type of welding torch where the oxygen and acetylene are both fed to the torch at the same pressure. The openings to the mixing chamber for each gas are equal in size, and the delivery of each gas is independently controlled.

balance, static — *See static balance.*

balance tab — An auxlliary control mounted on a primary control surface, which automatically moves in the direction opposite the primary control to provide an aerodynamic assist in the movement of the control. Sometimes called a servo tab.

balanced actuator — A hydraulic or pneumatic actuator having the same area on both sides of the piston. Fluid power into the actuator products the same amount of force in either direction of piston movement.

balanced amplifier — An electronic amplifier that has two output circuits that are equal, but opposite in phase. Also called a push-pull amplifier.

balanced control surface — A primary control surface with an overhang ahead of the hinge line to provide an aerodynamic assist to help the pilot in moving the control.

balancing — The act of performing a balance procedure using prescribed methods.

ball bearing – A form of anti-friction bearing consisting of grooved inner and outer races and one or more sets of steel balls held in a sheet metal retainer. Ball bearings can be designed to support thrust loads as well as radial loads.

ball bearing assembly – Consists of a grooved inner and outer races, one or more sets of balls, and, in bearings designed for disassembly, a bearing retainer. They are used for supercharger impeller shaft bearings and rocker arm bearings in some engines. Special deep-groove ball bearing are used in aircraft engines to transmit propeller thrust to the engine nose section.

ball check valve – A check valve in a fluid power system that uses a spring-loaded steel ball and a seat to allow flow in one direction only. The ball is forced tightly against its seat by fluid flowing into the valve from the end that contains the spring, thereby stopping the fluid flow through the valve. Fluid flowing into the valve from the ball end forces the ball off its seat, allowing flow through the valve.

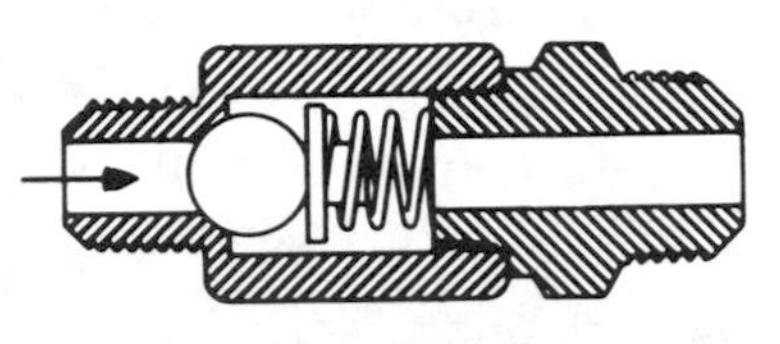

ball joint – A flexible expansion joint used in an aircraft engine exhaust system to allow relative movement of the parts as a result of their expansion and contraction.

ball peen hammer – A hammer with one side of its head shaped like a ball.

ballast –
[1]Permanently installed weight in an aircraft used to bring the center of gravity into the allowable range.
[2]A circuit element designed to stabilize current flow.

ballistic missile – A self-propelled long range missile which is guided by preset mechanisms as it goes upward, but is free falling as it comes down.

balloon – Lighter-than-air non-steerable aircraft that is not engine driven. Its rising capability comes from gases or hot air that is used to fill the bag.

balsa wood – The light, strong wood of the balsa, a tropical tree. It is sliced across its grain and sandwiched between two face sheets of thin metal or fiberglass to make rigid, lightweight panels.

balun – A type of transformer used to match a balanced antenna to an unbalanced transmission line.

banana plug — A mechanism that is used to make a temporary connection to an electrical circuit. The contacts of a banana plug are springs that have the general shape of a banana. These springs press out against the walls of the banana jack in order to make a low-resistance contact.

band — A range of frequencies.

band saw — A type of controllable speed power saw that is used to cut wood, plastics, or metal. The band saw blade is in the form of a narrow strip of steel with teeth along one edge. The ends of the blade are welded together to form a continuous loop, and the loop passes over two large wheels, one above the saw table and the other below the table.

band-pass filter — An electronic filter which passes a band of frequencies while rejecting all frequencies above or below the band.

band-reject filter — An electronic filter which rejects a specific band of frequencies while passing those above and below the band.

bandwidth — The difference between the maximum and minimum frequencies in a band.

bank — To incline or tilt an airplane along its longitudinal axis.

bank indicator — A flight instrument consisting of a curved glass tube filled with a liquid similar to kerosene and enclosing a round glass ball. It is mounted in the instrument panel to indicate to the pilot the relationship between the force of gravity and centrifugal force in a turn. If the bank angle is correct for the rate of turn, the ball will stay in the center of the tube. However if the angle of bank is too steep for the rate of turn, the ball will roll to the inside of the turn. If the angle of bank is not steep enough for the rate of turn, the ball will roll to the outside of the turn. A bank indicator is built into the face of a turn and slip indicator.

bar — A metric unit of pressure equal to 1,000,000 dynes per square centimeter. Pressure is often measured in meteorological services in millibars, which is 1/1,000 of a bar. The standard absolute pressure of the atmosphere at sea level is 1013.2 millibars.

bar folder — A forming machine used for making bends or folds along edges of metal sheets. It is best suited for folding small hems, flanges, seams, and edges. Most bar folders have a capacity for metal up to 22 gage in thickness and 42″ in length.

bar graph – A type of graph that is used to show relationships between different values. Each value in a bar graph is represented by a bar of an appropriate length. By comparing the lengths of the bars we can easily compare values.

bare conductor – An electrical conductor that is not protected with any type of insulating material.

barnstormers – Early aviation pioneers who traveled as entertainers to small towns. Some were clowns and characters, others more serious showed a promise for the future in aircraft transportation. Barnstorming included acrobatics and stunts such as picking up handkerchiefs off the ground with hooks attached to the wing tips etc. Many of the major airports in the United States were started by barnstormers.

barographs – Barographs are absolute-pressure measuring instruments often sealed and carried in an aircraft in order to make a permanent record of the altitude that has been reached by the aircraft.

barometer – Abbrev.: bar. An instrument used to measure atmospheric pressure thereby forecasting weather or measuring the height above sea level.

barometric pressure – Pressure existing above zero pressure, or a vacuum, normally measured in inches of mercury.

barrel – The part of a reciprocating engine cylinder made of a steel alloy forging with the inner surface hardened to resist wear of the piston and the piston rings which bear against it during operation. In some instances the barrel will have threads on the outside surface at one end so that it can be screwed into the cylinder head.

barrel nut – An internally threaded screw having a slotted head.

barrel roll – An airplane flight maneuver in which the airplane rolls around its longitudinal axis.

base –
[1]The center electrode of a transistor. The signal is normally applied to the base.
[2]The electrode in a bipolar transistor that is between the emitter and the collector.
[3]A bipolar transistor electrode that normally receives the signal.

base line – A line used as a basis for measuring.

base metal – The metal to which alloying agents are added.

baseball stitch – Type of hand stitching similar to that used to sew the cover on a baseball. It is used for hand sewing of aircraft fabric.

Basic fuel system – The basic parts of a fuel system include tanks, booster pumps, lines, selector valves, strainers, engine-driven pumps, and pressure gages.

basic load – The load on a structural member or part in any condition of static equilibrium of an airplane. When a specific basic load is expressed, the particular condition of equilibrium must be indicated in the context.

basic magneto – A high-voltage generating device in a reciprocating engine. It is adjusted to give maximum voltage at the time the points break and ignition occurs. It must also be synchronized accurately to the firing position of the engine.

basic size – The basic size is that size from which the limits of size are derived by the application of allowances and tolerances.

bastard file – A double-cut, metal-working file that has course cutting teeth. There are five grades of cuts from the coarsest to the finest are: coarse cut, bastard cut, second cut, smooth cut, and dead-smooth cut.

bathtub capacitor – A type of bathtub shaped capacitor that is sealed in a metal container.

battery – A device made up of a number of individual cells, used to store electricity by converting it into chemical energy. Electrons are caused to flow from one pole, the anode, to another pole, the cathode, by a chemically produced potential difference.

battery analyzer – A transformer rectifier unit used to charge nickel-cadmium batteries. The analyzer has a built-in load bank, timers, indicators, and controls for deep-cycling and recharging these batteries.

battery bus – The electrical tie point in an airplane where power from the battery is distributed to the various loads.

battery charger – *See battery charger/analyzer.*

battery charger/analyzer – A special power supply which converts alternating current into direct current for charging batteries. This is a special device with timer, load bank, and monitoring equipment for complete servicing of aircraft batteries.

battery ignition system – A type of ignition system that uses a battery as its source of energy, rather that a magneto. This system is similar to that used in an automobile. A cam driven by the engine opens a set of points to interrupt current in a primary circuit. The resulting collapsed magnetic field induces a high voltage in the secondary which is directed by a distributor to the proper cylinder.

bauxite – A clay like substance that is the source of aluminum. To extract the aluminum, the bauxite is changed into alumina (aluminum oxide). Then the alumina is reduced to metallic aluminum by an electrolytic process.

bay – Any specific compartment in the body of an aircraft. It may also refer to a portion of a truss, or fuselage, between adjacent bulkheads. struts or frame positions.

bayonet – Something that is detachable and can be easily put on or removed.

bayonet exhaust pipe – The elongated and flattened end of the exhaust pipe of a reciprocating engine. Its design is such that the exhaust noises will be minimized and exhaust valve warpage will be prevented by maintaining a relatively constant temperature at the exhaust ports of the cylinders.

bayonet exhaust stack – *See bayonet exhaust pipe.*

bayonet gage – A term used for a dipstick type gage that is used to measure the quantity of a liquid such as oil or hydraulic fluid.

bayonet thermocouple – A type of thermocouple used to indicate engine temperature. The bayonet probe fits into an adapter that is screwed into the cylinder chosen for installing the thermocouple.

bayonet thermocouple probe – A pickup for cylinder head temperature that presses into an adapter screwed into the side of a cylinder head. Used for measuring cylinder head temperature on an air-cooled aircraft engine. *See also bayonet thermocouple.*

bead –

[1]A trough-like impression formed in a sheet metal member for the purpose of stiffening the member.

[2]A bulge at the end of a tube.

[3]A raised rounded ridge that is formed near the end of a piece of rigid tubing. A hose is slipped over the end of the tube, and the hose clamp is installed between the end of the hose and the bead. The bead keeps the tube from being pulled out of the hose.

bead heel – The outer bead edge which fits against the wheel flange.

bead seat area – The highly stressed portion of a wheel where the bead of the tire seats against the wheel.

bead seat area of a wheel – The flat surface of an aircraft wheel on which the bead of the tire seats.

bead thermistor – A component in a fire detection system that signals the presence of a fire or an overheat condition. The beads in the detector are wetted with a eutectic salt which possesses the characteristic of suddenly lowering its electrical resistance as the sensing element reaches its alarm temperature. The lowered resistance starts the fire-warning procedure by turning on the fire-warning light and sounding the fire-warning bell.

bead toe – The inner bead edge closest to the aircraft tire center line.

bead welds – The part of a weld that joins edges of metal parts that have been heated and melted together to form one solid piece when solidified. The bead sticks up above the surface of metal that has been welded. Usually some additional metal is added to the weld, in the form of a wire or rod, to build up the weld seam to a greater thickness than the base metal. The characteristics of a bead are height uniformity, smooth and uniform ripple on its surface, and evenly blends into the base metal.

beaded coaxial cable – A type of coaxial transmission line in which the inner conductor is centered in the outer conductor by a series of beads made of insulating material.

beads – Steel wires embedded in rubber and wrapped in fabric. The beads anchor the carcass plies and provide firm mounting surfaces on the wheel. *See also bead seat area.*

beam – A supporting structural member in any construction designed to withstand loads in both shear and bending.

beam – A radio signal that is sent continuously in one direction.

beam power tube – An electron tube which utilizes directed electron beams to add to its power-handling capability. Beam power tubes are power amplifier tubes, contrary to it being a voltage amplifier.

beam radio antenna – A directional radio transmitting antenna that radiates its energy in a narrow beam that can be directed in a definite direction.

bearing –

1 An angular measurement of direction between a known point and the airplane in flight.

2 A surface that supports and reduces friction between moving parts.

bearing burnishing – An aircraft engine run-in process which creates a highly polished surface on new bearings and bushings installed in during overhaul. The burnishing is usually accomplished during the first periods of the engine run-in at comparatively slow engine speeds.

bearing cage – A thin sheet-metal separator that holds the bearing rollers equally spaced around the raced. The cage should not contact either of the races.

bearing cone – The assembly which consists of a tapered, hardened steel, cone-shaped bearing race which fits over the axle, and the rollers and the cage which holds the rollers in position.

bearing cup – The steel race of a roller bearing which is shrunk into the bearing cavity of the wheel.

bearing degausser – A device which removes magnetism from bearings and other engine components.

bearing failure – The failure of a riveted joint in which the failure is caused by the sheets tearing at the rivet holes, rather than the rivets shearing.

bearing field detector – A device which detects magnetism in rotating engine components.

bearing friction – Friction caused by a bearing.

bearing heater tank – An oil bath heater used to expand a bearing so that it can be hand fitted over its shaft journal.

bearing navigation – The horizontal direction of one object in relation to another object.

bearing pressurizing – The process of increasing air pressure in the bearing pockets by admitting compressor air.

bearing race – The hardened steel surface upon which anti-friction bearings ride.

bearing rollers – Hardened steel rollers that support the wheel. They roll between a hardened steel cone-shaped race on the axle and a hardened steel race, the cup, inside the wheel.

bearing scratch detector – A hand held ball bearing tipped tool which is passed over bearing surfaces. If the ball finds a depression the bearing is usually rejected.

bearing seal – *See carbon seal or labyrinth seal.*

bearing stack – A group of thrust type bearings placed one on top of the other to form a stack. Primarily used to allow propeller blades to rotate in the hub under high centrifugal loads.

bearing strength – The force required to pull a rivet through the edge of the sheet or to elongate the hole. The bearing strength of a material is affected by both its thickness and by the size of the rivet.

bearing sump – The compartment housing the engine main bearings. formed by bearing seals on either side of the bearing on the shaft. Seals are used to control the inward leakage of gas path air.

bearing support – The inner hub of a major engine case, which is supported by struts and houses a main bearing.

bearing surface – A surface that supports and reduces friction between moving parts or a moving load. Bearing surfaces are ordinarily treated in several ways to decrease the friction between the surface and the moving load.

bearing wheel – Airplane wheels are of the tapered roller type and consist of a bearing cone, rollers with a retaining cage, and a bearing cup, or outer race. Each wheel has the bearing cup., or race, pressed into place and is often supplied with a hub cap to keep dirt out of the outside bearing. Retainers and felt seals are supplied inboard of the inner bearing to prevent grease from reaching the brake lining.

bearings – Any surface which supports, or is supported by, another surface.

beat – A low-frequency vibration that is produced when two sources of vibration act on the same object at the same time. In a multi-engine airplane, if two engines have slightly different RPMs, the airframe vibrations caused by these engines will produce a very noticeable beat. This beat is caused by the difference in the frequency of the two vibrations.

beat-frequency oscillator – Abbrev.: BFO. A variable frequency electronic oscillator designed to produce a signal frequency which is mixed with another frequency in order to develop an intermediate frequency or an audio frequency that can be heard.

Beaufort scale – In meteorology, a scale used to describe wind force, ranging from 1 to 12, 0 (zero) represents less than 1 MPH, and 12 represents speeds of more than 72 MPH, or hurricane force.

beef up – To strengthen and reinforce parts.

beehive spring – A hard steel retaining spring used to hold a rivet set in a pneumatic rivet gun, which prevents the set from being driven out of the gun. It derives its name, beehive, from its shape.

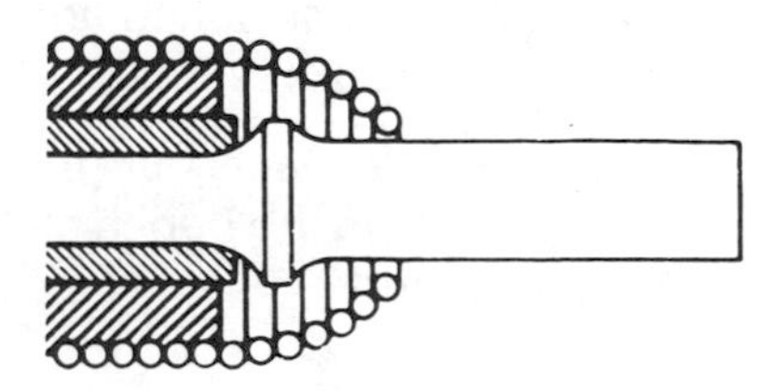

beep button — A switch on the collective control used to trim a helicopter turbine engine by increasing or decreasing the steady state RPM.

beeswax — A substance secreted by bees. It is used to coat rib lacing cord to protect it from moisture and prevent its slipping.

before bottom center — Abbrev.: BBC. The degree of crankshaft travel before the piston stops at the bottom of its stroke.

before top center — Abbrev.: BTC. The degree of crankshaft travel before the piston reaches the top of its stroke.

bel — A unit used to express the ratio of two values of power. The number of bels is the logarithm to the base 10 of the power ratio.

bell gear — The large stationary gear used in a spur gear-type planetary reduction gearing system.

bellcrank — A double lever in an aircraft control system used to change the direction of motion. Bellcranks are commonly used in an aileron system to change spanwise movement into chordwise movement to move the control surface.

Bellville washer — A special type of spring made in the form of a cupped steel washer. It has a great deal of compressive force, but a limited amount of travel.

bellmouth — A turbine engine air inlet duct having a flared, or a convergent shape used to direct air into a gas turbine engine. The shape of the bellmouth increases the efficiency of the incoming air to the engine.

bellows — Circular, pleated, or corrugated capsules or compartments used to measure pressure. They may be either evacuated or filled with a specific pressure of inert gas and exposed to the pressure to be measured. Their dimensional change is measured as the pressure surrounding them varies.

belt frame — A circumferential fuselage frame usually having a channel or hat section.

bench check — A functional check performed on a part which has been removed from an aircraft to determine its condition of serviceability. The equipment is set up on a test bench and operated to find out whether or not it is functioning as it should.

bench plate — A flat cast iron plate that is built into a bench that is used for working sheet metal. Holes in the bench plate support stakes that are used to form the sheet metal.

bench timing – A functional procedure to time a magneto by setting the breaker points, and for checking the rotor for the E-gap position.

bend allowance – The amount of material actually used in the bend of sheet metal. This amount of metal must be added to the overall length of the layout pattern to assure adequate metal for the bend. Bend allowance depends on four factors: the degree of bend, the radius of the bend, the thickness of the metal, and the type of metal used. The amount of material in the bend is usually found by using a bend allowance chart.

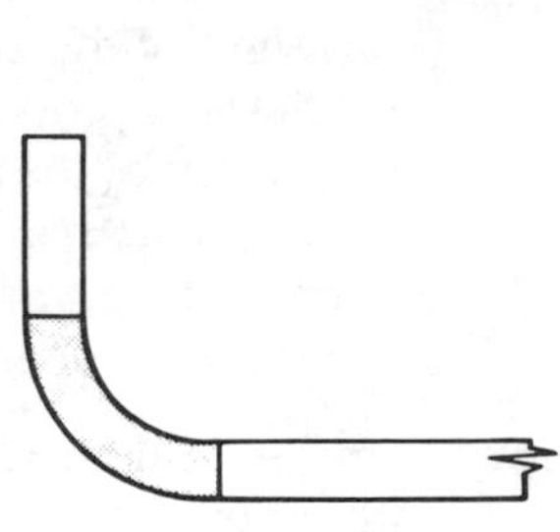

bend allowance chart – A chart used to save time in calculating the bend allowance. Formulas and charts for various angles, radii of bends, material thicknesses, and other factors have been established to make up the chart.

bend radius – The radius of the bend as measured on the inside of the curve.

bend tangent line – Abbrev.: BL. A line made on a sheet metal layout to indicate the beginning of the bend, and the line at which the metal stops curving. All the space between the bend tangent lines is the bend allowance.

bending – The stresses in an object caused by a load being applied to one end while the other is restrained. This results in a tensile load on one side and a compressive load on the other.

bending strength – The resistance of a material to curving under load and bending stresses.

Bendix fuel injection system – A continuous-flow fuel metering system which measures engine air consumption and uses airflow forces to control fuel flow to the engine.

benzene – A colorless, volatile, flammable, aromatic hydrocarbon liquid, (C_6H_6). Used as a solvent in aircraft finishing materials.

Bernoulli's principle – In physics, the interrelation between pressure, velocity, and gravitational effects in moving fluids. It states that for the steady flow of a frictionless and incompressible fluid, the total energy (consisting of the sum of the kinetic energy due to the velocity, the potential energy due to elevation in a gravitational field, and the pressure energy given by the pressure divided by the density) is a constant along the flow path. An increase in velocity at constant elevation must therefore be matched by a decrease in pressure. This principle is used to explain the lift of an airfoil, the theory of carburetors, etc.

beryllium – A hard metallic chemical element found in combination with other alloys.

beryllium bronze – An alloy of copper which is combined with approximately 3% beryllium.

best economy mixture – The fuel-air mixture used in reciprocating engines to achieve the greatest range of flight. It can only be used with reduced power, as it does not have the additional fuel needed for cooling.

best power mixture – The fuel-air mixture ratio used to allow the engine to produce its maximum power. The best power fuel-air mixture ratio, is richer than the ratio that is used for the best economy. It uses an excess of fuel to provide for cooling.

beta – The current gain of a transistor when it is connected as a grounded emitter amplifier. Beta is the ratio of the change in collector current to the change in base current when the collector voltage is held constant.

beta particle – A particle that is emitted from the nucleus of an atom during radioactive decay.

beta range – The pitch range of a turbopropeller system for ground handling and reversing whereby the propeller can be operated to provide either zero or negative thrust.

bevel – An angle other than a right angle.

bevel gears – A pair of toothed wheels having angled surfaces and whose shafts are not parallel. Bevel gears permit a shaft to drive another shaft that is not parallel to it.

bezel – The rim which holds the glass in an instrument case.

B-H curve – A curve that shows the association between the flux density (B) in a piece of magnetized material and the magnetizing force (H) that is needed to produce the flux density.

biannual – Occurring twice each year.

bias –
[1]A cut, fold, or seam made diagonally to the warp or fill threads of a fabric.
[2]An electrical reference used to establish the operating condition of a semiconductor device or an electron tube.

bias current – The current that flows in the emitter-base circuit of a transistor.

bias, forward – *See forward bias.*

bias, reverse – *See reverse bias.*

bias voltage – The DC voltage that is placed on the grid of an electron tube.

bias-cut surface tape – Aircraft covering surface tape that is cut at a 45° angle to the length of the tape from eselvage edge to selvage edge.

bicarbonate of soda – Common baking soda ($NaHCO_3$). It is used as a neutralizing agent for spilled battery acid.

bicycle gear – A landing gear that supports the main weight of the aircraft on wheels in line with each other along the length of the fuselage. The wings are supported by smaller outriggers near the wing tips.

bifilar – A system used for dampening rotor vibration that was developed by Sikorsky.

bifilar resistor – A type of resistor that is wound of wire which is doubled back on itself, in order to decrease the amount of inductance in the resistor.

bifilar transformer – A type of electrical transformer in which the primary and the secondary are wound side-by-side to increases the coefficient of coupling between the windings.

bifurcated duct – Split exhaust duct used on turbofan or lift fan engines.

bilge – The lowest part of an aircraft structure where water, dirt, and other debris accumulate.

bill of material – A list of the materials and parts necessary for the fabrication or assembly of a component or a system.

billet – An unfinished bar of metal less than 6″ × 6″ and approximately square.

bimetallic element – A device using two different metals joined together to produce either a mechanical bending or an electrical voltage as the temperature varies.

bimetallic strip – Two dissimilar metals such as chromel and constantan that are in close proximity to each other, used in fire detection systems.

binary number – A number in the binary number system which consists of only the two digits: zero and one.

binding post – A binding post is a special type of subassembly used for clamping or holding electrical conductors in a rigid position. It commonly consists of a screw having a collar head or body with one or more clamping screws.

binoculars – Field glasses. A form of hand-held optical instrument that is used to look at objects that are far away. Binoculars have a set of magnifying lenses for each eye. Prisms are used to get a high degree of magnification in a short physical length.

binomial – An algebraic expression that contains two terms connected by a plus or minus sign.

bioastronautics – The science that deals with the medical and biological aspects of astronautics.

biochemistry – Chemistry that deals with the chemical compounds and processes that are involved with living organisms.

biocidal action – The function of certain fuel additives which kill microbes and bacteria living in water in aircraft fuel tanks. This prevents scum which would promote corrosion in these tanks. Biocidal additives are also put in aircraft dope that is used on cotton or linen fabric to kill the bacteria that can destroy organic fabrics.

biocidal agent – A chemical combination that is destructive to certain types of living organisms.

biodegradable – A condition of a material that allows it to be broken down into simple products by the action of certain types of microorganisms.

biophysics – Interdisciplinary study of biological phenomena and problems using the principles and techniques of physics.

biplane – An airplane having two wings, one placed above the other.

bipolar transistor – The term used to describe either an NPN or a PNP transistor.

birch – One of several high grade woods used in the manufacturer of fixed-pitch wooden propellers.

bisector of a line – A position on a line that divides it into two segments of equal length.

bisector of an angle – A position on a line that divides an angle into two equal angles.

bismuth – A hard, brittle, grayish-white, trivalent, metallic chemical element that is also used as an alloying agent for changing the characteristic of certain metals. It is also used to dope silicon or germanium to make a P-type semiconductor material.

bistable – A condition that exists in a circuit in which either of two conditions may exist as a steady state.

bistable circuit – A circuit that has two stable conditions. Either condition may be chosen, and the circuit will operate in that condition until it is purposely changed.

bistable multivibrator – An oscillator circuit that uses two transistors of which only one transistor conducts at a time. When the first transistor stops conducting, the second transistor automatically starts to conduct.

bit – One unit of a binary number.

bitumen – An asphaltic residue which is the last of the products left in the fractional distillation of crude oil. Asphalt and tar are two commonly used bitumens.

bituminous paint – Heavy, thick, tar base paint used as an acid-resistant paint to reduce the corrosive action of the fumes and spilled electrolyte in battery compartments.

black box – A term used to describe any piece of electronic equipment that may be removed and replaced as a single unit.

black light – Ultraviolet light whose rays are in the lower end of the visible spectrum. While more or less invisible to the human eye, they excite, or make visible, certain materials such as fluorescent dyes.

bladder-type fuel cell – A neoprene impregnated fabric bag installed in a portion of the aircraft structure to form a cell and is used to hold fuel.

blade —
[1]A rotating airfoil used to create lift, as in compressor blades or to extract energy, as in turbine blades. Some manufacturers use "blade" for stationary airfoils. *See also vane.*
[2]A rotating airfoil shaped section used as a means of compressing air in a compressor or extracting energy from gases, as in a turbine.

blade alignment — An adjustment procedure, used on semirigid rotor systems, to place the blades in proper positions on the lead-lag axis of the rotor system. Blade alignment is sometimes called chordwise balance.

blade angle — The angle between the plane of propeller rotation and the face of the propeller blade.

blade angle check and adjusting — A method used to check the blade-angle setting at a predetermined blade station. The blade angle is checked using a device called a Universal Protractor.

blade antenna — A wide-band, quarter-wavelength antenna used on aircraft for communications or navigation in the ultra-high or very-high frequency bands.

blade back — The cambered side of a propeller blade that corresponds to the curved upper surface of an airfoil, similar to that of an aircraft wing. The opposite side of the blade face.

blade base — Same as the blade platform. The portion of the blade where the contoured section meets the root area.

blade beam — A paddle shaped lever having a slot shaped to fit the cross section of a propeller blade. Used for manually turning propeller blades. Blade beams are also called blade wrenches.

blade blending — A process used to remove small shallow scratches or dents of turbine blades using mild abrasive materials and sanding techniques. Bending requires maintaining the original contour and shape of the blade within prescribed limits.

blade butt — The root end of a propeller blade which fits into the hub of a propeller assembly.

blade chamber — The top or convex side of a rotating airfoil such as a compressor blade.

blade chord — The straight line between the blade leading and trailing edge.

blade chord line — An imaginary line drawn through the blade from the leading edge to the trailing edge.

blade climbing – A condition when one or more blades are not operating in the same plane of rotation during flight, which may not exist on ground operation.

blade coning – The acute angle between the helicopter rotor blade's spanwise axis and the plane of rotation. This is the result of lift versus gravity.

blade cross over – *See blade climbing.*

blade cuff – A metal, wood or plastic fairing that is installed around the shank of a propeller blade to carry the airfoil shape of the blade all of the way to the propeller hub. The airfoil shape of the cuff pulls cooling air into the engine nacelle.

blade dampener – A shock absorbing mechanism that is installed between a helicopter rotor blade and the hub to diminish or dampen blade movement on the lead-lag axis.

blade droop – The angle of the spanwise axis of the helicopter rotor at rest with only the forces of gravity acting on it.

blade face – The flat portion of a propeller blade, resembling the bottom portion of an airfoil. This is the portion of the blade that strikes the air first; thus its name.

blade fillet – The portion of the blade which is closest to the base or platform. Usually an area where the least damage is allowed.

blade flapping – Movement of helicopter rotor blades, about a horizontal hinge, in which the blades tend to rise and descend as they rotate. Blade flapping tends to minimize asymmetrical lift by increasing the angle of attack of the retreating blade while decreasing the angle of attack of the advancing blade.

blade grips – The part of a helicopter rotor hub into which the blades are attached by a lead-lag hinge pin. Blade grips are sometimes called blade forks.

Blade inspection method – Abbrev.: BIM. A system using an indicator and inert gas to detect rotor blade cracking. Used by Sikorsky Helicopter.

Blade inspection system – Abbrev.: BIS. A method used by Bell Helicopter to determine if rotor blades have cracked.

blade loading – The amount of weight each square foot of helicopter rotor blade supports. Blade loading is the ratio of the helicopter weight to the total area of the lifting area.

blade root — The portion of a propeller blade that fits into the propeller hub. The blade root is also called the blade butt.

blade section — A cross section of a propeller blade made at any point by a plane parallel to the axis of rotation of the propeller and tangent at the center of the section to an arc drawn with the axis of rotation as its center.

blade shank — The thick, rounded portion of a propeller blade near the hub.

blade span — The length of a blade from its tip to its root.

blade stall — A condition of the rotor blade in flight when it is operating at an angle of attack greater than the maximum angle of lift. This occurs at high forward speed to the retreating blade and to all blades during "settling with power".

blade station — Reference points on the blade measured in inches from the center of the propeller hub. Blade station measurements are used to identify locations along the blade of a propeller.

blade sweeping — An adjustment of the dynamic chordwise balance in which one or both blades are moved aft of the alignment point.

blade tabs — Fixed tabs mounted on the trailing edge of helicopter rotor blades for track adjustment.

blade tip — The part of a propeller blade that is the furthermost distance from the hub.

blade tracking —

[1]The process of determining the position of the tips of the propeller blades relative to each other.

[2]The mechanical procedure used to bring the blades of the rotor in satisfactory relationship with each other under dynamic conditions so that all blades rotate on a common plane.

blade twist — The variation in the angle of incidence of a blade between the root and the tip. The amount of thrust that is produced by a propeller blade is determined by the pitch angle of the blade at each blade station, and by the speed at which the blade is moving through the air. In order to maintain a constant amount of thrust along the blade, the blade angle must be twisted however, some degree of twist can also be caused by aerodynamic forces.

blade wrench — *See blade beam.*

blade-disk — A forged one-piece blade and disk as opposed to separate blades fitted into a disk. Either compressor or turbine.

blank — To cut out the surplus material from a part, prior to finishing.

blank blade — The identification of one blade of a helicopter during electronic balancing. It is the blade with the single intercepter.

blanket —
[1]A sheet of insulation material in the cabin and passenger compartments used to aid in suppressing noise.
[2]A shroud covering for airplane heat ducts.

blanket method of re-covering — The method of applying the fabric to an aircraft structure in which the fabric is wrapped around the structure and attached by sewing or by cementing it in place to either the trailing edge of a wing or to the longerons of a fuselage.

blanking — The process of removing metal by cutting small bits through the use of a tool with a stationary and a movable jaw.

bleed air — Compressed air removed from the compressor stages of a turbine engine by use of ducts and tubing. Bleed air can be used for deice, anti-ice, cabin pressurization, heating and cooling systems. *See also air bleed.*

bleed orifice — A calibrated orifice used to bleed down or adjust the pressure in a system.

bleed valve — In a turbine engine, a flapper valve, a popoff valve, or a bleed band designed to bleed off a portion of the compressor air to the atmosphere. Used to maintain blade angle of attack and provide stall-free engine acceleration and deceleration.

bleeder resistance — A permanently connected resistor connected across the output of a power supply and designed to "bleed off" a small portion of the current.

bleeder resistor — The resistor of a voltage divider through which the smallest amount of current flows. This resistor is generally closed so that the current through it is about 10% of the total circuit current.

bleeding —
[1]The act of removing air from a system.
[2]A maintenance procedure for purging the fuel system of air locks and to aid in flushing any traces of preservative oil from a pressure carburetor.
[3]A maintenance procedure in which air is removed from the hydraulic fluid that is in the brake system of an aircraft. If there is any air in the fluid in hydraulic brakes, the air will compress when the brakes are applied, and the brakes will feel "spongy", and their effectiveness will be reduced.

bleeding reds – Certain red pigments used in aircraft finishing materials which are soluble in the solvents used for their application. They will bleed up through any coat of finish put over them.

blemish – A defect or injury mark that damages an object or diminishes its value.

blending – A metal filing and stoning procedure to recontour damaged compressor and turbine blades back to an aerodynamic shape. A fine stone is used to smooth the reworked area into the original surface of the blade.

blimp – Early-day "B" class "limp", or non-rigid airships, i.e. no supporting structure for the gas bag.

blind flight – An early aviation term which was used before the current instrument flying term.

blind rivet – Special rivet designed to be used in sheet metal structure where it is not possible to use a bucking bar for riveting.

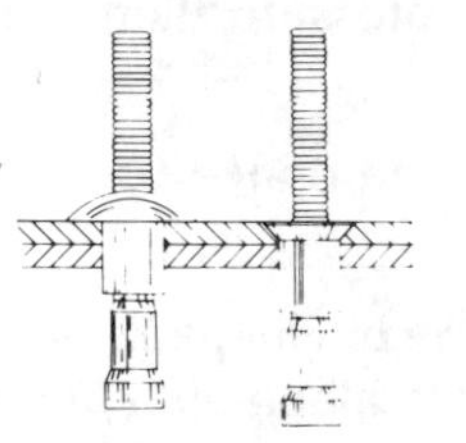

blind spot – Any area that is not visible.

blink Zyglo – A method of zyglo inspection wherein the part is cleaned and soaked with a fluorescent penetrant for an appropriate length of time. The part is rinsed and all of the penetrant is cleaned from its surface. The part is vibrated while it is being examined under a black light. If the vibration opens up a crack that has accepted some of the penetrant, the crack will show up as a blinking light.

blinker – Oxygen-flow indicator acting as a moveable shutter, opening and closing with each breath.

blister – An enclosed raised spot on the surface of a finish or a metal. It may be filled with vapor or with products of corrosion.

block – Securing and/or releasing of an airplane for flight. It includes ground crew personnel who aid the pilot in parking, mooring, or releasing by handling the wheelchocks, gear pins, etc. *See also block-to-block time.*

block diagram –
[1]A functional diagram of a system in which the units are represented by squares that describe the functions of the unit and show its relationship to the other units of the system. Arrows are used between the blocks to show the direction of the flow of energy or information within the system. Block diagrams do not show any of the actual components.
[2]Squares that describe the functions of units and show their relationship to other units of the system. Lines and arrows are used to connect the blocks to show the direction of the flow of energy or information within the system.

block heater – An electrical heater that is embedded in the die that is used for hot dimpling sheet metal.

block plane – A small hand-held carpenters tool used for smoothing the surface of wood.

block test – An operational test of an aircraft engine when the engine is installed in a test cell to determine the condition of the engine.

blocking capacitor – A capacitor that has a high impedance to DC and low frequency AC. However, it has a low impedance to the AC signal that is being passed through the circuit.

block-to-block time – The lapsed time between an airplane leaving the ramp for the purpose of flight and its returning after landing.

bloom – A bar of iron or steel ready for further working.

blow molding – A plastic molding process in which a hollow tube of thermoplastic material is heated inside a mold. Air pressure is applied to the inside of the tube, and the soft plastic material is forced out against the walls of the mold. The outside of the part takes the form of the inside of the mold.

blow-by, valve – *See valve blow-by.*

blowdown turbine – A form of power recovery device used on the Wright R-3350 engine that is driven by the exhaust gases from the engine, and coupled through a fluid coupling to the engine crankshaft. A blowdown turbine is also called a power-recovery turbine (PRT).

blower –
[1]Any mechanical device, such as a fan, that is used to move a column of air.
[2]An internal gear-driven supercharger in an aircraft reciprocating engine. Blowers are used to increase the pressure of the air after it has passed through the carburetor, and to improve the distribution of the fuel-air mixture to all of the cylinders evenly.

blower clutch – A unit in a two-speed supercharger system of a reciprocating engine, in which are single-stage, two-speed system can be driven at two different speeds by means of a clutch.

blower section – The blower section of an aircraft reciprocating engine crankcase houses the internal, gear-driven supercharger. *See also blower.*

blow-in doors – Spring-loaded doors located ahead of the first stage of the compressor. These doors are spring-loaded to hold them closed, but under conditions of low airspeed and high engine power, they open automatically to allow more air to enter the compressor. Blow-in doors help prevent compressor stall. Also known as auxiliary air-intake doors.

blown boundary layer control – A system used to decrease aerodynamic drag on the surface of a wing. Blown boundary layer control uses high-velocity air blown through ducts or jets to energize the boundary layer.

blow-out plug – A safety plug or disc on the outside skin of an aircraft fuselage near the installation of high pressure oxygen, CO_2, or other fire extinguisher agents. It is designed to rupture and discharge its contents overboard if, for any reason, the pressure of the gas in the cylinders rises to a dangerous value. Colored disks in the blow-out plugs identify the system that has been relieved in this manner.

blowtorch – A hand-held gasoline torch that uses compressed air to shoot out an intense hot flame used to melt metal or to burn off old paint from metal, etc.

blue arc – An instrument marking that indicates an operating range. For example, the blue arc might indicate the manifold pressure gage range in which an engine can be operated with the carburetor control set at automatic lean.

blueprint – A type of engineering drawing that is used to convey the construction or assembly of objects with the help of lines, notes, abbreviations, and symbols. Blueprints are made by placing a tracing of the drawing over a sheet of chemically treated paper and exposing it to a strong light for a short period of time. When the exposed paper is developed, it turns blue. The inked lines of the tracing now show as white lines on a blue background, thus its name blueprint. Blueprints have been replaced in many engineering departments by prints.

blush – The white or grayish cast which forms on a lacquer or dope film which has been applied under conditions of too high humidity. It is actually a nitrocellulose which has precipitated from the finish.

BMEP Indicator – An engine instrument which measures output shaft torque and converts it to brake mean effective pressure.

B-nut – A nut used in aircraft tubing used to connect a piece of flared tubing to a threaded fitting. B-nuts are used with a sleeve that is slipped over the tubing before the tubing is flared. The B-nut forces the sleeve tight against the flare which seals against the flare cone of the male fitting. The B-nut derives its nickname from its predecessor (no longer in use) which was called an A-nut.

board-foot – A unit of measurement commonly used to measure lumber. One board-foot is the amount of lumber in a piece of wood that is 1′ × 1′ × 1″.

bob weight – A mechanical weight in the elevator control system of some airplanes used to apply a nose-down force on the elevator control system. This force is counteracted by an aerodynamic force that is caused by the elevator trim tab. If the aircraft slows down enough that the aerodynamic force on the trim tab is lost, the bob weight forces the nose down, and the airplane picks up speed.

bogie landing gear – The landing gear of an aircraft that uses tandem wheels connected by a central strut. Aircraft having bogie landing gear are sometimes supported by outrigger wheels mounted far out on the wing when the aircraft is parked.

bogus parts – Parts that are not approved for use in aircraft. Bogus parts are often marked so as to appear to be authorized parts, but when they are installed in an airplane, its safety is compromised.

boiling point – Temperature at which a liquid changes to vapor. The boiling point of water, under standard conditions, is 212°F, or 100°C.

bolt — A bolt is an externally treaded fastener with an enlarged head on one end and threads on the other end.

bolt bosses — The enlarged portion of a casting or forging where the bolts pass through.

bomb tester — A spark plug tester in which the plug is exposed to approximately 200 PSI of air pressure. High voltage is applied to the center electrode cavity of the spark plug, and the electrodes are observed to see the type and amount of spark that is being produced. Plugs that can spark in this atmosphere are considered to be acceptable for use in the aircraft engine.

bond — An attachment of one material to another or of a finish to the metal or fabric.

bonded structure — A structure whose parts are joined together by chemical methods rather than mechanical fasteners. Honeycomb material, laminated fiberglass, and composite materials are examples of aircraft bonded structure.

Bonderizing — The registered trade name for a patented process of coating steel parts with a phosphate coating used to protect the parts from corrosion.

bonding —
[1] A procedure used in joining parts by using adhesives rather than any form of mechanical fastener.
[2] A method of electrically connecting all the components of an aircraft structure together so that static electricity cannot build up on one part of the structure to create a voltage that is high enough to allow it to jump to another part causing radio interference.

bonding agent — An adhesive used to bond structure parts together.

bonding jumper — A low-resistance wire or electrical connection used to electrically ground a component or structure to an airframe. Bonding jumpers carry the return current from an electrical component back to the battery. *See also bonding.*

bonnet assembly — Operating head of a fire extinguisher, which contains an electrically ignited powder charge used to rupture the disc and release the extinguishing agent.

boost —
[1] To assist.
[2] An older term synonymous with manifold pressure.

boost charge – A constant-voltage charge applied to low batteries installed in airplanes to restore their charge sufficiently to start the engine.

boost pump – An electrically driven fuel pump, usually of the centrifugal type and located in one of the fuel tanks. It is used to provide fuel to the engine for starting and providing fuel pressure in the event of failure of the engine driven pump. It also pressurizes the fuel lines to prevent vapor lock.

boost system – A hydraulically actuated system which aids the pilot in operation of the flight controls.

boost venturi – A small venturi whose discharge end is at the throat of the main venturi, and which surrounds the main discharge nozzle of a float-type carburetor. It increases the pressure drop for a given airflow.

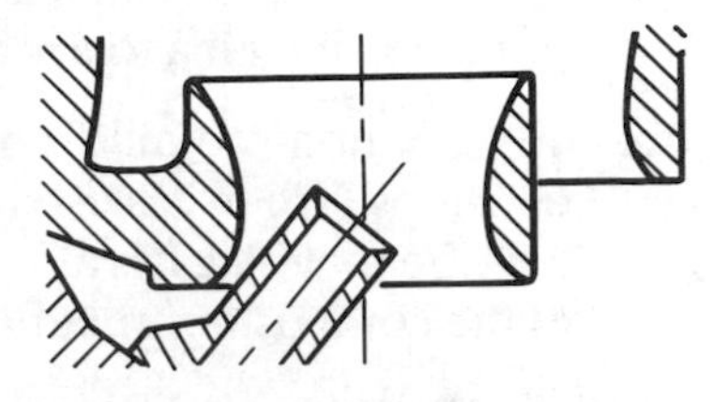

boosted brake – A form of brake power source using a master cylinder in which the hydraulic pressure from the aircraft hydraulic power system is used to aid the pilot in applying force to the master cylinder. This boost, or assistance, is automatically applied when the pressure required at the brake is greater than the pilot can produce with foot pressure alone.

boosted control system – *See boost system.*

booster coil – A transformer coil used with a vibrator to produce a high voltage at the spark plugs during starting.

booster magneto – A small high voltage magneto, usually turned by hand, used to produce a hot spark for starting reciprocating engines. The output for the booster magneto is fed into a trailing finger on the distributor, which fires the cylinder following the one in position for normal ignition.

boot – A telescoping type of rubber seal usually placed over a cable which passes through a bulkhead separating a pressurized and non-pressurized section.

bootstrapping –
[1]Technique with which something is brought into the desired state through its own action.
[2]Indication of unregulated power charge that results in continual drift of manifold pressure on a turbosupercharged engine; an undesirable cycle.

bore –
[1]Diameter of an engine cylinder.
[2]The internal diameter of a pipe, cylinder, or hole for shafting.

borescope – An optical tool with which a visual inspection can be made inside an area which is otherwise impossible to see. It consists of a light, mirrors, and lenses.

boric acid – A white crystal that can be dissolved in water to make a weak acid solution. Boric acid is used to neutralize spilled electrolyte from nickel-cadmium batteries.

boring – A process of increasing the size of a hole in a piece of material by cutting it with a rotary cutting tool.

boron – A non-metallic chemical element having three valence electrons. When it is used to dope silicon or germanium, it produces a P-type material. It is also used to add stiffness and strength to some of the composite structural materials use in modern aircraft.

boss – An enlarged or thickened part of a forging or casting to provide additional material for strength at its attaching point.

bottle bar – A special bucking bar recessed to hold a rivet set. It is used in reverse riveting.

bottled gas – Any of the gases kept under pressure (acetylene, propane, oxygen, and nitrogen) that are in heavy steel containers.

bottom dead center – Abbrev.: BDC. The crankshaft position when the piston is at the bottom extreme of its stroke, and the crank pin is below and directly in line with the wrist pin and the center of the crankshaft.

bottoming reamer – A reamer used to smooth and enlarge holes to exact size, but having no taper. A bottoming reamer completes the reaming of blind holes.

bottoming tap – A tap used to cut full threads at the bottom of a blind hole. The bottoming tap is not tapered. It is used after the hole has been partially tapped with a tapered tap.

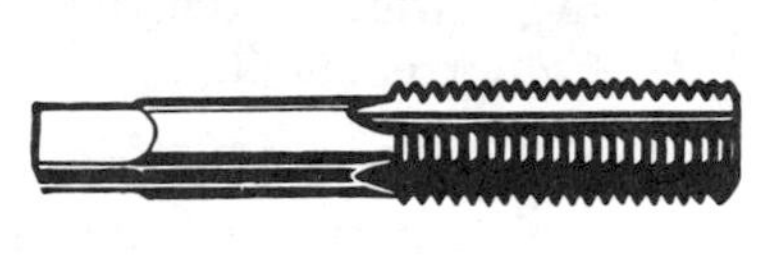

bounce – A condition where the breaker points of a magneto tend to bounce after they close. This is caused by a weak breaker spring.

boundary layer – The layer of air immediately adjacent to the surface of an airfoil. Its flow, rather than being laminar, is essentially random, or circulatory, and produces a great deal of aerodynamic drag.

boundary layer control – A method of removing random flowing air from the immediate surface of an airfoil caused by the turbulent flow of the boundary layer. Boundary layer control is obtained by either blowing it away or sucking or vacuuming it off through tiny holes inside the structure.

Bourdon tube – The mechanism in a pressure gage consisting of a flat or elliptical cross-sectioned tube bent into a curve or spiral. When pressure is applied, the tube attempts to straighten. The amount the tube straightens is proportional to the amount of pressure inside the tube, and as it straightens, it moves a pointer across the instrument dial.

bow or camber – The amount that a side of a surface deviates from being straight.

bow wave – A shock wave that forms immediately ahead of an aircraft that is flying at a speed faster than the speed of sound.

Bowden cable – A control system which uses a spring steel wire, enclosed inside a helically wound wire casing, used to transmit both pushing and pulling motion to the device that is being actuated.

bowline knot – A type of knot used to tie down an aircraft and to start the rib-stitching when attaching fabric to an aircraft structure. A properly tied bowline knot will not slip, and is easy to untie.

box brake – A metal-forming machine that is similar to a leaf (cornice) brake. It is used to form all four sides of a box by allowing the sides that have been bent up to fit between the fingers of the clamp while the last bends are being made. A box brake is also called a finger brake.

box spar – A type of design for wood spars in the shape of a box or a square. The top and bottom of the box are called the caps of the spar, and the sides of the box are called the webs of the spar.

box wrench – A wrench with an enclosed end that has six, eight, or twelve points. It can be used to tighten or loosen nuts and bolts, and can easily fit into close spaces and can be used to apply a greater amount of force than can be applied with an open-end wrench.

boxbeam wing – A type of wing construction made in the form of a box, that uses two main longitudinal members with connecting bulkheads to furnish additional contour and strength.

boxing of paint – A mixing procedure in which the paint is poured back and forth between two containers until the pigment and the vehicle are completely mixed.

Boyle's law – A gas law which states that at a constant temperature the volume of the gas will vary inversely as its temperature changes.

brace – A device that holds parts together, or in place. Something that gives support.

braced wing – A wing that requires external bracing; definitely not self supporting; used on some liaison aircraft. *See cantilever for contrasting definition.*

brad – A thin wire nail or spike that has a small-diameter, and a small barrel-shaped head.

braid – The rubber-coated, woven metal fabric reinforcing cord used to encase hydraulic flex hoses.

braided shield – A covering of braided metal over one or more insulated conductors to form shielded cable. This braid intercepts the magnetic field that is produced by the alternating current and keeps the field from causing radio interference. *See also shielded cable.*

brake –

[1]A device inside an aircraft wheel that is used to apply friction to the wheel to slow or stop its rotation. Wheel brakes slow the aircraft down during taxiing and landing. Types of brakes used on aircraft are in four general categories: shoe, expander tube, single disc, and multiple disc brakes.

[2]A metal-working shop tool that is used to make straight bends across sheets of metal. Brakes can be adjusted to make bends with the proper bend radius and bends that have the correct number of degrees. *See also box brake, cornice brake, finger brake, and press brake.*

brake back plate – A retainer plate to which the wheel cylinder and the brake shoes attach.

brake caliper – The clamp in a disc brake system which holds the brake linings. When pressure is applied to the brake, the calipers apply pressure to the linings to produce the braking action.

brake horsepower – The actual horsepower delivered to the propeller shaft of an aircraft engine.

brake line – Made on flat sheet stock which is set even with the nose of the radius bar of a cornice brake and serves as a guide in bending.

brake lining – A material with a high coefficient of friction and the ability to withstand large amounts of heat. It acts as the wearing surface in aircraft brakes.

brake mean-effective pressure – Abbrev.: BMEP. A computed value of the average pressure that exists in the cylinder of an engine during the power stroke. BMEP is measured in pounds per square inch, and is used to compute the amount of power the engine is developing.

brake or sight line – The mark on a flat sheet of metal which is set even with the nose of the radius bar of a cornice brake and serves as a guide in bending. The brake line can be located by measuring out one radius from the bend tangent line closest to the end which is to be inserted under the nose of the brake. *See also brake.*

brake specific fuel consumption: – Abbrev.: BSFC. The number of pounds of fuel burned per hour to produce one horsepower in a reciprocating engine.

brass – A metal alloy consisting essentially of 67% copper and 33% zinc.

Brayton cycle – The name given to the thermodynamic cycle of a gas turbine engine to produce thrust. This is a varying volume constant pressure cycle of events and is commonly called the constant-pressure cycle. A more recent term is continuous combustion cycle because of the four continuous and constant events which are the intake, compression, expansion (including power), and exhaust.

Brayton cycle of energy release – The constant pressure cycle of energy release that is used to describe the action of a gas turbine engine. The reason for this is that in the gas turbine engine pressure is fairly constant across the combustion section as volume increases and gas velocities increase.

brazier-head rivet – A form of aircraft rivet with a large thin head. Its specification was AN455. These rivets have been superseded by AN470 (MS 20470) universal head rivets.

brazing – Refers to a group of metal-joining processes in which the bonding material is a nonferrous metal or alloy with a melting point higher than 800°F. It is a method of joining two pieces of metal by wetting their surface with a molten alloy of copper, zinc and tin. The brazed joint has more strength than a soldered joint but less than a welded joint.

break line – A line used in drawings to indicate that a portion of the object is not shown.

break-away point – *See shear point.*

break-before-make switch – A type of double-throw switch that breaks one circuit before it makes contact with the other circuit.

breakdown voltage — Breakdown voltage in a capacitor is that voltage at which the dielectric is ruptured, or the voltage level in a gas tube at which the gas becomes ionized and starts to conduct.

breaker assembly — A mechanism used in high-tension magneto ignition systems, automatically open and close the primary circuit at the proper time in relation to piston position in the cylinder to which an ignition spark is being furnished. The interruption of the primary current flow is accomplished through a pair of breaker contact points. *See also breaker contact.*

breaker contact — A pair of electrical contacts that are opened and closed by a cam in the magneto for the purpose of timing the ignition of a reciprocating engine.

breaker point bounce — A condition caused by a weak breaker point spring. It is a fault in which the breaker points in an aircraft magneto bounce open rather than remaining closed when the cam follower moves off of the cam lobe.

breaker points — Interrupter contacts in the primary circuit of a magneto or battery ignition system. They are opened by a cam the instant the highest current flows in the primary circuit, thus producing the maximum rate of collapse of the primary field.

breakers — Extra layers of reinforcing nylon chord fabric are placed under the tread rubber to protect casing plies and strengthen tread area. Breakers are considered an integral part of the carcass construction.

breast drill — A type of drill designed to hold a larger size twist drill than the hand drill, and is used to drill relatively large holes in wood. A breast plate affixed at the upper end of the drill permits the use of body weight to increase the pressure on the drill.

breather — A vent in an engine oil system that keeps pressure within the tank the same as the atmospheric pressure. *See also engine breather.*

breather pressure system — The breather pressurizing system ensures a proper oil spray pattern from the main bearing oil jets and furnishes a pressure head to the scavenge system.

breather pressurizing valve — An aneroid-operated valve and a spring-loaded blowoff valve. Pressurization is provided by compressor air which leaks by the seals and enter the oil system. At sea level pressure the breather pressurizing valve is open. It closes gradually with increasing altitude and maintains an oil system pressure sufficient to assure jet oil flows similar to those at sea level.

breech chamber – A chamber in which a combustible material is ignited to produce pressure for starting small turbine engines.

bridge circuit – An electrical circuit that contains four impedances connected in series in such a way that their schematic diagram forms a square. One pair of diagonally opposite corners is connected to an input device, and the other two corners are connected to the output device. The bridge rectifier differs from the full wave rectifier in that a bridge rectifier does not require a center-tapped transformer, but does require two additional diodes.

bridge rectifier – An electrical rectifier circuit using four diodes arranged in a bridge circuit to change AC to DC.

brine – A solution of table salt (sodium chloride) and water, used as a quenching agent in the heat treatment of metal. Quenching steel parts in a brine solution are harder than parts that are quenched in either water or oil.

Brinell hardness test – A test used to determine the hardness of a metal by forcing a hardened steel sphere into the surface with a given force and holding it for a specified time. The diameter of the indention which is measured with a special microscope, is directly related to the hardness of the material.

brinelling – Indentations in bearing races usually caused by high static loads or application of force during installation or removal. They are usually rounded or spherical, due to the impressions left by contacting balls or the rollers of the bearing.

British thermal unit – Abbrev.: Btu. The amount of heat required to raise the temperature of one pound of water 1°F.

brittleness – The property of a metal to break when bent, deformed, or hammered. It is the resistance to change in the relative position of the molecules within the material.

broaching – The process of removing metal by pushing or pulling a cutting tool, called a broach, along the surface.

broken-line – A graph using sharp, abrupt changes in the information line.

broken-line graph – A type of graph that represents the way in which values change. The horizontal axis of the graph represents one value, and the vertical axis represents the another value that is changing. Straight lines are used to connect points that show true values at each plotted point.

Bromochlorodifluoromethane (Halon 1211) – Chemical formula ($CBrClF_2$). A liquefied gas with a UL toxicity rating of 5, used as a fire extinguishing agent. It is colorless, noncorrosive and evaporates rapidly leaving no residue whatever. It does not freeze or cause cold burns and will not harm fabrics, metals, or other materials it contacts. Halon 1211 acts rapidly on fires by producing a heavy blanketing mist that eliminates air from the fire, and interferes chemically with the combustion process.

Bromotrifluoromethane (Halon 1301) – Chemical formula (CF_3Br). A liquefied gas with a UL toxicity rating of 6, used as a fire extinguishing agent. It has all the characteristics of Halon 1211. The significant difference between the two is that Halon 1211 throws a spray similar to CO_2 while Halon 1301 has a vapor spray that is more difficult to direct.

bronze – An alloy basically of copper and tin. Used for bearing surfaces.

brush – A component device in an electric generator or motor designed to provide an electrical contact between a stationary conductor and a rotating element. Brushes are made of a carbon compound which contact each segment of a rotating commutator. The brush picks up the voltage generated by the rotating armature and directs it to an external circuit.

brush – A device composed of bristles, or hairs fastened to a wood or plastic handle. Brushes are used to apply paint or other substances to a surface.

brush guard – A protective device used to guard the tail rotor blades of a helicopter from damage during ground operations.

brushing – A motion that barely touches an object in passing.

bubble – A small space filled with a volume of air or gas entrapped in a liquid.

bubble octant – A celestial navigation instrument, like the sextant, that uses a bubble level in the octant to provide an artificial horizon that allows a navigator to find the angle between a line tangent to the earth's surface (the horizon) and a line to the stars that are used for navigation.

bucker – The person holding the bucking bar used to upset a rivet.

bucket – A slang term used for a turbine blade.

bucket root – A method of turbine disk blade retention in which the blade root has a stop made on one end of the root so that the blade can be inserted and removed in one direction only, while on the opposite end is a tang. This tang is bent to secure the blade in the disk.

bucket wheel – Slang for turbine wheel in turbine engines.

bucking – The coordinated process between the bucking bar holder and the pneumatic rivet gun operator. whereby a shop head is formed on a solid rivet. A special hardened steel bar is held against the rivet shank, which has been inserted into a hole drilled in metal to be joined, while the pneumatic hammer is held at the rivet head during the hammering.

bucking bar – A tool made of alloy steel stock which is held against the shank end of a rivet while the shop head is being formed.

buckle – A bend or kink in or on a surface of a metal structure. Caused by the failure of the part under a compressive load.

buckled areas – Localized areas in a turbine engine combustion chamber liner in which small areas have been heated to an extent to cause the area to buckle.

buffer –
[1]Any device used to absorb shock.
[2]Used to isolate an input or to strengthen a signal in a digital electronic circuit that has one input and one output, with the output having the same condition as the input.

buffer amplifier – An amplifier in a transmitter circuit designed to isolate the oscillator section from the power section thus preventing a frequency shift or otherwise operate improperly.

buffet – A series of waves or blows caused by imbalance such as can occur with flight controls. The aircraft feels as though it were hit with a series of blows, shocks, or waves.

buffeting – Erratic movement of aircraft controls caused by the turbulent flow of air over the surfaces.

bug – An unexpected malfunction.

bug light – A tool made up of an electrical wire, flashlight battery and bulb used to check the continuity of an electrical circuit.

build-up and vent valve – A manually operated valve on a liquid oxygen converter. In the build-up position, pressure is allowed to reach a preset value and excess pressure is vented to atmosphere. In the vent position, gas is vented to atmosphere without pressure buildup.

bulb angle – An extruded angle of metal with a rounded edge resembling a bulb on one of the legs.

bulb root – A means by which turbine engine rotor blades are attached to the rotor hub. The base of the blade is cylindrical and larger than the rest of the blade. This fits into a mated hole in the rotor hub.

bulb temperature – A unit of a carburetor induction system that monitors the air inlet temperature to be sure the inlet temperature does not exceed the maximum value specified by the engine manufacturer.

bulbed Cherrylock® rivet – A special form of blind rivet manufactured by the Cherry division of Townsend, Inc., in which the stem is locked into the hollow shank by a special locking collar which swages into a groove in the stem.

bulb-fit – A design of compressor blade attachment to the disk shaped in a bulb fashion.

bulkhead – A structural partition in a fuselage or wing. Bulkheads usually divide the fuselage or wing into bays, and provide additional strength to the structure.

bumping – The shaping or forming of sheet metal by hammering or pounding. During this process, the metal is supported by a dolly, a sandbag, or die.

bundled cable – Any number of individually insulated electrical wires tied together with lacing cord or with special plastic wire-wrapping straps.

bungee – A shock-absorber cord made from natural rubber strands encased in a braided cover of woven cotton cords treated to resist oxidation and wear.

bungee cord – An elastic cord encased in a braided cloth cover that holds and protects the rubber, yet allows the rubber to stretch. Bungee cords are used in some of the simpler aircraft landing gears to assist in retracting the landing gear and to absorb shock.

buoyancy – The upward pressure that is produced on an object when it is placed in a fluid.

burble – A breakdown of the laminar airflow over an airfoil caused by too high an angle of attack. The result is an increase in drag and a loss of lift.

burble point – The angle of attack at which burbling first occurs on an airfoil. *See also burble.*

burn down coat – A coat of dope with some of its thinner replaced with retarder, sprayed on a blushed area to attack and reflow the surface and remove the blush.

burned areas – Localized areas of a turbine engine combustion chamber liner that have been heated to an extent to cause visible damage.

burned surface – A condition resulting from high surface temperatures with relatively low pressures and accompanied by heat discoloration. This condition may or may not otherwise mark the surface. The cause is usually improper clearance or insufficient lubrication. Areas effected are bearings or journals.

burner – *See combustor.*

burner cans – Any number of individual combustion chambers in a combustion section of a gas turbine engine.

burner compartment – A section of the cowling behind which the burner section of a turbine engine is located.

burner pressure – Symbol: Pb. Static pressure signal used as a measure of mass airflow through the engine and sent to the fuel control unit for fuel scheduling purposes.

burning –
[1]Surface damage due to excessive heat. It is usually caused by improper fit, defective lubrication, or over temperature operation.
[2]The combustion process that occurs when fuel is mixed with air and ignited.

burning in – An electronic components manufacturing process in which the equipment is operated for a specified period of time in order to stabilize the operating characteristics of the components.

burning point – The lowest temperature at which a petroleum product in an open container will continue to burn when it is ignited by an open flame held near its surface. Among these are alcohol, benzol, kerosene, and gasoline.

burnish – To polish a metal surface by rubbing with a smooth, extremely hard tool, called a burnishing tool. a lubricant is usually required. Small scratches can be smoothed by burnishing.

burnishing – Polishing of one surface by sliding contact with a smooth, harder surface. Usually no displacement nor removal of metal.

burr – A sharp or roughened projection of metal usually resulting from machine processing.

burring – *See burr.*

burst RPM – The RPM at which the blades of a turbine motor will separate from the rotor due to excessive centrifugal loads.

bus — A main electrical power circuit to which a number of component circuits connect.

bus bar — An electrical power distribution point to which several circuits may be connected. It is often a solid metal strip having a number of terminals installed on it.

bushing — A removable cylindrical lining for an opening used to minimize resistance and serve as a guide.

butt fusion — A method of joining two pieces of thermoplastic material. Butt fusion is done by heating the ends of the two pieces until they are in a molten state and forcing them together before they cool and harden.

butt joint — A welded joint made by pacing two pieces of material edge to edge, so that there is no overlapping, making but a single thickness of metal, and then welded. The types of butt joints are: flanged, plain, single bevel, and double bevel.

butt rib — The last rib at the inboard end of an airfoil. The rib on a wing which is closest to the wing attachment fittings.

butterfly tail — A type of design which combines the vertical and horizontal surfaces of the empennage. The shape is that of a "V".

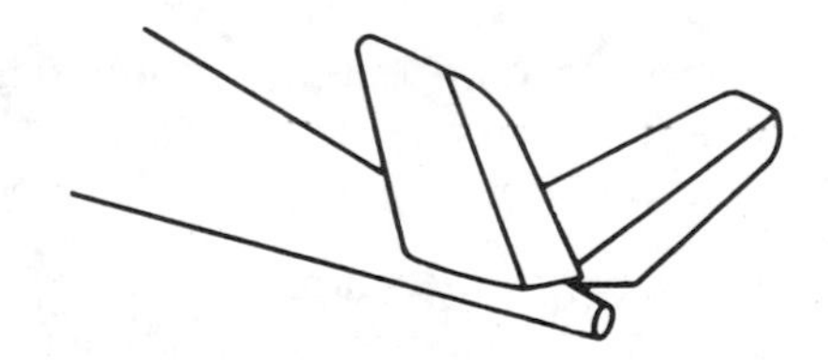

butterfly valve — A damper or valve consisting of a disk turning on its diametral axis to control the flow of fluid in a round tube.

buttock line — A measurement of width, left or right of, and parallel to, the vertical center line. Also called BL or butt line.

Butyl — The trade name of a synthetic rubber product made by the polymerization of isobutylene. It withstands such potent chemicals as Skydrol hydraulic fluid.

butyrate dope — A finish for aircraft fabric consisting of a film base of cellulose fibers dissolved in acetic and butyric acids, with the necessary plasticizers, solvents, and thinners.

Buys-Ballot's law – A law of meteorology that communicate to us that if we stand in the northern hemisphere with the wind striking us at our back, the center of the low-pressure area, around which the wind is blowing, is ahead of us, and to our left.

buzz – An airflow instability which occurs when a shock wave is alternately swallowed and regurgitated by the inlet. At its worst, the condition can cause violent fluctuations in pressure through the inlet, which may result in damage to the inlet structure or, possibly, to the engine itself.

bypass capacitor – A capacitor that provides a low-impedance path for alternating current to bypass a circuit component, when the component is being used to produce a DC voltage drop.

bypass duct – Cold airstream duct. Also called fan exhaust duct on a turbofan engine.

bypass jacket – An annular by-pass around an oil cooler through which oil flows when it does not need cooling.

bypass jet – A form of turbojet engine in which a portion of the compressor air is bypassed around the combustion chamber and into the tailpipe.

bypass ratio –
[1]The ratio of the mass airflow in pounds per second through the fan section of a turbofan engine to the mass airflow that passes through the gas generator portion of the engine.
[2]Ratio between fan mass airflow (lb/sec.) and core engine mass airflow lb/sec).

bypass turbojet engine – Forerunner of the bypass fan engine, whereby the low pressure compressor discharge is divided in two; one portion of air to enter a bypass duct and the other portion of air to enter the high pressure compressor inlet.

bypass relief valve – A pop-off valve used to relieve excessive pressures at oil filters, oil coolers, etc. Set to crack at a predetermined PSID.

bypass valve – A valve whose function is to maintain a constant system pressure. When the system pressure is exceeded by a predetermined amount, the valve will allow excess pressures to bypass the system thereby not allowing the system to rupture due to excess pressure.

byte – A computer term used to describe a group of binary digits consisting of eight bits.

C

C battery – A small, low-voltage battery.

cabane – An arrangement of struts used to support a wing above the fuselage of an airplane. Such an airplane wing attachment is called a parasol.

cabin – That portion of an aircraft used for passengers and/or cargo.

cabin altitude – Cabin pressure in terms of equivalent altitude above sea level.

cabin differential pressure – The difference between the pressure inside a cabin and the outside air pressure. The maximum cabin differential pressure is determined by the aircraft structural strength.

cabin pressure regulator – A means of controlling cabin pressure by regulating the outflow of air from the cabin.

cabin pressurization safety valve – A combination pressure and vacuum relief and dump valve used to prevent cabin pressurization exceeding safe limits.

cabin supercharger – Mechanical air pumps used to provide the air pressure for cabin pressurization.

cabinet file – A coarse file that is flat on one face, and is rounded on the other face, used for metalworking and woodworking. It is often called a half-round file.

cable –

[1]In common usage, any heavy conductor.

[2]In electronics, two or more conductive paths bound into a single package.

[3]A group of insulated electric conductors, usually covered with rubber or plastic to form a flexible transmission line.

[4]A stranded wire generally composed of a number of wires that are enclosed in single bundle or group.

cable control – The system of operating aircraft controls by the use of high-strength flexible steel cables.

cable drum – A cylindrically shaped spool around which a control cable is wound to increase the amount of cable that is moved each time the handle is turned.

cable guard — A pin installed in the flange of a control cable pulley bracket to prevent the cable jumping out of the pulley grooves.

cable rigging tension chart — Charts showing the relationship between control cable tension and temperature.

cadmium — A bluish-white, malleable, ductile, toxic, metallic chemical element. *See also cadmium plating.*

cadmium cell — A basic unit of the nickel-cadmium battery. It consists of positive and negative plates, separators, electrolyte, cell vent, and cell container. The positive plates are made from porous plaque on which nickel-hydroxide is deposited. The negative plats are made from similar plaques on which cadmium-hydroxide is deposited. The voltage that is produced by a cadmium cell at 20°C is 1.0186 v.

cadmium plating — A thin coating of cadmium metal electroplated on a steel part to protect the steel from corrosion. This is accomplished by the cadmium serving as the anode in a corrosive action.

caging device — A mechanism used in a gyroscopic instrument to erect the rotor of a gyro to its normal operating position prior to flight or after tumbling.

caging system — *See caging device.*

calcium carbide — A combination of calcium and carbon which is made to react chemically with water to produce acetylene gas.

calendar month — The measure of time used by the FAA for inspections and certification purposes. A calendar month ends at midnight of the last day of the month, regardless of the day it began.

calendering — Process of dipping cotton yarn or fabric into a hot solution of caustic soda to shrink the material and give it greater strength and luster.

calibrate — A procedure in which the indication of an instrument is compared with a standard value in order to fix, check, or correct the graduations of a measuring device.

calibrated airspeed — Indicated airspeed of an aircraft, corrected for position and instrument error. Calibrated airspeed is equal to true airspeed in standard atmosphere at sea level.

calibrated orifice — A hole with a specific internal diameter used to measure or control the amount of flow through it.

calibration — Testing the accuracy of a measuring instrument or scale by comparing it with a known standard.

calibration card – A card mounted on an instrument panel near an instrument to show the errors in an instrument in order for the pilot to apply an appropriate correction.

calibration curve, instrument – A curve on a graph that is plotted to show the instrument errors at different points on the scale of the instrument. The pilot uses the curve to interpolate the error at points between those that have been plotted.

callouts – Numbers or names used to identify components or parts in an aircraft drawing. Callouts are placed near the part that is being identified, connected by a thin leader line.

call-up – The act of calling on the radio from the aircraft to a facility, place or service.

calorie – The amount of heat required to raise the temperature of one gram of water 1°C.

calorimeter – An apparatus for measuring specific heat.

cam – An eccentric plate or shaft used to impart motion to a follower riding on its surface or edge.

cam dwell – The cam dwell is the number of degrees the cam rotates between the time the breaker points close and the time they open.

cam lobe – An eccentric used to change rotary motion into linear motion. For example, the cam followers that operate valves ride on the cam lobes and as the cam shaft rotates, the cam lobes move the cam followers up and down in a direction that is perpendicular to the axis of the cam shaft.

cam nose – The peak, or the highest point on a cam that pushes up on the cam follower.

cam pawl – A special device that allows a wheel or gear to turn in one direction but prevents its turning in the opposite direction.

cam plate – A radial-engine driven disc or plate with lobes machined onto its circumference. Cam followers ride on the lobes and open the engine valves through a system of push rods and rocker arms.

cam ring – An open cam plate driven by teeth around its circumference.

camber –

[1]The curvature of an airfoil above and below the chord line surface. An airfoil is often described as having an upper and lower camber.

[2]The mean camber of an airfoil section is a line that is drawn through a series of points that are located midway between the upper and the lower camber.

[3]The amount of angle the wheels of an aircraft are from the vertical. If the wheel tilts outward, the camber is positive, and if it tilts inward, the camber is negative.

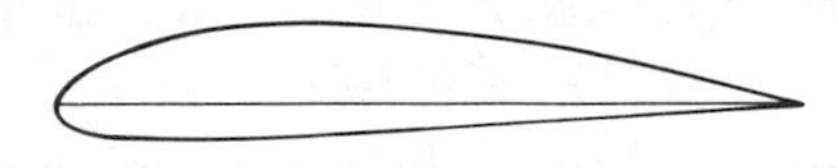

cambric – A finely woven cotton or linen material.

cam-ground piston – An aircraft engine piston ground in such a way that its diameter parallel to the wrist pin boss is less than its diameter perpendicular to the boss. When the piston reaches its operating temperature, the difference in mass has caused the piston to expand to a perfect circular form.

camlock fastener – A patented cowling fastener in which a hard steel pin is turned in a special cam-shaped receptacle.

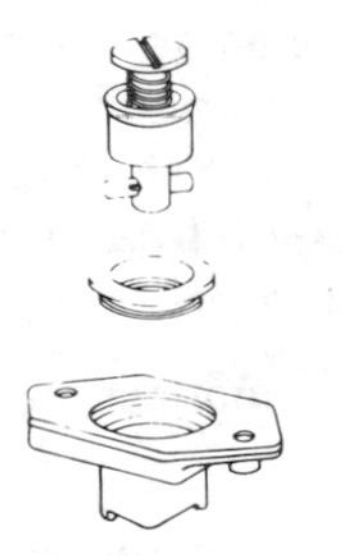

cams compensated – A cam which is designed to operate with a specific engine and has one lobe for each cylinder to be fired by the magneto. The cam lobes are machine ground at unequal intervals to compensate for the top-dead-center variations of each piston position. *See also compensated cam.*

camshaft – A long shaft running parallel to the crankshaft of an inline or horizontally opposed reciprocating engine. Lobes are ground at intervals along its length to operate the valves through push rods and rocker arms.

can or multichamber – *See can-annular combustor; can-type combustor.*

can tap valve – A valve fastened onto a small can of Freon refrigerant. It punctures the can seal and controls the flow of refrigerant.

can-annular combustor – Can annular combustion chambers arranged radially around the axis of the engine. The combustion chambers are enclosed in a movable steel shroud, which covers the entire burner section; designed for more complete cooling and mixing of fuel and air. The burners are interconnected by projecting flame tubes which facilitate the engine-starting process.

canard – An aircraft with its horizontal stabilizing and control surfaces in front of the wings.

candela – A unit of incandescent intensity.

candlepower – The luminous intensity of light produced by a candle expressed by the number of candles.

cannibalize – To remove serviceable parts from a non-flying aircraft for use on another machine.

cannon connector – A connector used to attach the battery to the aircraft electrical system. It is a high-current connector in which the cables are held onto the battery by pressure applied by a hand screw.

canopy –

1. A transparent cover for the airplane's cockpit. It provides streamlining and protection for the pilot against the elements. Sometimes called cockpit canopy.
2. The large cloth umbrella shaped body of a parachute.

canted bulkhead – A bulkhead or wall that is not vertically straight.

cantilever – A beam or other member that is supported at or near one end only, without external bracing.

cantilever wing – A cantilever wing uses no external wing struts in which all of its support is obtained inside the wing itself.

can-type combustor – A combustor, or burner section of a gas turbine engine that is made up of eight to ten individual burner cans. These cans are long cylinders that consist of an outer case or housing within which there is a perforated stainless steel combustion chamber liner or inner liner. The can-type combustors are arranged radially around the axis of the engine. Compressed air from the compressor flows through the cans and fuel is sprayed into them and burned to add energy to the air. Cooling air flows through holes in the inner liners to keep the temperature of the liners low enough that they will not be damaged.

canvas – A heavy, woven coarse cloth that is generally made of cotton.

cap – The longitudinal members at the top and bottom of a beam which resist most of the bending loads by their strength in compression and tension.

cap screw – A type of threaded fastener. The head of the cap screw, which is turned with a wrench, clamps two pieces of material together.

cap strip – Cap strips are extrusions, formed angles, or milled sections to which the web is attached. These members carry the loads caused by the wing bending and also provide a foundation for attaching the skin.

capacitance – The ability of an insulator to store electrical energy in the form of electrostatic fields, expressed in farads. The amount of electricity a capacitor can store depends on several factors, including the type of material of the dielectric. It is directly proportional to the plate area and inversely proportional to the distance between the plates. The formula for capacitance is $C = Q/E$, in which C is the capacitance in farads, Q is the quantity (amount) of charge in coulombs, and E is the electrical pressure in volts.

capacitance bridge – A null-type precision measuring instrument used to measure capacitance.

capacitance-type fuel gauging system – A fuel quantity indicating system using the fuel as the dielectric of a capacitor. It actually measures the weight of the fuel instead of its volume.

capacitive electrical load – An electrical load that produces more capacitive reactance than there is inductive reactance in the circuit.

capacitive reactance – Abbrev.: Xc. The opposition to the flow of alternating current electricity caused by the capacitance in a circuit, and is measured in ohms. Capacitive reactance is evaluated by the capacitance of the circuit, and by the frequency of the AC.

capacitive time constant – The amount of time, determined by the resistance of the circuit, and by the capacitance of the capacitor measured in seconds, that is needed for the voltage across a capacitor to rise to 63% of the source voltage.

capacitor – A device used to store electrical energy in the form of electrostatic fields. A capacitor is essentially two conductors separated by an insulator.

capacitor-discharge ignition system – A high-energy ignition system for turbine engines in which a large quantity of electrons are stored in a capacitor and are released to provide a spark of exceptional intensity.

capacitor-input filter –
[1]A network consisting of a capacitor and inductor, used to smooth the ripple output of a rectifier.
[2]A form of electronic filter used to smooth out the pulsations in the output of an electrical power supply. A capacitor-input filter is installed in parallel with the rectifier output, and an inductor is installed in series with the rectifier output.

capacitor-start induction motor – An AC motor whose rotor is excited by voltage induced from the field windings. A second winding whose phase is shifted by a capacitor is used to provide a rotating field for starting. When the motor gets up to speed, a centrifugal switch opens the circuit in which the capacitor is situated.

capacitor-type ignition system – A form of gas turbine ignition system consisting of two identical independent ignition units operating from a common low-voltage DC electrical power source, the aircraft battery. A high voltage, supplied by the ignition exciter unit, charges a storage capacitor with a charge, up to 4 joules, which generates an arc across a wide igniter spark gap to ignite the fuel for engine starting.

capacity –
[1]The ability to hold or contain in a holding space.
[2]In electricity, the quantity of electrons that can be stored in a capacitor. This depends on the area of the plates and the thickness and dielectric constant of the insulator.

cape chisel – A cold chisel used when cutting square corners or slots.

capillary action – An action causing a liquid to be drawn up into extremely tiny tubes or between close-fitting parts.

capillary tube – A tube with a very small bore, used to transmit pressure of fluid or gas to an indicating gage, or to meter a fluid.

capstan – A spool shaped device in the control system of an aircraft similar to a grooved drum-like wheel. A control cable is wound around the capstan, and the ends of the guide are attached to the aileron, the elevator, or the rudder control cable.

capstan screw – A special purpose machine screw with holes across the head to accommodate a bar that can be passed through these holes and used to turn the screw.

captive balloon – Aerial observation platforms anchored to the ground used during early American wars.

captive screw – A screw whose section of threads on its shank have been cut away. A captive screw has the ability to turn in the body in which it is mounted, but it will not drop out when it is unscrewed from the part it is holding.

carbide drill – A specially manufactured cutting drill that has its cutting edges surfaced with tungsten carbide, tantalum carbide, or titanium carbide.

carbide tool – A metal-cutting machine tool whose cutting faces have been surfaced with either tungsten carbide, tantalum carbide, or titanium carbide.

carbo-blast – A grit blast. Field cleaning agent, a lignocellulose material consisting of ground up walnut shells and apricot pits. *See also field cleaning.*

Carboloy – The name of certain cutting tools and dies having tungsten carbide bonded to their cutting surface.

carbon – Nonmetallic element which is a part of all organic compounds. It ranges in appearance from black, fluffy soot, to hard, transparent diamond.

carbon arc – An electric arc, produced by a welding machine, that jumps between two carbon electrodes, or from a carbon electrode to a metal electrode. A carbon arc makes an intensely bright light and it produces enough heat (approximately 10,000°F) to melt metals for welding or cutting.

carbon arc lamp – A type of electrical lamp in which an electric arc produces a high-intensity light.

carbon black – A soft and fluffy carbon deposit. Carbon black is produced by the incomplete burning of acetylene gas when the flame does not have enough oxygen for complete combustion.

carbon brake – One of the newest developments in aircraft brakes required for extremely high energy dissipation. Both the rotating and stationary discs are made of pure carbon.

carbon composition resistor – The most widely used electronic component today. It is a resistor formed by embedding wire leads in a cylindrical slug of carbon and filler material. The whole is then usually covered with an epoxy or other plastic insulating jacket.

carbon deposits – Residue from overheated oil or incompletely burned gasoline. It forms as a hard, black crust inside the engine.

carbon dioxide – Abbrev.: CO_2. A colorless, odorless, nonflammable gas often used as a fire extinguishing agent in aircraft.

carbon fouling – A carbon deposit soot that forms as a result of overly rich, idle fuel/air mixtures. The carbon settles on the inside of combustion chambers and spark plugs because the heat of the engine and the turbulence in the combustion chamber are slight.

carbon knock – The preignition of the fuel/air charge inside the cylinder of a reciprocating engine before the engine is ready for ignition to occur.

carbon microphone – A microphone used in telephones and some types of radio transmitters. It consists of a flexible diaphragm of carbon granules that is acted on by sound waves pressing against it thereby changing its resistance.

carbon monoxide – Abbrev.: CO. A colorless, odorless, highly toxic gas that forms from incomplete combustion of a hydrocarbon fuel.

carbon monoxide detector – A device used to detect the presence and concentration of carbon monoxide gas.

carbon oil seal – A type of oil seal used in gas turbine engines. These seals are usually spring loaded and are similar in material and application to the carbon brushes used in electrical motors. Carbon seals rest against a surface provided to create a sealed bearing cavity or void; thus, the oil is prevented from leaking out along the shaft into the compressor airflow or the turbine sections.

carbon pile resistor – A type of variable resistor that is used in some electrical equipment. A carbon pile is made of a stack of thin, pure carbon disks. Its resistance is changed by varying the amount of pressure that is held on the stack.

carbon pile voltage regulator – A voltage regulator that depends on the resistance of a number of carbon disk arranged in a pile or stack. The resistance of the carbon stack varies inversely with the pressure applied. When the stack is compressed under appreciable pressure, the resistance in the stack is less. Pressure on the carbon pile depends upon two opposing forces: a spring and an electromagnet. The spring compresses the carbon pile, and the electromagnet exerts a pull which decreases the pressure. When the generator voltage varies, the pull of the electromagnet varies thereby increasing or decreasing the pressure on the disks. This change allows a change in the generator output voltage.

carbon resistor – An electrical component that is used to put a controlled amount of resistance in an electrical circuit. Carbon resistors are composed of a rod of compressed graphite and binding material, with wire leads called "pigtail" leads, attached to each end of the resistor. Colored bands marked on the resistor indicates its resistance value.

carbon seal –
[1]A heat-resistant device used in turbine engines to seal the lubricating oil in the bearing cavity.
[2]A ring of carbon material which rides on a highly polished metal surface used to prevent lubricating oil from seeping into the gas path. Located especially at main bearing locations.

carbon steel – A group of iron alloys having carbon as the principal alloying agent. Low-carbon steel contains less than 0.20% carbon and is not as strong as high-carbon steel that contains up to 0.95% carbon where greater strength is necessary.

carbon tetrachloride (Halon 104) – Once used as a fire extinguishing agent. A chemical formula CCl4, a liquid with a UL toxicity rating of 3. When used as a fire extinguishing agent it becomes very toxic and harmful to humans and other animals.

carbon tracking – A fine track of carbon which is deposited inside the magneto, distributor or in the terminal cavity of a spark plug as a result of a flashover. It acts as an electrical conductor to ground, or to another electrical lead.

carbonaceous – Containing carbon.

carbon-film resistor – An electrical resistor that is composed of a thin film of carbon on a ceramic cylinder. Wires connected to each end of the carbon film allows the resistor to be connected to an electrical circuit.

carbon-zinc cell – A portable primary cell consisting of a carbon rod placed in a can made of zinc filled with a paste of ammonium chloride. The chemical reaction between the paste and zinc causes electrons to leave the zinc can and travel through an external circuit to the carbon rod. The common flashlight battery is of this type and produces 1.5 volts.

Carborundum – A manufactured aluminum oxide abrasive similar to natural emery. It is used for grinding wheels and for abrasive papers.

carburetor –
[1]Pressure: A hydromechanical device employing a closed feed system from the fuel pump to the discharge nozzle. It meters fuel through fixed jets according to the mass airflow through the throttle body and discharges it under a positive pressure. Pressure carburetors are distinctly different from float-type carburetors, as they do not incorporate a vented float chamber or suction pickup from a discharge nozzle located in the venturi tube.
[2]Float-type: Consists essentially of a main air passage through which the engine draws its supply of air, a mechanism to control the quantity of fuel discharged in relation to the flow of air, and a means of regulating the quantity of fuel/air mixture delivered to the engine cylinders.

carburetor air temperature – The temperature of the induction air before it enters the carburetor. Controlling the temperature of the air as it enters the carburetor keeps the fuel/air mixture temperature high enough to prevent water condensing out of the air and freezing and low enough to prevent detonation.

carburetor float – A float mechanism within a float chamber is provided between the fuel supply and the metering system of a carburetor. The float chamber provides a nearly constant level of fuel to the main discharge nozzle. The float is connected to a needle valve and seat which meters the correct amount of fuel to the induction system according to the demand.

carburetor heater – A heater muff or jacket installed around the exhaust manifold through which induction air is drawn to warm it before it enters the carburetor. This heat prevents the formation of carburetor ice.

carburetor ice – Ice which forms inside the carburetor due to the temperature drop caused by the vaporization of the fuel. Induction system icing is an operational hazard because it can cut off the flow of the fuel/air charge or vary the fuel/air ratio.

carburetor maintenance – A form of maintenance that may include idle speed adjustment, removal and installation, adjusting idle mixtures, rigging, and inspection.

carburizing – A form of case hardening of steel in which carbon is infused into the surface of the steel to increase its hardness.

carburizing flame – A flame used in oxyacetylene welding in which there is an excess of acetylene gas. Also called reducing flame. This type of flame introduces carbon into the steel. It can be recognized by the greenish-white brushlike second cone at the tip of the first cone. The outer flame is slightly luminous and has about the same appearance as an acetylene flame burning freely in air alone.

cardinal headings – Headings along the four main points of a compass: North, South, East, and West.

cardioid microphone – A microphone with the ability to pick up sounds ahead of it, rejecting sounds behind it.

cargo – Freight that is transported in an airplane.

cargo aircraft – An airplane whose main function is to carry freight.

carrier frequency – The high frequency alternating current that produces the electromagnetic waves that radiate from a radio transmitting antenna.

carrier wave – High-frequency alternating current which can be modulated to carry intelligence by propagation as a radio wave.

cartridge fuse – A type of fuse used to protect an electrical circuit from an excess of current. It consist of a fusible link held between metal rings, or caps, that screw onto each end of an insulating tube.

cartridge starter – A starting device for reciprocating engines, using solid fuel pellets which are electrically ignited. The pressure is used to move a piston to start the engine rotating.

cartridge-pneumatic starter – A combination air-turbine starter and cartridge starter. It can be operated by bleed air or by an explosive charge, both of which exhaust through a turbine wheel connected to a reduction gearbox. Its purpose is to start main engines.

cartridge-type filter – A disposable filter element of paper, cellulose or the like for both fuel and oil systems.

cascade electrical circuits – A system of connecting multiple levels of electrical circuits so that the output of one level feeds the input of the next level.

cascade thrust reverser – *See aerodynamic blockage thrust reverser.*

cascade transformer — A device that can be used in an electrical circuit to get a high voltage. A system of connecting multiple levels of electrical step up transformers so that the output of one level steps up the next level transformer cascading until the required high-output voltage is obtained.

cascade vane — An air turning vane. One common use is in thrust reversers.

case hardening — A form of heat treatment of a metal in which the surface is made extremely hard and brittle while the core of the metal retains its toughness.

case pressure — A low pressure maintained inside the case of a hydraulic pump. In the event of a damaged seal fluid will be forced out of the pump rather than allowing air to be drawn in.

casein glue — A form of powdered glue made from milk. Casein glues are widely used in wood aircraft repair work. For aircraft use, casein glues should contain suitable preservatives such as the chlorinated phenols and their sodium salts, to increase their resistance to organic deterioration under high humidity exposures.

casing — The rubber and fabric body of a pneumatic tire. The casing is the same as the carcass of the tire. It is composed of diagonal layers of rubber-coated fabric cord (running opposite angles to one another), providing the strength of a tire.

casing plies, aircraft tires — Diagonal layers of rubber-coated nylon cord fabric (running at opposite angles to one another) provide the strength of a tire.

cast iron — Iron that contains 6-8% carbon and silicon. Cast iron is a hard unmalleable pig iron made by casting.

cast-aluminum alloy — Aluminum alloy which has been heated to its molten state and poured into a mold to give it the desired shape.

castellated nut — *See castle nuts.*

casting — Objects formed by pouring molten metal into molds. castings usually have less strength than forgings or extrusions.

castle nuts — Commonly used general purpose hexagonal nuts for aircraft or engine use. They are shaped to resemble a castle. with the slots between the "turrets" for locking the nut to the bolt with cotter pins.

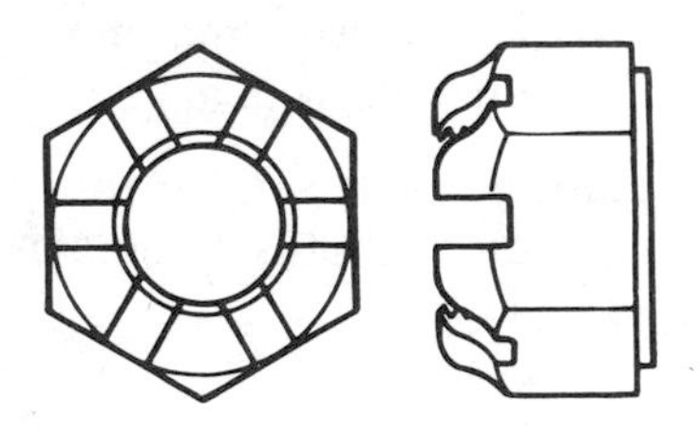

catwalk – A narrow path used for walking.

catalyst – A material which is used to bring about a change, or speeds up the rate of change of a chemical action, but does not actually enter into the change itself.

catalytic cracking – A method of refining petroleum products in which catalytic cracking is used to change high-boiling-point hydrocarbons into low-boiling-point hydrocarbons.

catalyzed material – A material whose cure is initiated by the addition of a catalyst.

catapult – A mechanism that is used to launch an object into the air. Catapults are used to launch heavily-loaded aircraft from the decks of aircraft carriers at a high rate of speed.

category – With respect to the certification of aircraft, this is a grouping of aircraft based upon intended use or operating limitations. Examples include: transport, normal, utility, acrobatic, limited, restricted, and provisional.

category II operation – With respect to the operation of aircraft, this is a straight-in ILS approach to the runway of an airport, under a category II ILS instrument approach procedure issued by the Administrator or other appropriate authority.

catenary curve – A curve that is formed by a flexible cord, or rope, that is suspended between two points at the same level.

catenary thermal shield – A curved sheet metal section between turbine wheels of a particular set. It serves as a heat barrier between the gas path and the inner portion of a drum type turbine wheel.

cathedral – The downslope of the wings from the fuselage. It is the opposite of dihedral. Airplanes that imploy cathedral have an increase in maneuverability but a decrease in stability.

cathode –

[1]The negative terminal of a semiconductor diode or the element in an electron tube, from which the electrons are emitted.

[2]An active element in an electrochemical cell that loses oxygen in the chemical action which causes electrons to flow.

cathode of a semiconductor diode – That end of a semiconductor diode that is made of N-type material.

cathode protection – Another name for sacrificial corrosion. A material more anodic than the material being protected is attached to or plated on the material, which then becomes the cathode and is not corroded.

cathode ray oscilloscope – An electrical measuring instrument in which the readout is on the surface of a tube similar to that in a television set. Electrons are made to strike the inside of the tube where they cause the coating of the tube to glow. Recurring voltage changes are displayed on this tube in the form of a green line.

cathode-ray tube – Abbrev.: CRT. A special type of electron tube in which a stream of electrons (cathode rays) from an electron gun impinges upon a fluorescent screen, thus producing a bright spot on the screen. The electron beam is deflected electrically or magnetically to produce patterns on the screen.

cation – A positive charged ion that moves toward the cathode in the process of electrolysis.

caustic material – Any substance having the ability of burning, corroding, or eroding other substances by chemical action.

caustic soda – A common name for sodium hydroxide.

cavitation – Cavitation is a partial vacuum of an area of low pressure behind an object that is moving in a fluid.

cavity – A hole or hollow place within a body or structure.

C-clamp – A metal clamp in the general shape of the letter C. It is used to exert pressure and to temporarily hold objects together.

c-d inlet or exhaust – *See convergent-divergent.*

Ceconite – A fabric woven from polyester fibers.

ceiling – The height above the ground of the base of the clouds.

ceiling balloon — A small, black, helium-filled balloon that is used to find the height of the bottom of a ceiling of clouds.

ceiling light — A light that is used by weather observers to measure the height of the bottom of a layer of clouds at night.

cellular combustor — *See can-type combustor.*

celluloid — The registered trade name of a thermoplastic material consisting essentially of cellulose nitrate and camphor.

cellulose — A material that is obtained from natural fibrous plants such as wood.

cellulose acetate butyrate — A compound formed by the action of acetic and butyric acid on cellulose.

cellulose acetate butyrate dope — A form of aircraft dope having a cellulose acetate butyrate film base and suitable plasticizers, along with the necessary solvents and diluents. Butyrate dope has a better tautening effect on fabric and is less flammable than nitrate dope.

cellulose nitrate — A compound formed by treating cellulose with a mixture of nitric and sulfuric acids.

cellulose nitrate dope — Aircraft dope consisting of a nitrocellulose film base with the appropriate plasticizers, thinners, and solvents. It has excellent encapsulating properties, but its high flammability has caused its decrease in popularity as a finish for fabric covered aircraft.

Celsius — Temperature measurement that sets 0° at freezing, and 100° as the boiling point of water.

center — A point equally distanced from all points.

center console — The space between the pilot and copilot where the power lever control system is positioned on most multi-engine type airplanes.

center drill — A combination of twist drill and a 60° countersink. Used to center a hole and a countersink in a piece of metal.

center line — Alternate long and short dashes indicating the center of an object or part of an object. Used in aircraft drawings.

center of airfoil moments — The point about which the basic airfoil moment coefficients are given, usually the aerodynamic center of 1/4 of the mean aerodynamic chord.

center of gravity — Abbrev.: CG. A point within an aircraft at which all of the weight may be considered to be concentrated.

center of gravity envelope — A graphic depiction of the fore-and-aft range of center of gravity limits, showing the way these limits vary with the gross weight of the aircraft.

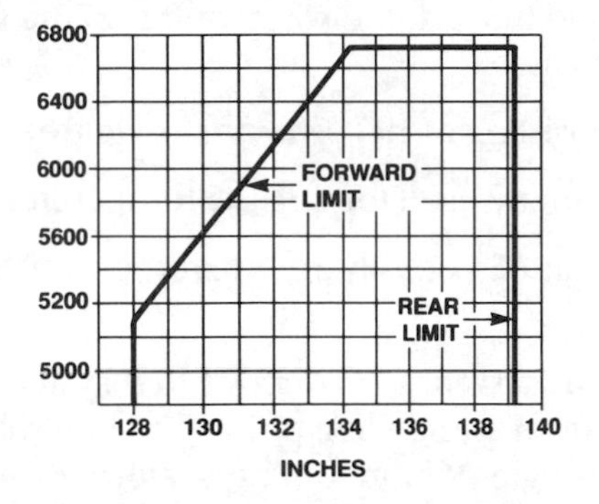

center of gravity limits — The extreme forward and rearward positions at which the center of gravity of an aircraft may be located.

center of gravity range — The distance between the forward and rearward center of gravity limits, as specified on the Type Certificate Data Sheet for the aircraft.

center of lift — The resultant of all of the centers of pressures of an airfoil.

center of mass — The location within an aircraft at which its entire mass can be considered to be in equilibrium.

center of pressure — The point on the chordline of the airfoil where all of the aerodynamic forces are concentrated.

center of pressure coefficient — The ratio of the distance of the center of pressure from the leading edge to the chord length.

center of thrust — The resultant of all of the thrust forces of the propellers and/or the exhaust jet stream.

center punch — A punch having a somewhat blunt point, used to form an indentation in sheet metal which may be used to start the twist drill.

center-of-rotation line — The line on a drawing about which an object will rotate.

center-tapped winding — A winding on an electrical transformer that has a connection (tap) located in its electrical center. It is used to divide the winding in half with each half having opposite polarities.

centervent system — Use of the main rotorshaft as an air-oil separator in place of a driven centrifugal device. After separation, oil is scavenged back to the oil reservoir and air, which was entrained in the oil, is vented through the rotorshaft into the gas path in the area of the turbine wheel.

centigrade — Consisting of 100 divisions or degrees.

centigrade — Formerly used for Celsius temperature. *See also Celsius.*

centistoke — A unit of viscosity measurement of both fuels and oils. 1⁄100 of a "stoke".

central refueling provisions — Aircraft fuel system in which all tanks may be filled from one fueling point. The central refueling system is also called a single-point refueling system, or a pressure refueling system.

central refueling system — *See central refueling provision.*

centrifugal brake — A friction brake which is used to apply friction to stop if the unit rotating turns at a speed that is faster than is permitted.

centrifugal breather — *See rotary breather.*

centrifugal clutch — A type of friction clutch which engages when a drive wheel reaches a predetermined speed. The clutch is engaged by centrifugal force that acts on a flyweight mechanism.

centrifugal filter — A filtering element that separates contaminants from a fluid by centrifugal action. It throws contaminants by rotary motion into traps that hold them until they can be removed.

centrifugal flow compressor — An impeller shaped device which receives air at its center and slings air outward at high velocity into a diffuser for increased pressure. Sometimes referred to as a radial outflow compressor.

centrifugal force — The outward pull on a body as it rotates or spins.

centrifugal moment — A force that tries to cause a rotation caused by the amount of centrifugal force acting on an object.

centrifugal oil filter — A rotary filtering element used to throw contaminants outward into sediment traps.

centrifugal pump — Any pump that uses a high-speed impeller to throw the fluid outward by centrifugal action.

centrifugal switch – An electrical switch that is mounted inside of a rotating induction motor of a capacitor-starter. The switch is actuated by centrifugal force disconnecting the starter winding when the rotor is turning at a predetermined speed.

centrifugal tachometer – A mechanical tachometer that measures the speed of rotating shaft. Flyweights are mounted on a collar around the rotating shaft in such away that centrifugal force pulls the flyweights away from the shaft. As the flyweights move away from the shaft, the collar moves up the shaft causing a pointer to move over a dial registering the shaft speed.

centrifugal twisting force – The centrifugal forces acting on a propeller blade. The twisting force is present in all rotating propellers and always acts to send the blades toward a lower pitch position.

centrifugal twisting moment – The tendency of a propeller blade to twist on its axis due to the centrifugal forces acting on the blade. The twisting moment is present in all rotating propellers and always acts to send the blades toward a lower pitch position.

centrifugal-type pump – Pump that uses a high-speed impeller to throw the fluid outward at a high velocity.

centrifuge – A device that is used to separate a liquid mixture or a suspension into its various components that have different specific gravities.

centrifuge action – A force which tends to separate particles according to their density, or to pull an object apart by rotating it rapidly about its center.

centrifuging – A method of separating particles that have different densities by spinning them around in a centrifuge.

centripetal force – The force within a body that opposes the centrifugal force as the body rotates or spins.

centroid – The center of mass of a body or a point about which all of its mass is concentrated.

ceramic – Clay-like material composed primarily of magnesium and aluminum oxide, which may be molded and fired to produce an excellent insulating material.

ceramic magnet – A permanent magnet made by compressing a mixture of ceramic material and sintered magnetic particles.

certificate – An official FAA document authorizing a privilege, fact, or legal concept.

certificated — An object or person which has been granted a certificate of approval. In aviation, these are normally issued by the FAA.

certificated aircraft — An aircraft designed to meet minimum specifications and requirements specified by the FAA. When these conditions are met, an Approved Type Certificate is issued for the aircraft. In order for the aircraft to maintain the certificate it must be maintained in such a way that it continues to meet these specifications, to be considered legally airworthy.

certificated technician — A person who holds a valid technician's certificate issued by the FAA with either an Airframe, Powerplant, or both ratings.

cesium — A soft, ductile, bluish-gray metallic chemical element that is used in the manufacture of photoelectric cells.

cesium-barium 137 — A radioactive substance used to coat ignition system air-gap points to synchronize discharge of current to the igniter plug.

CFR engine — An engine which is used by the Cooperative Fuel Research to determine the octane rating of a hydrocarbon fuel. A CFR engine has a variable compression ratio, and it can cause any of the fuels that are being tested to detonate. When the correct percentages of iso-octane is obtained, an octane number is given to the fuel.

chadwick balancer — A term used to describe electronic balancing or tracking of rotor blades. It is actually the name of the manufacturer of the balancing and tracking equipment.

chafe — To wear away by a rubbing action.

chafers, tires — Layers of fabric and rubber that protect the tire carcass from damage during mounting and demounting. They insulate the carcass from brake heat and provide a good seal against movement during dynamic operations.

chaff — A metallic material, such as aluminum, which is used to confuse enemy radar. Chaff is usually ejected from an airplane and is picked up by the radar as a large object.

chaffed surface — A transfer of metal from one surface to another resulting from a slight relative movement between two surfaces under high contact pressure. The surface of each part reveals metal removed and metal added.

chafing — Rubbing action between adjacent or contacting parts under light pressure which results in wear.

chafing strip – *See chafing tape.*

chafing tape – Cloth or paper tape placed over any metal seam or protruding screw head that is to be covered with fabric. It is used to protect the fabric from wear.

chain gear – A gear or sprocket that is used to transmit motion from one shaft to another shaft connected by a roller chain similar to that used in a bicycle.

chain hoist – A mechanism that is used in a shop to lift heavy weights. A chain hoist uses an endless loop of chain to drive a geared wheel which also supports and lifts the load as it is pulled up by the geared wheel.

chain reaction – A self-sustaining action in which one event causes other events of the same kind to happen.

chamfer – A bevel cut on the edge of a piece of material.

chamfered point of a threaded fastener – The point of a bolt or a screw that is formed in the shape of a cone with its top cut off. The chamfered point allows easy entry into the hole for starting.

chamfered tooth – The tooth of the gear on the rotating magnet or the distributor gear which is beveled to identify it for use when timing the magneto.

chamois – A piece of soft leather used to filter gasoline. Gasoline will pass through a chamois, but water will not. Gasoline that has been filtered through a chamois can be considered to be free from water.

chandelle – An abrupt climbing turn to approximately a stall in which the momentum of the airplane is used to obtain a higher rate of climb than would be possible in unaccelerated flight. The purpose of this maneuver is to gain altitude at the same time that the direction of flight is changed.

channel – A metal structural member either extruded or bent into a U-shape.

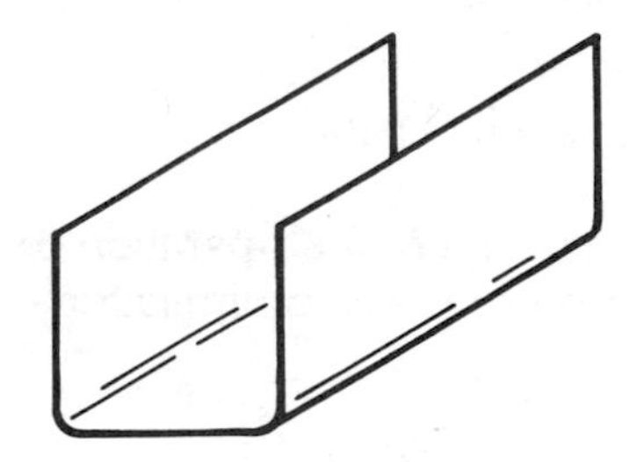

channel iron – Extruded steel either extruded or bent into a U-shape.

channel section – A form of structural material that has the cross sectional shape of a channel or the letter U.

characteristic curves –
[1]A series of graphically presented curves which describe in mathematical terms the characteristics of lift and drag produced by an airfoil section.
[2]A graph which shows the performance of an electron tube or a transistor under various operating conditions.

characteristic potential difference – The theoretical potential difference produced by a chemical cell using specific pole materials.

charcoal – Black porous carbon.

charge –
[1]A quantity of electricity. If the charged material holds a greater number of electrons than normal, it is said to be negatively charged. If the material has a deficiency of electrons, it is positively charged.
[2]The physical condition that gives rise to an electric field.

charging a battery – Preparation of battery for service by passing low DC voltage through the battery through the positive terminal. A lead acid battery should be recharged when one cell reads 1.240 or below on the hydrometer. A fully charged battery reads about 1.300 on the hydrometer.

charging current – A current passed through a secondary cell which restores the active material on the plates to a condition that allows them to change chemical energy into electrical energy.

charging stand – A handy and compact arrangement of air conditioning service equipment, containing a vacuum pump, manifold set, and a method of measuring and dispensing the refrigerant.

Charles' law – A law of physics which states that if a gas is held at a constant pressure, it will expand in direct relationship to the increase in its absolute temperature.

chart –
[1]A pictorial presentation of data.
[2]A graph.
[3]A graphic representation of the operation of engine performance, fuel consumption, horsepower, or limitation of some specific unit.

chart, navigation – A special map that is used for aerial navigation that gives the location, and the necessary information about all of the navigation aids. A chart shows the grids of latitude and longitude, and provides a surface for plotting courses and locating fixes.

chasing threads – Cutting screw threads by moving a tool along the axis of the work to be threaded.

chassis – An aluminum, copper, or plated steel body around which any unit is built. It serves as the support for the airplane, engine parts, electronic components, etc., and is often used as a voltage reference point.

chattering brakes – A heavy vibration in the brakes produced by the brake friction as the disks rotate. Chattering is caused by glazed discs.

check flight – An operational check out of all of the aircraft systems during a test flight to check the aircraft performance after major re-work or repairs.

check list – A sequential systematic list of specific procedures to be followed when performing any complex operation. For example, check lists are used in the performance of preflight inspections, and 100-hour and annual inspections of aircraft to be sure that all required operations are completed.

check nut – Thin nut jammed against the regular nut to prevent its loosening.

check, propeller – An operational check of a newly installed propeller on the engine, and during and after the engine has been ground operated.

check valve – A valve which allows free flow of fluid in one direction, but no flow or restricted flow in the opposite direction.

checkpoint – A navigation location identified either visually, or electronically.

cheesecloth – Lightweight cotton gauze that has no sizing in it. It is used as a straining element to remove lumps and contaminants from liquids, or as a polishing cloth.

chemical bond – The joining of two or more parts or pieces by molecular attraction of an adhesive agent which wets the parts.

chemical compound – Substance formed by the chemical reaction between two or more chemical elements.

chemical element – A fundamental substance that consists of atoms of only one kind. Examples of chemical elements include oxygen, carbon, gold, silver, and hydrogen.

chemical energy – Energy stored in chemicals due to their attraction to or reaction with other chemicals.

chemical etching –
[1]A process in which small cracks in aluminum may be detected by application of a caustic soda solution.
[2]A chemical process use to etch (roughen) the surface of metal in preparation for priming or painting.

chemical fire extinguisher – Any type of fire extinguisher that extinguishes fire by expelling the fire extinguishing agent and blankets the fire to keep oxygen away from it.

chemical milling – A chemical etching process used to machine large sheets of metal. Chemical milling economically reduces the weight of the aircraft, and produces a lightweight skin that has all of the needed strength and rigidity than can be done with conventional machining or by using riveted-on stiffeners.

chemical reaction – A chemical alteration in a substance to form a chemical compound. This is always accompanied by an energy change.

chemical salt – The result of the combination of an alkali with an acid. Salts are generally porous and powdery in appearance and are the visible evidence of corrosion in a metal.

cherry picker – A hydraulically operated boom with a man-carrying basket on its end. A person can be lifted in the basket in order to work at a high location point of large airplanes.

Cherry rivet® – A form of blind rivet patented and manufactured by the Cherry Rivet division of Townsend, Inc. Its upset head is formed by pulling its tapered stem through its hollow shank.

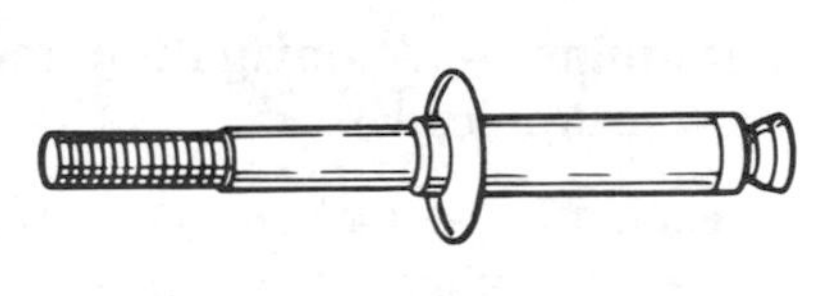

chevron seal – A single-direction seal in a hydraulic or pneumatic actuator. It derives its name from its V-shaped (a chevron) cross section.

chilled iron – Cast iron that has been cast in a steel mold. The casting is quickly cooled by the steel mold so that it retains most of the carbon, and retains a high degree of hardness.

chin – An aircraft structural part that sticks out from the bottom of the forward part of the fuselage.

chine – The longitudinal member on the side of a float or seaplane hull where the bottom and the side meet.

chine tire – A nose wheel tire that has a deflector molded into its sidewall. Chine tires are mounted on the nose wheel of jet aircraft and prevent water, ice, or snow and slush from getting into the intake of the engines by throwing the water and slush outward and away from the engines.

chinook – A name given to a warm, dry wind that blows down the eastern slopes of the Rocky Mountains in the United States from Canada. The moisture in the air of a chinook is almost completely lost as it blows up the western slopes of the mountains, and it is dry and warm as it blows down the eastern slopes.

chip –
[1]A small fragment of metal removed from a surface by cutting with a tool.
[2]An electronic component containing an integrated circuit.

chip detector – An electrical metal detection warning system. A magnetic sump or drain plug with an electrode at its center, and with ground potential at its casing. When ferrous particles bridge the gap, the current path is completed illuminating a warning light in the cockpit.

chipping – Breaking away of pieces of material by excessive stress or by careless handling.

chisel – A hard steel cutting tool used to shear metal when it is hammered.

chlorobromomethane (Halon 1011) – A chemical formula – CH_2CLBr. A liquefied gas, with a UL toxicity rating of 3. Commonly referred to as CB, chlorobromomethane is more toxic than CO_2. It is corrosive to aluminum, magnesium, steel and brass. It is not recommended for aircraft use.

chock – Block of material wedged under the tires of an aircraft to act as a safety for the brakes.

choke –
[1]An electrical inductor that is used to appose the flow of pulsating DC electricity. Chokes are used with capacitors to make filter circuits that smooth out the voltage changes and make pulsating direct current into smooth flowing DC.
[2]An inductor used to smooth the pulsations in rectified AC.

choke bore – A method of boring the cylinder of an aircraft engine in which the top, that portion affected by the mass of the cylinder head, has a diameter slightly less than that of the main bore of the barrel. When the cylinder reaches operating temperature, the mass of the head has caused the bore to be straight throughout its length.

choke coil – An inductance coil designed to provide a high reactance to certain frequencies, and generally used to block or reduce currents at these frequencies.

choked – A condition of a turbojet engine where airflow from a convergent nozzle is at Mn = 1.0 and cannot be further accelerated regardless of pressure applied. Occurs normally at the turbine nozzle and exhaust nozzle, but is a cause of stall conditions in the compressor.

choked cylinder bore – The cylinder of a reciprocating engine whose bore is slightly smaller in the part of the cylinder that is screwed into the cast aluminum head than it is in the center of the cylinder barrel. The cylinder head expands at normal operating temperature enough that the bore straightens out and has the same diameter throughout.

choked nozzle – A jet engine nozzle whose flow rate has reached the speed of sound.

choke-input filter – A form of filter that is used with an electronic power supply to change pulsating direct current into smooth DC.

choo-choo – A mild compressor surge condition caused by insufficient compression ratio across the compressor.

chopper – Slang for helicopter.

chord –
[1]An imaginary straight line which passes through an airfoil or wing section from the leading edge to the trailing edge.
[2]A straight line that passes through the circle and touches the circumference at two points. Also called the diameter of the circle.
[3]An imaginary line that is drawn through an airfoil from its leading edge to its trailing edge. The chord, or chord line, is used as a reference (a datum line) for laying out the curve of the airfoil.

chord length – The length of the projection of the airfoil area by the span.

chord line – *See chord.*

chordwise – Passing from the leading edge to the trailing edge of an airfoil.

chrome molybdenum steel – An alloy steel containing chromium and molybdenum. The most generally used steel for aircraft structure and is the SAE 4100 series. Has high strength, good toughness, and is highly weldable.

chrome nickel molybdenum steel – A steel which has been alloyed with chromium, nickel, and molybdenum.

chrome pickling – A method used to convert the surface of magnesium to form a hard oxide film to protect it from corrosion. This is accomplished by soaking the magnesium in a solution of potassium dichromate.

chrome plated cylinder – Hard chrome plating applied to the inside walls of and aircraft cylinder to form a hard, wear resistant surface.

chrome plating – An electroplating process transferring chromium to the surface of the steel. Either hard chrome or decorative chrome may be applied.

chrome plating – A type of treatment for cylinder walls of reciprocating engines. It hardens the walls and helps lubricate them. Worn cylinder barrels may be ground so that their bore is straight and round. Then hard chromium is electroplated on the cylinder walls to a depth that brings the diameter of the cylinder bore back to its original dimensions. The surface of the chrome plating on the cylinder walls resembles a maze of spider webs. There are thousands of tiny, interconnected cracks in its surface. The electroplating current is then reversed, and these tiny cracks open up enough that they can hold oil. Porous chrome plating provides a hard, wear-resistant surface for the piston rings to ride on. The oil that is trapped in the tiny grooves lubricates the wall to minimize piston ring and cylinder wall wear, and to help the rings seal.

chrome vanadium steel – A steel alloyed with chromium and vanadium. It is the SAE 6100 series and is used extensively in the manufacture of technicians' hand tools.

chrome-alumel – Bimetallic metal used in the exhaust temperature indicating system. Alumel contains an excess of free electrons which when heated move into the chromel lead. This current flow is read as an indication of temperature.

Chromel – An alloy of nickel and chromium highly resistant to oxidation, and has a high electrical resistance.

chromic acid — An acid similar to sulfuric acid except for the substitution of chromium for the sulfur. It is used as an etchant for preparing aluminum alloys for finishing and as a corrosion inhibitor.

chromic acid etch — A solution of sodium dichromate, nitric acid, and water. This is used to etch or roughen.

chromium — A hard, brittle, white metallic chemical element which is highly resistant to corrosion. Used for plating metal to harden its surface or to protect it from rust or corrosion.

chronometric tachometer — An instrument that is used to measure the speed in revolutions per minute of the crankshaft of an aircraft reciprocating engine. The chronometric tachometer repeatedly counts the number of revolutions in a given period of time and displays the average speed on its dial.

chuck — A special clamp-like device on a lathe or drill that is used to hold the material that is being worked. Chucks have three or more jaws that are used to clamp and hold the material.

chugging — Low frequency oscillations of airflow within a turbine engine. A mild, audible stall condition which can usually be controlled by proper throttle movement.

chute — An inclined trough or channel that is used to allow objects or materials to be sent from one level or place to another.

cigarene — *See cigarette.*

cigarette — A ceramic or synthetic rubber insulator used at the end of an ignition lead to insulate it from the shielded barrel of a spark plug.

circle — A closed plane curve in which all points along the curve are an equal distance from a point within the curve called the center.

circle graph — A graph using a circle divided like a pie to convey data.

circuit — The complete path in which electrical current flows. It must contain a source of electrical energy, a load to absorb this energy, and conductors to carry the electron flow.

circuit breaker — A circuit protecting device which opens the circuit in the case of excess current flow. It differs from a fuse in that it can be reset without having to be replaced.

circuit diagram — An electrical drawing that uses conventional symbols to show how the components in an electrical system are interconnected.

circuit protector — A device which will open an electrical circuit in the event of an excessive current flow.

circular inch — The area of a circle whose diameter is 1″.

circular mil — A measurement of area equal to that of a circle having a diameter of 1/1,000″, 1 mil, or 0.001″.

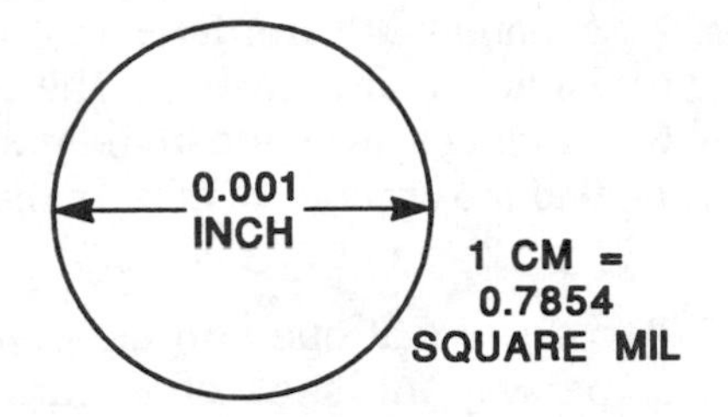

circular motion — The motion of an object along a curved path in which the object stays a constant distance from the center of the motion.

circular saw — Powered saws which use a circular blade that is driven by an electric motor.

circular slide rule — A slide rule having scales arranged in circles on the surface of a disk. Transparent runners, attached at the center of the disk may be moved over the scale to add or subtract portions of the scales to perform the various mathematical operations.

circumference of a circle — The linear distance around a circle. The circumference of a circle is always 3.1416 times the length of the diameter of the circle.

circumferential frame — A circular or oval frame; also called belt or transverse frame. It gives shape to a fuselage or nacelle.

circumscribed circle — A circle that is drawn around the outside of a another figure in such a way that all of the points touch the circumference of the circle.

cistern — A container that is used to store a liquid.

CIT sensor — A device which sends an inlet duct temperature signal to the fuel control as a scheduling parameter.

civil aircraft — Aircraft other than public aircraft.

clad aluminum — Aluminum alloy which has a coating of pure aluminum rolled onto both sides for corrosion protection.

cladding — A method of protecting aluminum alloys from corrosion by rolling a coating of pure aluminum onto the surface of the alloy. This is done in the rolling mill and it reduces the strength of the material somewhat.

clamp — Any of a variety of devices used to exert pressure and to temporarily hold objects together.

clamp-on ammeter — A hand-held ammeter that clamps around a current-carrying wire that is to be measured. The changing magnetic field around the wire induces a voltage in the jaws of the ammeter which is proportional to the amount of current that is flowing in the line.

clamshell doors — Two doors that open on the opposite sides of the center line the same way the shell of a clam opens. *See also clamshell thrust reverser.*

clamshell thrust reverser — A thrust reverser, clamshell door system that fits in the exhaust system of a turbojet engine. When the reverser is deployed for thrust reversing, the doors move into position to block the normal tailpipe and duct the exhaust gases around so that they flow forward to oppose the forward movement of the aircraft. *See also mechanical blockage thrust reverser.*

clapper — *See mid-span shroud.*

class — With respect to the certification of aircraft, this is a broad grouping of aircraft having similar characteristics of propulsion, flight, or landing.

class of thread — Classes of threads are distinguished from each other by the amount of tolerance and/or allowance specified. Classes 1A, 2A, and 3A apply to external threads, and classes 1B, 2B, and 3B apply to internal threads. Classes 2 and 3 apply to both external and internal threads.

clean and true — A term used in valve seat grinding whereby the rough stone is used until the seat is true or exactly matches the valve guide and until all pits, scores, or burned areas are removed.

cleanout —
[1]Term used for cleaning out or cutting away the damaged area in preparing it for the repair.
[2]Estimating the amount of structure which must be removed, prior to repair.

clear ice — Transparent or glaze ice which forms on the surface of an aircraft while flying through freezing rain.

clearance – The clear space or distance between two mechanical objects or moving parts.

clearance volume – The volume of the cylinder of a reciprocating aircraft engine with the piston at the top of its stroke.

clearing engine – Purging the combustion chambers of unburned fuel by rotating the engine with the starter. The air flow caused by the compressor will carry off dangerous accumulations of fuel vapors and vaporize the liquid fuel present.

Cleco fastener – Spring-type fastener used to hold metal sheets together until drilling or riveting procedures are accomplished.

clevis – The forked end of a push-pull tube which is usually fastened to a bell crank in a control assembly.

clevis bolt – A special-purpose bolt whose round head is slotted or recessed to accept a screwdriver. The threaded portion of the shank is very short and it is used only for shear loads.

clevis pin – *See flathead pin.*

climb indicator – A rate of pressure change indicator, used to furnish the pilot with information regarding his rate of vertical ascent or descent.

climbing blade – A condition when one or more blades are not operating in the same plane of rotation during flight. This may not occur during ground operation.

clinometer – Closed-end curved glass tube filled with a liquid similar to kerosene and enclosing a round glass ball. It may be used as a leveling device or in a turn and slip indicator to indicate the relationship between the force of gravity and centrifugal force in a turn.

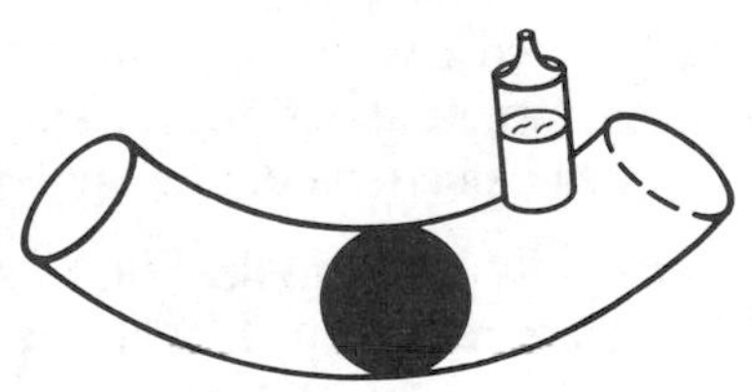

clip – A small attachment device used to join parts in aircraft construction.

clockwise rotation – The direction in which the hands of a clock rotate.

closed angle – The angle formed in sheet metal after it has been bent more than 90°. For example, if a piece is bent through 135°, it forms a 45° closed angle.

close-tolerance bolt – A hex-head aircraft bolt whose shank has been centerless ground to a tolerance of +0.000 – 0.0005". It is identified by a triangle on its head enclosing the material identification mark.

cloud point – The temperature of an oil at which its wax content, normally held in solution, begins to solidify and separate into tiny crystals, causing the oil to appear cloudy or hazy.

clove hitch – A type of knot used for making individual spot ties for securing electrical wire bundles. In this use, the clove hitch is locked with a square knot.

club propeller – A short, stubby propeller used for testing engines after the engine has been reconditioned.

clubhead – Rivet that does not formcorrectly during the bucking process and can be corrected by rapidly moving the bucking bar across the rivet head in a direction opposite that of the malformed travel. This corrective action can only be accomplished while the gun is in action.

cluster weld – A welded joint made at the intersection of a number of tubes which meet at a common point.

clutch – A device used to connect and disconnect a driving and driven part of a system, such as a transmission and main rotor of a helicopter.

coalescent bag – A bag in the water separator of an air-cycle air conditioning system on which the moisture that condenses from the air may coalesce.

coast-down check – Time a turbine engine takes to motor down to a complete stop from idle speed after the fuel is shut off. A maintenance test cell check of engine performance.

coating – The application of some material such as a metal, organic compound, etc. to a surface.

coaxial cable – A transmission line in which the center conductor is surrounded by an insulator and a braided outer conductor. All of this is enclosed in a weatherproof outer insulator.

cobalt chloride – An additive to silica gel dehydrator plugs which serves as an indicator of the amount of moisture absorbed by the plug. A dry dehydrator with this additive will be bright blue, but, if it has been exposed to excessive moisture, it will turn pink.

cobalt chromium steel – A steel alloy containing cobalt and chromium. Used in exhaust valves.

cobalt-based alloy – A cobalt, tungsten, molybdenum alloy of extreme high temperature strength in the family of turbine super alloy. Extremely expensive and used almost exclusively in the hot section.

cockpit – The pilot's compartment of an aircraft.

code markings – Aircraft fluid lines are often identified by markers of color codes, words, and geometric symbols. These identify each line's function, content, direction of fluid flow, and primary hazard.

coefficient – A dimensionless number expressing degree of magnitude.

coefficient of expansion – A dimensionless number relating to the amount of dimensional change of a material with a change in temperature.

cohesion – The act or process of holding together tightly.

coil – A conductor consisting of turns of wire in which the magnetic field around one turn cuts across the other turns, increasing the inductive effect of the wire.

coil assembly – The magneto coil assembly consists of a soft iron core around which is wound the primary and secondary coil with the secondary coil wound on top of the primary coil.

coil booster – *See booster coil.*

coin dimpling – Performed by a special machine which has, in addition to the usual dies, a "coining ram". This ram applies an opposing pressure to the edges of the hole so that the metal is made to flow into all the sharp contours of the die giving the dimple greater accuracy and improving the fit.

coin pressing – A dimpling process using a countersunk rivet as the male dimpling die, placing the female die in the usual position and backing it with a bucking bar. The rivet is then struck with a pneumatic hammer.

coke – A solid, carbon-like residue left by mineral oil after the removal of the volatile material by heat.

coking – Carbon build-up from decomposition of oil in vent lines. This build up can over a period of time cause a restriction of flow.

cold – The absence of heat.

cold dimpling – Accomplished while the material is at room temperature by either the coin ram or coin dimpling method.

cold flow – Term used to describe deep and permanent impressions or cracks caused by hose clamp pressure.

cold heading – Forcing metal to flow cold into dies to form thicker sections and more or less intricate shapes. The operation is performed in specialized machines where the metal, in the form of a wire or bar stock, may be upset or headed in certain sections to a larger size and, if desired, may be extruded in other sections to a smaller diameter than the stock wire.

cold section – The air compression sections of the engine.

cold spark plug – A spark plug in which the nose insulator provides a short path for heat to travel from the center electrode to the shell. Cold spark plugs are used in high-compression engines to minimize the danger of pre-ignition.

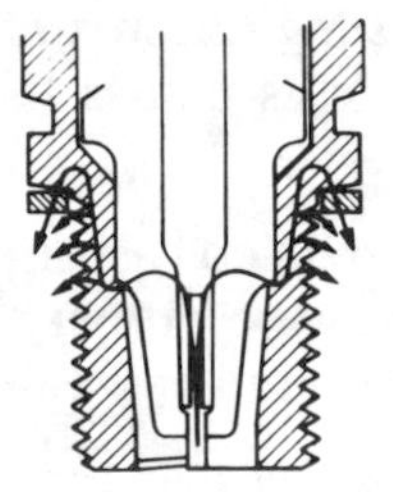

cold swaging process – A method of reducing or forming steel or other material while cold by drawing to a point or reducing the diameter as may be required.

cold tank system – A lubrication system wherein the oil cooler is located in the scavenge oil subsystem. The oil passes through the cooler and returns to the tank cooled. *See also hot tank system.*

cold working – Any mechanical process which will increase the hardness of a metal. This may be done by repeatedly hammering the material, passing it through rollers, or pulling it through dies.

cold-rolled steel – Steel which has been cold-worked by passing through a series of compression rollers or dies.

cold-starting oil relief valve – A by-pass relief valve in a main oil system which acts as an emergency pop-off valve when cold oil causes excessive system pressure. Used in systems having no oil pressure regulating relief valve.

collapsed surface – A dimensional change with neither removal of material nor an abrupt change of surface and usually affecting large sections of the object. Causes are excessive pressure or forces and improper abusive engine operation. Parts effected usually include valves, piston rings and springs.

collar – A collar is a raised ring or flange of material on the head or shank of a fastener.

collective pitch control – The control in a helicopter in which the pitch of all the rotor blades may be changed at the same time.

collector –
[1]The electrode in a transistor through which conventional current leaves the transistor.
[2]The exhaust cone collector in a turbine engine collects the exhaust gases discharged from the engine turbine buckets and gradually converts them into a solid jet. In performing this, the velocity of the gases is decreased slightly and the pressure increased. It also helps to direct the flow of hot gases rearward and prevents turbulence and at the same time imparts a high final exit velocity to the gases.

collector ring – A corrosion-resistant steel assembly which collects the exhaust gases from the cylinders of a radial engine and routes them overboard.

color code – A means of identifying an object by the use of various combinations of colors.

color wheel – A means of visualizing the color which will result when the basic colors are mixed.

combination compressor – A compressor design which utilizes an axial compressor and a centrifugal compressor (usually attached together) to compress incoming air prior to combustion.

combination inertia starter – An inertia starter for reciprocating aircraft engines which may be energized by either an electric motor or by a band crank.

combustion – A chemical process in which a material is united with oxygen at such a rapid rate that light and heat are released.

combustion chamber – Section of the engine into which fuel is injected and burned.

combustion liner – The perforated and louvered inner section of the combustor in which fuel burning is controlled.

combustion liner louvre – Small slots in the liner to direct cooling airflow and provide the inner walls with a cooling air blanket.

combustion section – The combustion section is located directly between the compressor and the turbine sections. It contains a casing, a perforated inner liner, a fuel injection system, some means for initial ignition, and a fuel drainage system to drain off unburned fuel after engine shutdown. The combustion section houses the combustion process which raises the temperature of the air passing though the engine. This process releases energy contained in the air/fuel mixture.

combustion starter – *See fuel-air combustion starter.*

combustor – The section of the engine into which fuel is injected and burned.

combustor – Point within the engine where combustion occurs. *See also combustor types: annular, can-annular, can, and scroll.*

combustor efficiency – A measure of the percentage of fuel burned completely or Btu's of heat attained as opposed to the Btu potential of fuel introduced. Typical figures are in the 99% range.

combustor outlet duct – *See transition duct.*

combustor turbine – *See free turbine.*

coming-in speed – The speed of a magneto which is just sufficient to produce the voltage required to fire all of the spark plugs consistently.

commercial fastener – A fastener manufactured to published standards and stocked by manufacturers or distributors. The material, dimensions and finish of commercial fasteners conform to the quality level generally recognized by manufacturers and users as commercial quality.

commercial operator – A person who, for compensation or hire, engages in the carriage by aircraft in air commerce of persons or property other than as an air carrier.

commercial ratings – *See engine ratings.*

commutator – The copper bars on the end of a generator armature to which the rotating coils are attached. AC is generated in the armature, and the brushes riding on the commutator act as a mechanical switch to convert it into DC.

comparator – A device for inspecting parts by comparing them with a greatly enlarged standard chart.

compartment – A separate and enclosed space in an aircraft structure.

compass –
[1]A device for determining direction measured from magnetic north.
[2]A drafting instrument used to create circles or arcs.

compass correction card – A card mounted near the compass in full sight of the pilot to indicate the difference between compass heading and magnetic heading.

compass locator – A low-frequency nondirectional beacon co-located with the marker beacons used to help establish the pilot on the localizer for an ILS approach.

compass north –
[1]The north to which a compass actually points. Its field is produced by the combination of the earth's magnetic field and the local magnetic fields within the aircraft.
[2]The direction of magnetic north corrected for local deviation errors.

compass rose – A circle marked out on a flat part of an airport away from magnetic interference. It is marked every thirty degrees of magnetic direction. The airplane is taxied onto this rose, and the magnetic compass is adjusted to agree with the heading of each of the marks.

compass swinging – The process of aligning the aircraft on a series of known magnetic headings and adjusting the compensating magnets to bring the compass heading as near the magnetic heading as possible.

compensated cam – The magneto cam used on high-performance radial engines. One lobe is provided for each cylinder, and the lobes are ground in such a way that the magneto points will open when the piston is a given linear distance from the top of the cylinder, rather than a given angular distance. This compensates for the relationship of the master rod pistons and those connected to the crankshaft through the link rods.

compensated relief valve – An oil pressure relief valve with a thermostatic valve to decrease the regulated oil pressure when the oil warms up. High pressure is allowed to force the cold oil through the engine, but the pressure is automatically decreased when the oil warms up.

compensating cam – A cam used in conjunction with the collective pitch control to add the correct amount of engine power for the pitch of the rotor. Used on turbine powered helicopters.

compensating port – A port inside a brake master cylinder which vents the wheel cylinder to the reservoir when the brake is not applied. It prevents fluid expansion due to heat causing the brakes to drag.

compensating winding – A series winding in a high-output generator. wound between the main pole and the interpoles to aid in brushless commutation and in overcoming armature reaction.

compiler – A compiler is a special computer program that converts a high-level computer language that is easy for a programmer to use into machine language that can be used by the computer.

complex circuit – A circuit consisting of a number of components, some arranged in series and others in parallel.

compliance – To accomplish as required by regulation or directive.

component – Any one of several parts in a combination of parts to make up a unit or whole.

composite fan blades – An advanced technology blade design, not yet in current use. Its composition is an epoxy-resin material and graphite fiber. It is stronger than fiberglass and 20-30% lighter than metals of the same strength.

compound – A new entity formed by a union of elements or parts. The constituents lose their original identity and assume the characteristics of the compound.

compound 314 – A chemical preparation which is applied to the propeller blade to prevent the formation of ice during flight.

compound curve – Curvature of a metal surface in more than one plane.

compound-wound generator – A generator which has both a series and a shunt field.

compressibility burble – A region of disturbed flow produced by and aft of a shock wave.

compressibility of air – Refers to the idea that air acts as an incompressible fluid at subsonic flow rates.

compression – The resultant of two forces which act along the same line and also act toward each other.

compression fastener – A fastener the primary function of which is to resist forces which tend to compress it.

compression member or strut – A heavy member, usually of tubular steel, which separates the spars in a Pratt truss wing and is used to carry only compression loads.

compression ratio –
[1]The ratio of the volume of an engine cylinder with the piston at top center to the volume when the piston is at the bottom center.
[2]Sometimes used to refer to compressor pressure ratio; not entirely correct because compression ratio infers a ratio of volumes as in a piston engine.

compression rib – A heavy-duty rib specially made with heavy cap strips and extra strength webs. A compression rib is designed to withstand compression loads between the wing spars.

compression rings – The top piston rings used to provide a seal for the gases in the cylinder, and to transfer heat from the piston into the cylinder walls.

compression wave – More familiarly known as shock wave. *See shock wave.*

compressive load – A load or a force which tends to compress or squeeze an object together.

compressive strength – The ability of a body to resist a force that tends to shorten, compress, or squeeze it.

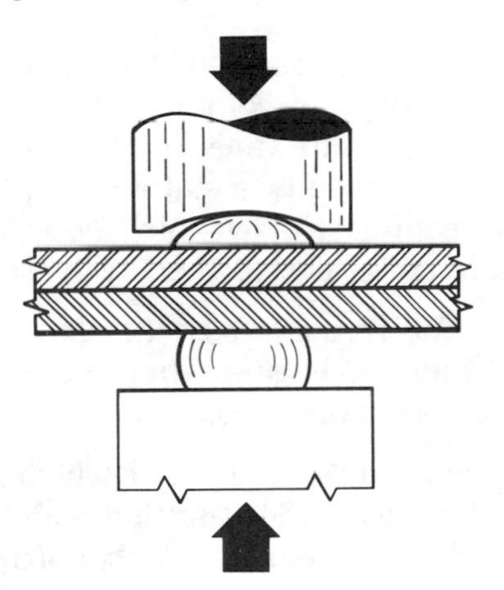

compressive stress – The basic stress which tends to shorten an object by pressing its ends together.

compressor – The section of a turbine engine which increases the pressure and density of the air which flows through the engine.

compressor bleed air – Air taken out of the compressor section of a turbine engine to prevent stall and operate certain components.

compressor case — The outer compressor housing, usually split front to rear or top to bottom. This case provides support to stator vanes.

compressor discharge pressure — Symbol: Pt4. Pressure signal taken at the compressor exit and sent to the fuel control unit for fuel scheduling purposes.

compressor disk — The compressor inner section to which the blades are attached. A disk assembly is made up of one segment per stage and bolted together to form one large rotating piece. *See also compressor drum.*

compressor efficiency — A measure of aerodynamic efficiency, one important factor of which is the ability to compress air to the maximum pressure ratio with the minimum temperature rise.

compressor front frame — The compressor inlet case.

compressor hub — The front and rear portion of the compressor to which the compressor shafts attach.

compressor pressure ratio — The ratio of compressor discharge pressure to compressor inlet pressure.

compressor stage —
[1]Each section of a compressor in which the air pressure is progressively increased. A stage of compression consists of one row of blades and one row of stator vanes in an axial flow compressor.
[2]A rotor blade set followed by a stator vane set. Simply stated, the rotating airfoils create air velocity which then changes to pressure in the numerous diverging ducts formed by the stator vanes.

compressor stall — The abrupt loss of the efficiency of the axial flow compressor in a turbine engine when the angle of attack of the compressor blades becomes excessive.

compressor stall-margin curve — A curve which shows a relationship between the compression ratio and mass air flow which must be maintained for a particular engine. If either of the factors goes out of limits, a compressor stall results.

compressor surge —
[1]A severe compressor stall across the entire compressor which can result in severe damage if not quickly corrected. This condition occurs from a complete stoppage of airflow or a reversal of airflow.
[2]An operating region of violent pulsating air flow usually outside of the operating limits of the engine by virtue of the flow control settings. A primary cause is compressor blade stall. Surge may result in flameout and, in severe cases, structural damage.

concentration cell corrosion – A type of corrosion in which the electrode potential difference is caused by a difference in ion concentration of the electroyte instead of a difference in galvanic composition within the metal.

concentric – Having a common center.

concentric shafts – Two shafts having a common axis, one inside the other.

condensation – The process of changing a vapor into a liquid.

condenser –
[1]Another name for capacitor. *See capacitor.*
[2]The component in a vapor cycle air conditioning system where heat energy is given up to the air and the refrigerant vapor is changed into a liquid.

condition lever – A turboprop cockpit lever. On some aircraft, it serves as prop control lever for flight (alpha range). On other engines, it serves only as a fuel shut-off lever.

conductance – The ability of a substance to conduct electricity.

conductivity – The characteristic of a material which makes it possible for it to transmit heat or electrical energy by conduction.

conductor – A material whose outer ring electrons are loosely bonded. Therefore, a relatively low voltage will cause a flow of these electrons.

conduit – A duct or tube enclosing electrical wires or cable.

conformity – Meeting all of the requirements of its original or properly altered conditions as specified in the Type Certificate Data Sheets and the manufacturer's specifications.

congealed oil – Oil that has solidified because of cold or contaminants.

conical – Cone-shaped.

coning – The upward bending of the blades of a helicopter rotor in flight.

coning angle – The angle formed by the rotor blades and the axis of rotation of a helicopter rotor system. The magnitude of the angle is determined by the relationship between the centrifugal force and the lift produced by the blades.

connecting rod – The component in an internal combustion engine which connects the piston to the crankshaft.

connector — Device used to join two pieces of wire, tubing, or hose to a component.

console — The pedestal or panel in an aircraft cockpit in which the operating controls are located.

constant — A value, used in a mathematical computation, that is the same every time. For instance, the relationship between the circumference of a circle and its diameter is a constant, 3.1416 (pi, π).

constant current charge — A method of charging a battery in which the voltage is adjusted as the charge progresses to keep the current constant.

constant displacement pump — A pump which displaces a constant amount of fluid each time it turns. The faster it turns, the more it puts out.

constant pressure cycle — *See Brayton cycle.*

constant section — That part of the fuselage of an aircraft which has a uniform cross-sectional shape.

constant voltage charge — A method of charging a battery in which the voltage across the battery remains constant. The current is high at the start of the charge, but tapers off to a low value as the charge progresses.

constantan — A copper-nickel alloy used as a negative lead in thermocouples for reciprocating engines.

constant-speed drive — Abbrev.: CSD. A hydraulic transmission which may be controlled either electrically or mechanically. It is used for alternators and enables the alternator to produce the same frequency regardless of the engine's variation of speed from idle to maximum RPMs.

constant-speed propeller — A controllable-pitch propeller whose pitch is automatically varied in flight by a governor to maintain a constant RPM in spite of varying air loads.

constrained-gap igniter — A type of turbine igniter plug that has the center electrode recessed in the insulator in order to cause the spark to arc well past the tip of the igniter.

construction theory of aircraft engine — *See theory.*

contact cement — A syrupy adhesive applied to both surfaces which bonds on contact.

contaminant — Anything that pollutes or defiles a fluid.

contamination — The entry of foreign materials into the fuel, oil, hydraulic, or other system.

continuity — The condition of being unbroken or uninterrupted.

continuity light — A simple test device in which a light indicates continuity in an electrical circuit while an open circuit will prevent the light burning.

continuous airworthiness program — A maintenance program consisting of the inspection and maintenance necessary to maintain an aircraft or a fleet of aircraft in airworthy condition and is usually used on large or turbine powered aircraft.

continuous curved-line graphs — A graph utilizing a smooth and even line to convey data.

continuous gusset — A brace used to strengthen corners in a structure. It runs the full width of the structure.

continuous ignition system — A secondary, lower power, ignition system installed along with the main system. It is used to fire one igniter plug during takeoff, landing, and in bad weather, for relight purposes in case of flameout.

continuous wave — Abbrev.: cw. An rf carrier wave whose successive oscillations are identical in magnitude and frequency.

continuous-element-type detector — A type of fire detection system, consisting of a stainless steel tube containing a discrete element which has been processed to absorb gas in proportion to the operating temperature set point. When the temperature rises due to fire, overheating, etc., the gas is released from the element, causing a pressure increase in the tube. This mechanically actuates a diaphragm switch, activating the warning lights and an alarm bell.

continuous-flow oxygen system — Any oxygen system which provides a continuous flow of oxygen at a rate constant for any given altitude.

continuous-loop fire detector system – A form of fire detection system utilizing a continuous loop consisting of two conductors separated by a thermistor material. At normal temperatures, the thermistor is an insulator, but in the presence of a fire or overheat condition, the thermistor becomes conductive and signals the presence of a fire.

contour – The outline of a figure.

contour template – A tool used to measure or duplicate the contour of a surface.

contract – To become reduced in size by squeezing or drawing together.

contrail – *See condensation trail.*

contrarotating propellers – Two propellers mounted on concentric shafts which turn in opposite directions. This type of rotation cancels the torque forces.

control – The act of regulating, directing, or coordinating any device or activity.

control cable – Specially designed steel cable connected to linkages used in flight control systems and engine controls.

control circuit – Any one of a variety of circuits designed to exercise control of an operating device, to perform counting, timing, switching, and other operations.

control column – The unit in the cockpit controls in an airplane on which a wheel is mounted. The ailerons are controlled by rotation of the wheel, and the elevators are controlled by its in-and-out movement.

control grid – The electrode in a vacuum tube to which the signal is applied.

control locking devices – Devices used to secure control surfaces in their neutral positions when the aircraft is parked.

control rod – A rigid, tubular rod used to actuate control surfaces. They are often called push-pull rods or torque tubes.

control snubber – Method of protecting control surfaces equipped with a hydraulic booster unit.

control stick – A vertical stick in the cockpit of an airplane used to move the elevators by fore-and-aft movement or the ailerons by side-to-side movement.

control surface – Any of the major or flight controls such as the ailerons, elevator, and rudder.

control wheel – Hand-operated wheel in the cockpit of an airplane used to actuate the elevators by in-and-out movement and the ailerons by rotation of the wheel.

controllability – The quality of the response of an aircraft to the pilot's commands while maneuvering the device.

controllable-pitch propeller – A propeller whose pitch may be changed in flight.

convection – Transfer of heat energy from one place to another by circulatory movement of a mass of fluid.

convection cooling – Refers to internal cooling air which escapes through small holes and slots, as opposed to transpiration cooling through porous walls.

conventional – Conforming to formal or accepted standards in drawings; rules.

conventional current –
1 Current flowing in an electrical circuit from positive to negative, outside the power source.
2 Also known as Franklin current. This theory is not in common use today.

conventional landing gear – A type of landing gear with wheels attached to a strut assembly located forward of the center of gravity and either a skid or wheel assembly at the tail.

convergent duct – A duct which decreases in a cross-sectional area front to back as in a subsonic aircraft exhaust duct.

convergent-divergent exhaust – Afterburner design, a supersonic exhaust duct. The forward section is convergent to increase gas pressure. The aft section is divergent to increase gas velocity to supersonic speed. This arrangement is necessary in order for the aircraft to attain supersonic speed.

convergent-divergent inlet – A supersonic engine inlet duct. The forward section is convergent to increase air pressure and reduce air velocity to subsonic speed. The aft section is divergent to increase air pressure still further and slow airflow to approximately Mach 0.5 before entering the engine.

conversion coating — A chemical solution used to form a dense, non-porous oxide or phosphate film on the surface of aluminum or magnesium alloys.

converter — A circuit in the control box of an anti-skid system using AC wheel speed sensors. It converts changes in AC frequency into changes in DC voltage.

convex — Having a surface that curves outward.

convey — Communicate, transmit.

cooling fins — Ribs projecting from the surface of a component to increase its area so that heat may be more easily transferred into the airstream flowing over the fins.

coordinated bucking — The process of allowing the bucking bar to vibrate in unison with the rivet gun set.

copilot — An assistant pilot who aids or relieves the pilot.

copper crush gasket — A copper gasket with a fiber core which is allowed to be crushed and thereby take the shape of mating surfaces for the purpose of affecting a leakage-free seal.

copper steel — When any minimum copper content is specified, the steel is classed as copper steel. The copper is added to enhance corrosion resistance of the steel.

cord body — *See casing plies.*

core speed sensor — Same as tachometer generator (N_2 speed on a dual spool engine).

cornice brake — A large sheet metal forming tool used to make straight bends; often called a leaf brake.

corona — The discharge of electricity from a wire when it has a high potential.

Coriolis force — The force that is produced when a particle moves along a path in a plane while the plane itself is rotating.

corrolation box — A cam to add power to a reciprocating engine used on a helicopter as the collective pitch control is raised.

corrosion — An electrochemical process in which a metal is transformed into chemical compounds which are powdery and have little mechanical strength.

course deviation indicator — Abbrev.: CDI. The instrument used for flying along a VOR. Also called a left-right indicator.

covalent bond — A bond between two atoms which comes about when valence electrons are "shared" by atoms.

cowl flaps — Movable doors on the air exit of an aircraft engine cowling. The cylinder head temperature may be controlled by varying the amount the flaps are opened.

cowl panels — The detachable covering of those areas into which access must be gained regularly.

cowl support ring — A large ring attached to a radial engine mount to provide firm support for cowl panels and also for attachment of cowl flaps.

cowling — A removable cover or housing placed over or around an aircraft component or section, especially an engine.

cowling, NACA — A cowling enclosing a radial air-cooled engine consisting of a hood, O-ring, and a portion of the body behind the engine so arranged that the cooling air smoothly enters the hood at the front and leaves through a smooth annular slot between the body and the rear of the hood; the whole forming a relatively low-drag body with a passage through a portion of it for the cooling air.

crack — A partial separation of material usually caused by vibration, overloading, internal stresses, defective assemblies, fatigue, or too rapid changes in temperature.

cradle — A support with pads used for supporting fuselage and wings during assembly, disassembly, or repairs.

crankcase — The housing that encloses the different mechanisms of an engine and is the foundation of the engine.

crankpin — That part of a crankshaft to which the connecting rods attach.

crankshaft — A shaft with a series of throws used for transforming the reciprocating motion of the crankshaft. This, in turn, turns the propeller.

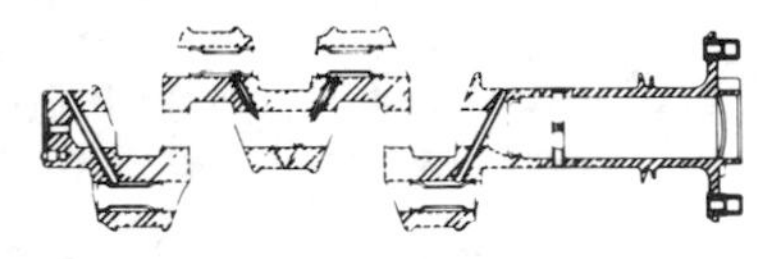

crater — A small pool of molten metal in the flame or arc during the process of welding.

craze — Hairline cracks in plastic due to age, stress, and exposure to sun.

creep — A condition of permanent elongation in rotating airfoils from thermal stress and centrifugal loadings.

crest – That surface of the thread which joins the flanks of the thread and is farthest from the cylinder or cone from which the thread projects.

crest clearance – As in a thread assembly, the distance, measured perpendicular to the axis, between the crest of a thread and the root of its mating thread.

crest truncation – The crest truncation of a thread is the distance measured perpendicular to the axis, between the sharp crest (or crest apex) and the cylinder or cone which bounds the crest.

crew member – A person assigned to perform duty in an aircraft during flight time.

cross member – A structural member which joins two longerons, or other lengthwise structural members. It carries loads other than the primary loads.

cross modulation – The modulation of a desired signal by an unwanted signal resulting in two signals in the output.

cross section – The representation on a drawing of the interior of a part cut at right angles to an axis.

cross sectional area – The area of the plane section of an object cut at right angles to its length.

crossover – A condition which exists in a helicopter rotor system in which the climbing and diving blades cross.

crossover tube – *See interconnector.*

crude petroleum – Unrefined petroleum, in a raw or natural condition, before being prepared for use.

cruise – A moderate speed of travel at optimum speed for sustained flight.

cruise control – Engine operation procedures that allow the best efficiency for power and fuel consumption during cruising.

cruise power – 60-70% of maximum continuous power; used for fuel economy and engine life during cruising.

crystal –
[1]A thin piece of piezoelectric material having a specific resonant frequency used to control the frequency of an oscillator.
[2]A small piece of galena, or lead sulfide, which will allow electron flow in one direction only.

crystal lattice — The basic pattern in which atoms are arranged in a specific manner which is repeated throughout the solid.

crystal microphone — A microphone making use of the piezoelectric properties of a crystal, acted on by the pressure of sound waves.

crystalline — A substance in which the atoms or molecules are arranged in a definite pattern, tending to develop definitely oriented plane surfaces.

Cuno filter — The proprietary name of a fluid filter made up of a stack of discs separated aby scraper blades. Contaminants collect on the edge of the discs and are periodically scraped out and collected in the bottom of the filter case.

cure — A chemical change which takes place in a finishing system which produces the desired surface.

cure time — The time required for a resin to complete its solidification.

current — The flow of electricity. Technically, it is electrons that flow, but more commonly, this is called current-flow. Current-flow is measured in amperes.

current limiter — A device which limits the generator out put to a level within that rated by the generator manufacturer.

current-fed antenna — A half-wave antenna fed in its center.

curvature — A curving or being curved.

curvic coupling — A circular set of gear-like teeth on each of two mating flanges which provide a positive engagement when meshed together and bolted. Used to attach together turbine wheels, compressor disks, etc.

cusp — The indentation on each side at the floor level when a fuselage shape has a "figure eight" shape. The cusp design is used to avoid unneeded width in this area.

customer bleed air — Air extracted from the engine (usually at the diffuser) to provide air for aircraft systems such as air conditioning, fuel tank pressurization, engine starting, etc.

cut off — To sever an object, or to stop a flow.

cut thread — A thread produced by removing material from the surface with a form cutting tool.

cutout switch — An elecltrical switch that interrupts the power to a motor or actuator when the limit of its desired travel is reached.

cuts out – A colloquial term describing the intermittent operation of a magneto or ignition system.

cutting edge – Edge of a tool or device used to remove material when some type of force is applied.

cutting plane – A line on an aircraft drawing used to indicate the surface of an auxiliary view.

cyanoacrylate – A single component polyester-type resin that hardens by exposure to ambient moisture and surface alkalinity.

cycle – A complete series of events or operations that recur regularly. The series ends at the same condition as it started, so the next series of events can immediately take place.

cyclic pitch control – The control in the cockpit of a helicopter with which the pilot can change the pitch of the rotor blades at a specific point in their rotation. The resulting change imparts lateral, forward, or backward movement to the helicopter.

cycling – The operation of a unit, such as the landing gear retraction system, through its full range of operation.

cycling switch – A switch which opens and closes the circuit, permitting a unit to cycle on and off.

cylinder –
1. A geometric shape having ends of a circular form and its sides parallel.
2. That component of a reciprocating engine in which the fuel is burned to increase the pressure which does the work.

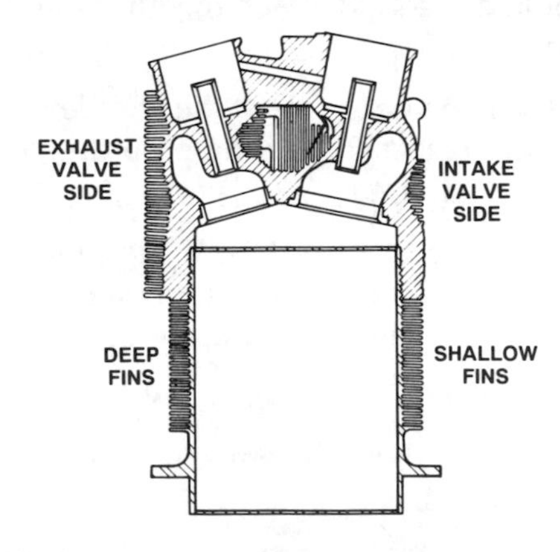

cylinder barrel – That portion of the cylinder or a reciprocating engine in which the piston moves up and down.

cylinder bore – The diameter of the cylinder barrel.

cylinder fins – Metal fins on a cylinder that increase the area of metal exposed to air which allows the heat to radiate out into the air for cooling.

cylinder flange – The base of a cylinder which incorporates a machined mounting flange by which the cylinder assembly is attached to the crankcase.

cylinder head – Closed end of the combustion chamber of a reciprocating engine.

cylinder head temperature – The temperature of the cast-aluminum head of a reciprocating aircraft engine cylinder.

cylinder pads – The machined surfaces on the crankcase of an aircraft engine on which the cylinders are mounted.

cylinder skirt – That portion of a cylinder of a reciprocating engine that extends below the mounting flange.

D

Dacron® – Polyester fibers made by E.I. DuPont de Nemours & Company.

dampen – To deaden, depress, reduce, or lessen.

damper – A device used to limit movement.

damper vane – A vane in a fuel flowmeter used to dampen fluctuations caused by erratic flow.

damper-type combustion air fuel valve – An automatically actuated damper-type valve located in the combustion air inlet of an aircraft heater. The valve is held open by fusible links which melt and allow the valve to shut off the combustion air in case of a fire or over-temperature condition.

damping action – An action which smooths out pulsations in the flow of an oscillation.

damping tube – A short length of tubing with an extremely small inside diameter inside a manifold pressure gage. It prevents a pressure surge caused by the engine backfiring, damaging the mechanism of the instrument.

danger area – A specified or specific area that is designated dangerous constituting a potential hazardous situation to persons or property.

d'Arsonval meter movement – Most commonly used meter movement in DC measuring instruments. A movable coil on which a pointer is mounted rotates in a permanent magnetic field. The amount of current in the coil determines the strength of the electromagnet; thus, the amount of pointer deflection.

dash numbers – Numbers following, and separated from, a part number by a dash identifying the components of the part.

dashpot – A mechanical damper used to cushion or slow down movement by restricting the flow of a viscous fluid.

data plate – A permanent identification plate affixed to an aircraft, engine, or component.

data stamp – Information stamped on units or components providing information on the correct name, part number, date of manufacture, and cure date, if effective.

data-plate speed — The speed at which the manufacturer determines "rated power" of an engine and stamps this value on a data plate affixed to the engine. The engine is required to perform within a certain range of this value throughout its service life.

datum — An arbitrary reference line from which all measurements are made when determining the moments used for weight and balance computations.

datum line — *See datum.*

dead — In electricity, having no potential or current flow.

dead center — *See bottom dead center and top dead center.*

dead engine — An engine that has been shut down during flight.

de-aeration — A process of removing air from a liquid.

de-aerator chamber — In hydraulics, an area of an oil tank where de-aeration takes place.

de-aerator tray — A container which collects the return oil from the oil system of a turbine engine and allows the air bubbles to separate out of the oil before it returns to the system.

debarkation — The unloading of passengers and cargo.

debooster — A unit used in the brake system which gives faster application and release of the brakes. Used to reduce system pressure to brake pressure and to permit rapid release of the brakes.

decades — A series of quantities in multiples of 10.

decalage — The difference in the angle of incidence of the two wings of a biplane. Decalage is positive if the angle of incidence of the upper wing is greater than that of the lower wing.

decarbonizers — Potent solvents used to soften the bond of carbon to a metal part.

deceleration — The rate of decrease in velocity.

deceleration check — A check made on an engine while retarding the throttle from the acceleration check. The RPM should decrease smoothly and evenly with little or no tendency for the engine to afterfire.

decibel —

[1]A measure of sound intensity equal to 1/10 of a bel.

[2]A unit used to express the ratio of the amounts of electrical or acoustical power and equal to 10 times the logarithm of this ratio.

decimal – A proper fraction in which the denominator is a power of ten.

decomposing – A term that describes material such as water being broken down into its basic elements by the process of electrolysis.

decouple – To release or disconnect a unit.

deep cycling – A treatment of nickel cadmium batteries in which the battery is completely discharged, the cells shorted out and allowed to "rest". The battery is then recharged to 140% of its ampere-hour capacity.

defect – Any imperfection, fault, flaw, or blemish which may require repair or replacement of a part.

deflation – Decreasing the amount of air held by an object.

deflecting-beam torque wrench – A form of hand-operated torque wrench in which the amount of torque applied to a bolt is indicated by the amount the beam is bent. The indication is read against a fixed scale on the handle of the wrench.

deflection – The movement of an electron beam up and down or sideways in response to an electric or magnetic field in a cathode-ray tube.

degeneration – Feedback of a portion of the output of a circuit to the input in such a direction that it reduces the magnitude of the input; also called negative feedback. Degeneration reduces distortion, increases stability, and improves frequency response.

degreaser – A solvent used for removing oil or grease from a part.

dehydrator plug – A plastic plug with threads to screw into a spark plug opening of an aircraft engine cylinder. These plugs are filled with silica-gel and an indicator to remove moisture from the air inside the cylinder and indicate the condition of preservation of the cylinder.

deicer – A system which removes ice from an aircraft structure after it has formed.

deicer boots – Inflatable rubber boots attached to the leading edge of an airfoil. They may be sequentially inflated and deflated to break away ice that has formed over their surface.

deicer tubes – The inflatable tubes in the deicer boot.

deicing – Removing ice after it has formed.

delaminated –
[1]A condition caused by exfoliation corrosion in which the layers of grain structure in an extrusion separate from one another.
[2]Separation of the core and face sheets of a bonded structure along a bond line.

deLavel nozzle – Same as convergent divergent nozzle.

delta – Greek letter (Δ) used in weight and balance computations to indicate a change.

delta connection – A method of connecting three components to form a three-sided circuit, usually drawn as a triangle, hence the term delta.

delta hinge – The hinge located at the root end of the rotor blade with its axis parallel to the plane of rotation of the rotor which allows the blade to flap equalizing lift between the upwind and downwind sides of the rotor disc.

delta winding – The connection of the windings of three-phase AC machines in which the ends of all three windings are connected together to form a loop or a single path through the three windings.

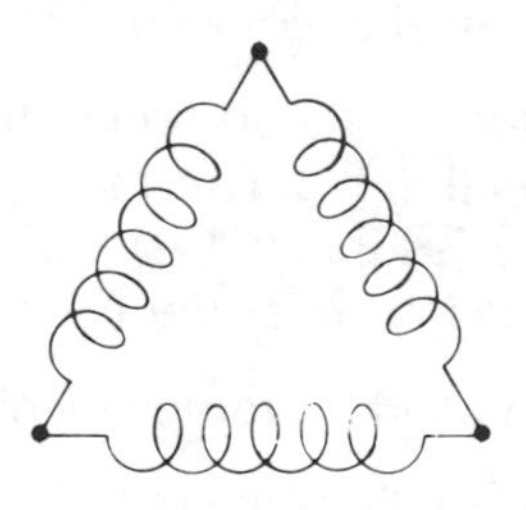

delta wing – The triangular planform of a wing of a supersonic aircraft.

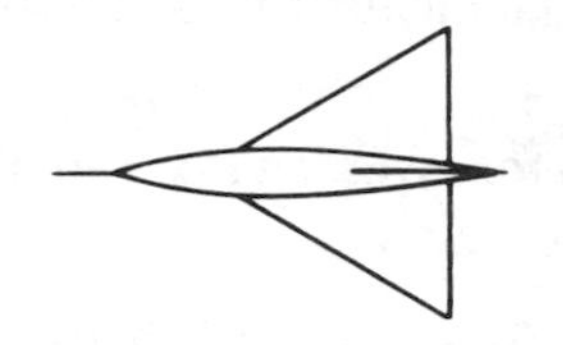

demand oxygen system – Any oxygen system in which the delivery rate of oxygen is determined by the requirement of the individual user.

demodulation – The recovery of the af signal from an rf carrier wave. Also called detection.

demulsibility – The measure of the ability of an oil to separate from water.

denominator – The part of a fraction that is below the line indicating division. It is an indication of the number of parts into which a number is divided.

density – The mass of a substance per unit of its volume. The weight per unit volume expressed in pounds per cubic foot.

density altitude – That altitude in standard air which corresponds with existing air density. It may be found by correcting pressure altitude for non-standard temperature.

dent – A depression in a surface usually caused by the part being struck with an object.

dented surface – A smooth depression of the surface without removal of material, usually affecting small areas, and usually with abrupt changes in surfaces, distinguishing it from a collapsed surface. Cams, tappet rollers, ball and roller bearings are the parts most often involved.

depletion area – That area on both sides of the junction of a semiconductor, which varies its characteristics between acting as a conductor and an insulator.

depolarization – The absorption of generated gases in a chemical cell especially during the "rest" periods. This may cause an apparent rejuvenation of the cell.

depreservation – A procedure that includes a special inspection and cleaning of aircraft parts removed from storage prior to being installed on the aircraft or engine.

depth micrometer – A form of micrometer caliper used to measure the depth of a recess.

derichment – An automatic leaning of the fuel-air mixture ratio to a ratio that will produce maximum power regardless of the heat released. Derichment occurs when the anti-detonation injection system injects liquid into the cylinders to remove this excess heat.

descent – A reduction in altitude.

desiccant –
[1]Any form of absorbent material.
[2]A material used in a receiver-dryer to absorb moisture from the refrigerant.

desiccant bags – Cloth containers of a silica gel desiccant packed with an engine or component that is placed in long-time storage.

design load – The load for which a member is designed. It is usually obtained by multiplying a basic load by a specified design load factor.

design size – That size from which the limits of size are derived by the application of tolerances. When there is no allowance, the design size is the same as the basic size.

designated – Being given the legal right and authority to perform certain specified functions by the FAA.

detachable – A unit or part which can be unfastened and separated.

detail drawing – An aircraft drawing which describes a single part in detail.

detail view – An auxiliary view incorporated into an aircraft drawing to show additional details of a part.

detailed inspection item – An inspection item of a progressive inspection that requires close and careful inspection, may involve disassembly to inspect and could even be to overhaul a component or part.

detector – That portion of an electronic circuit which demodulates or detects the signal.

detergent oil – A mineral oil to which ash-forming additives have been added to increase its resistance to oxidation. Because of its tendency to loosen carbon deposits, it is not used in aircraft engines.

deterioration – To become worse.

detonation – The almost instantaneous release of heat energy from fuel in an aircraft engine caused by the fuel-air mixture reaching its critical pressure and temperature. It is an explosion rather than a smooth burning process.

Deutsch rivet – A high-strength blind rivet.

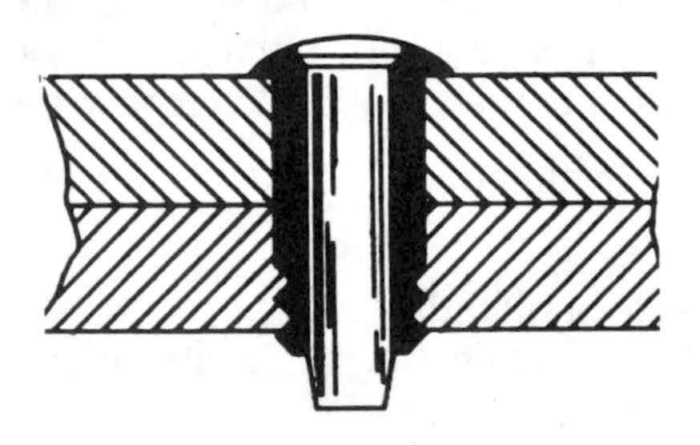

developed width – Width of the flat layout of a sheet metal part.

developer – A powder sprayed on a surface which has been treated with a penetrating dye. The powder acts as a blotter, pulling penetrant out of any crack, exposing the crack.

deviation – A compass error which is caused by the compass magnets attempting to align with extraneous magnetic fields in the airplane. Deviation error changes with the airplane's heading rather than the geographic location.

dew point – The temperature to which a body of air must be lowered before it will no longer hold water in a vapor state.

Dewar flask – A double-wall vacuum chamber used for the storage of liquid oxygen. Similar in principle to a Thermos® bottle.

diagram – A graphic representation of an assembly or system.

dial indicator – A precision linear measuring instrument whose indication is much amplified and is read on a circular dial.

dial-indicating torque wrench – A hand-operated torque wrench of the deflecting beam-type in which a dial indicator measures the deflection of the beam and reads directly in foot-pounds, inch-pounds, or meter-kilograms of torque.

diamagnetic material – A material having extremely low magnetic permeability and considered to be nonmagnetic.

diameter – The length of a chord passing through the center of a circular body.

diamond dressing tool – An industrial diamond mounted in a tool and held in contact with a grinding wheel to true the wheel.

diaphragm switch – A switch whose position is controlled by movement of a diaphragm.

diaphragm-controlled – A mechanical movement controlled by the action of a pressure or suction applied to a diaphragm.

dibromodifluoromethane – A fire extinguishing agent. Noncorrosive to aluminum, brass, and steel, it is less toxic than CO_2. It is one of the more effective fire extinguishing agents available.

die –
[1]A tool used to shape or form metal or other materials.
[2]A cutting tool used to cut threads on the outside of round stock.

die casting – Method by which molten metal is forced into suitable permanent molds by hydraulic pressure in order to improve the grain structure of the resulting casting.

dielectric — A material which will not conduct electricity.

dielectric constant — Symbol: k. The characteristic of an insulator that determines the mount of electrical energy that can be stored in electrostatic fields.

dielectric qualities — These conditions are normally expressed in terms of dielectric strength and the dielectric constant.

dielectric strength — A measure of the ability of a dielectric to prevent being punctured by electrical stresses.

diesel engine — An internal combustion reciprocating engine whose ignition is achieved by spraying a highly atomized, accurately metered charge of fuel into the air which has been heated by compression.

dieseling — The continued firing of a reciprocating engine after the ignition has been turned off. Ignition is caused by incandescent particles in the combustion chamber.

differential aileron travel — The increased travel of the aileron moving up over that of the aileron moving down. The up aileron produces extra parasite drag to compensate for the additional induced drag caused by the down aileron. This balancing of the drag forces helps minimize adverse yaw.

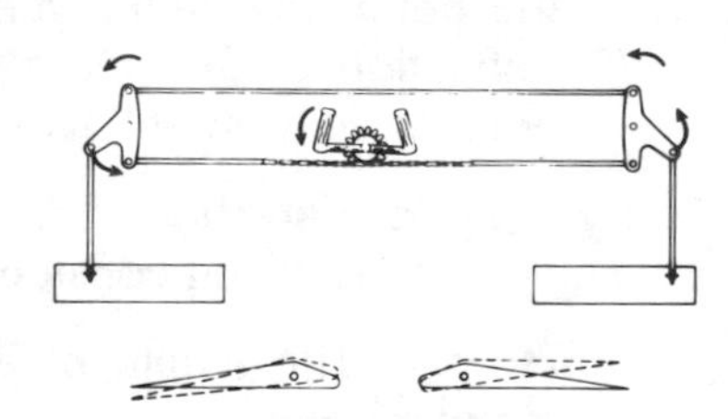

differential compression check — A test of the condition of an engine cylinder in which the amount of leakage past the piston rings and valves is determined by measuring the pressure drop across a calibrated orifice in the tester.

differential pressure — A difference between two pressures. The measurement of airspeed is an example of the use of a differential pressure.

differential pressure range of pressurization — The range of cabin pressurization in which a constant differential is maintained between cabin pressure and the outside air pressure. This mode of pressurization is used after the cabin reaches its maximum allowable differential pressure.

differential pressure switch – A diaphragm and electrical micro-switch arrangement which receives two pressure senses. If the difference in the two pressures exceeds a certain value, the microswitch completes a warning light in the contamination light circuits.

differential voltage reverse-current relay – Abbrev.: DVRCR. A form of reverse current relay used for high-output systems, in which the relay closes when the generator voltage is at a specific value above the voltage of the battery.

differentiating circuit – A circuit which produces an output voltage proportional to the rate of change of the input.

diffuser –
[1]A duct used on a centrifugal flow turbine engine to reduce the velocity of the air and increase its pressure.
[2]The divergent section of the gas turbine engine used to convert velocity energy of compressor discharge air to pressure energy.

diffuser vane – A turning or cascading vane in a centrifugal flow engine diffuser used to change air direction from radial as it leaves the impeller to an axial direction into the combustor.

digital readouts – Presentation of information by an instrument in a digital form such as light emitting diodes or drums, rather than by the move movement of a pointer over a numbered dial.

dihedral – The positive acute angle between the lateral axis of an airplane and a line through the center of a wing or horizontal stabilizer.

dikes – A common expression for diagonal cutting pliers.

diluter-demand oxygen system – An oxygen system which delivers the oxygen mixed or diluted with air to maintain a constant oxygen partial pressure as the altitude changes.

dilution air – The portion of combustion secondary air used to control the gas temperature just prior to its entry into the turbine nozzle area.

dimension – A measurement of length, width, thickness, size, or degree listed within a drawing.

dimension line – A light solid line, broken at the mid-point for insertion of measurement indications and having opposite pointing arrowheads at each end to show origin and termination of a measurement.

dimensional inspection – The physical measurement of a part against a recognized standard to determine the amount of wear or deformation.

dimming relay – A relay that allows a light or lights in a circuit to be dimmed.

dimming rheostat – A rheostat used to control the degree of brilliance of a lighting circuit.

dimpling – A process that is used to indent the hole into which a flush rivet is to be installed. Some metals, such as the harder aluminum alloys, cannot be dimpled while the metal is cold because it is likely to crack. This type of metal must be hot dimpled. Hot dimpling equipment consists of a pair of electrically heated dies that have a pilot to go through the rivet hole. The pilot is passed through the hole and the heated dies are pressed together. The dies heat the metal enough to soften it and force it into the shape of the die. This is the proper shape for the rivet.

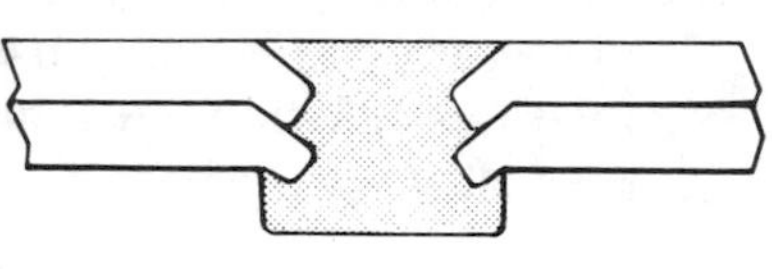

diode – An electron check valve. A device which allows a flow of electrons in one direction but not the opposite.

dipole antenna – A center-fed halfwave antenna.

dipping – A method of paint application where the part is dipped into a tank of finishing material.

dipstick – A bayonet-type gage used to determine the quantity of fluid in a reservoir.

direct current – Abbrev.: DC. A flow of electrons in one direction throughout a circuit.

direct-cranking electric starter – A high-torque, direct-current electric motor used to rotate a reciprocating aircraft engine for starting.

distance measuring equipment – Abbrev.: DME. Electronic navigation equipment which measures the time required for a signal to travel from the airplane to a ground station and return to the airplane. This time is translated into nautical miles to the station.

distillate fuel – A liquid hydrocarbon fuel which has been condensed from vapors of crude petroleum.

distortion – Undesirable change in the waveform of the output of a circuit compared with the input.

distributor – That part of a high tension magneto which distributes the high voltage to each spark plug at the proper time. Distributors for low tension ignition systems distribute the low voltage to the transformers at each spark plug at their proper time.

distributor block – A dielectric block in a magneto which contains stationary electrodes to pick up the voltage from a rotating distributor brush or finger and deliver it to the proper ignition lead.

distributor brush – A carbon brush used on a low-tension magneto to distribute the voltage to the distributor block as it rotates.

distributor finger – A rotating conductor in the distributor of a high tension magneto that delivers voltage to the distributor electrodes.

distributor valve – A device that controls the inflation sequence of deicer tubes.

dive – A steep, sudden descent in flight.

dive flaps – Devices used on an airplane to produce drag without an attendant increase in lift. They are also known as speed brakes.

divergent duct – A cone-shaped passage or channel in which a gas may be made to flow from its smallest area to its largest area with a resulting velocity decrease and pressure increase.

dividend – A number to be divided.

dividers – A measuring tool having two movable legs, each with sharp points. Dividers may be used to transfer distances, to divide straight or uniformly curved lines into an equal number of parts.

diving blade – A blade track of a helicopter's main rotor that lowers with an increase in RPM.

divisor – The number by which a dividend is divided.

dock – An enclosed work area where airplanes can be placed for repairs.

docking – Placing an airplane in a hangar where dock platforms are used to facilitate maintenance.

doghouse – A mark on a turn and slip indicator which resembles a doghouse. It is located one needle width away from the center, and when the pointer aligns with it, a standard rate of turn is being made.

dolly – Any of several devices used in lifting or carrying heavy aircraft components.

dolly block – Variously shaped anvils used to form and finish sheet metal parts.

domain – Magnetic fields. Spheres of magnetic influence around molecules of metals containing iron.

donor — An impurity used in a semiconductor to provide free electrons as current carriers. A semiconductor with a donor impurity is of t he n type.

dope — The finishing material used on fabric surfaces, serving to tauten and strengthen the fabric as well as making it weatherproof.

dope proofing — Treatment of painted surfaces to be covered with aircraft fabric with a material to prevent the solvents in the dope lifting the finish.

dope roping — A condition in the application of dope in which the surface dries while the dope is being brushed. This results in a stringy, uneven surface.

doped-in panel — A repair to aircraft fabric covering by installing a new panel of fabric extending from one wing rib to another, and from the leading edge to the trailing edge. The panel is doped rather than sewn in place, but it is attached to the structure with rib-stitching.

doped-on fabric repair — The repair of small damage to a fabric covered aircraft by doping a patch directly to the fabric covering, using no other attachment.

dope-proof paint — A finish applied over a varnished surface to prevent the solvents in the dope coming in contact with the varnish and lifting it.

doppler effect — The change in frequency of energy waves as their source moves toward or away from the receiver or observer.

dorsal — To the rear and on top of an object.

dorsal fin — A vertical aerodynamic surface extending from the top of the fuselage to the vertical fin. The purpose of a dorsal fin is to increase the directional stability of the aircraft.

double flare — A flare made in aluminum or steel tubing in which the tubing is doubled back on itself to produce two thicknesses of metal in the flare.

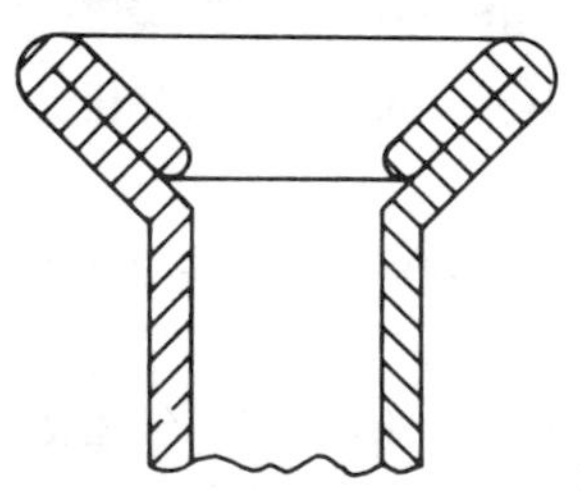

double spread – The spreading of an adhesive equally divided between the two surfaces to be joined.

double-acting actuator – A linear actuator which is moved in both directions by fluid power.

double-acting hand pump – Hydraulic hand pump that moves fluid with both the forward and rearward movement of the handle.

double-backed tape – An adhesive tape with adhesive on both sides.

double-loop rib-stitching – Attachment of fabric to the aircraft structure using a double loop of rib-stitch cord at each stitch.

double-pole, double-throw switch – Abbrev.: DPDT. An electrical switch that establishes three conditions in two circuits.

double-pole, single-throw switch – Abbrev.: DPST. An electrical switch that establishes two conditions in two circuits.

doubler – A piece of sheet metal placed against an aircraft skin where some component is attached. It provides stiffness or additonal strength.

double-row radial engine – A radial engine having two rows of cylinders and using two master rods attached to single crankshaft having two throws.

double-tapered wing – A wing where both the chord and the thickness ratio vary along the span.

dovetail fit – A shape similar to a cabinet maker's "dovetail cut". Primarily used to fit compressor blades into a compressor disk.

dow 19 treatment – An acid treatment for magnesium alloy parts which reduces an oxide film on the surface which inhibits the formation of corrosion.

Dow metal – A series of magnesium alloys produced by the Dow Chemical Corporation.

dowel – A short rod of either wood or metal used to hold objects together.

downdraft carburetor – A carburetor mounted on top of the engine in which the flow of air into the engine is downward through the venturi.

downlocks – Mechanical locks that hold a retractable landing gear in the "on" position, preventing its retracting when the hydraulic pressure is released.

downtime – The time an airplane is out of commission.

downwash –
[1]Air deflected perpendicular to the motion of the airfoil.
[2]Air that has been accelerated downward by the action of the main rotor of a helicopter.

downwash angle – The angle the air is deflected downward by an airfoil. It is the difference between the angle of air approaching the airfoil and the air leaving it.

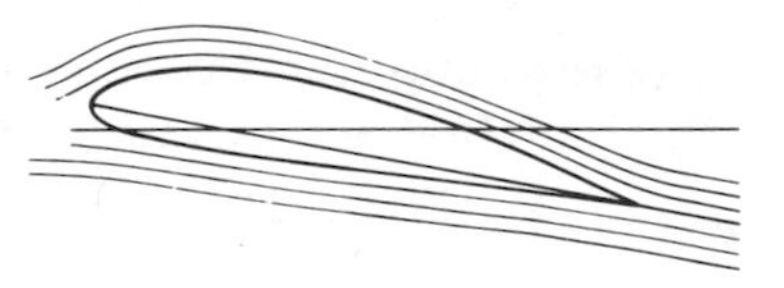

draftsman – A person who makes aircraft drawings.

drag – An aerodynamic force on a body acting parallel and opposite to the relative wind.

drag brace –
[1]In the configuration of the main landing gears; acts as side support for shock strut.
[2]An adjustable brace used to position the main rotor in a fixed position preventing movement of the blade at the attached point on semi-rigid rotors.

drag coefficient – One of the aerodynamic characteristics of an airfoil section which illustrates the increase in induced drag as the angle of attack is increased.

drag hinge – The hinge on a helicopter rotor blade that is parallel to the axis of rotation of the blade. It allows the blade to move back and forth on a horizontal plane, minimizing the blade vibrations.

drag wire – A diagonal, load-carrying member of a Pratt truss wing. It runs from the front spar inboard to the rear spar outboard and carries tensile loads tending to drag back on the wing.

dragging brakes – Brakes which have not fully released and which maintain some friction as the wheel rolls. Dragging brakes cause serious overheating.

drain – The electrode in a field effect transistor (FET) that corresponds to a collector of the ordinary transistor.

drain can – A container to catch fuel rained from the main fuel manifold filter shutdown of a turbine engine.

drain hole – Hole placed in the lower surface of a wing or other sealed components to provide ventilation and allow the drainage of any accumulated moisture.

drain plug – A removable plug located at the lowest point of a system used or drainage purposes.

drain valve – A spring-opened and burner pressure-closed mechanical valve located in the lower portion of the combustor outer case, installed to drain off puddled fuel after an aborted start.

draw filing – A method of hand filing in which the file is grasped with both hands and moved crosswise over the work. Draw filing produces an exceptionally smooth surface.

draw set – A riveting tool used to force sheets of metal together before they are riveted.

drawing – A graphic method of conveying information.

drawing number – Number assigned to each aircraft drawing and located in the lower right-hand corner of the title block. It identifies the drawing and is usually the part number of the component or part described by the drawing.

drift angle – The horizontal angle between the longitudinal axis of an aircraft and its path relative to the ground.

drift magnet – A small permanent magnet in a fixed coil ratiometer indicator, used to drift or pull the pointer off scale when the instrument is not energized.

drift punch – A pin punch with a long straight shank.

drill –
[1]A rotary cutting tool that is driven with a drill motor or a drill press.
[2]To sink a hole with a drill, usually a twist drill.
[3]A pointed cutting tool rotated under pressure.

drill bushing – Hardened steel sleeves inserted in jigs, fixtures, or templates for the purpose of providing a guide for drills so that holes will be straight and in the proper location.

drill chuck – The clamp on the spindle of a drill motor or drill press into which the drill is fastened.

drill jig – A device which holds parts or units in the proper position while holes are being drilled.

drill press – A power-driven drilling device that includes a table for holding the material, a chuck for holding the drill, a motor for driving the chuck, and a means of feeding the drill into the material.

drilling burrs – Sharp ragged particles of metal left by the drill when a hole is made.

drip valve – *See dump valve.*

drip-stick gage – A visual means of checking the fuel level from beneath the wing of a large jet transport aircraft. The stick is released and pulled downward until fuel drips from its end, signifying that the inside is even with the top of the fuel in the tank. The fuel quantity is read where the drip-stick enters the wing.

drive coupling – A coupling between the accessory section of an engine and the component which is driven. It is used to absorb torsional shock or to serve as a safety link which will shear in case the component seizes.

drive fit – A fit between mating parts in which the part to be inserted into the hole is larger than the hole, and the parts must be driven, or forced together. Also known as an interference fit.

drive screws – Plain-headed, self-tapping screws used for attaching name plates to castings or to plug holes in tubular structure through which rust-preventative oil has been forced.

driver head – A head, on a bolt or screw, designed for driving the fastener by means of a tool other than a wrench, such as a screwdriver.

droop –
1 Refers to the RPM loss which occurs when a fuel control flyweight governor speeder spring is extended and weaker. It takes less flyweight force to come to equilibrium with the weaker spring force and consequently slightly less speed will result.
2 The inability of the engine power to increase as the rotor pitch is increased, causing the rotor to slow down.

droop restraint – A device used to limit the droop of the main rotor blades at low RPM.

drop forging – A process of forcing semi-molten metal to flow under the pressure of repeated hammer blows into a mold or die.

drop hammer – Large, heavy, hammer-type metal-forming machine which uses sets of matched dies to form compound curved sheet metal parts. The metal is placed over the female die and the male die is dropped into it forcing the metal to conform to the shape of the dies.

drop tank – An externally mounted fuel tank that is designed to be dropped in flight.

drop-forged part – A steel part which has been formed by the drop-forging process.

droplets – Tiny drops of liquid.

dropping resistor — A resistor used to decrease the voltage in a circuit.

dry air — Air that contains no water vapor. Dry air weighs 0.07651 pounds per cubic foot under standard sea level atmospheric conditions of 59°F (15°C) and a barometric pressure of 14.69 PSI, or 29.92 inches of mercury.

dry air pump — An engine-driven air pump using carbon vanes, which does not require any lubricating oil in the pump for sealing or cooling.

dry operation — Operation of an aircraft engine equipped with a water injection system, but operating without the benefit of water injection.

dry rot — Condition of wood attacked by fungus causing brittleness and decay.

dry wash — Aircraft cleaning method where cleaning material is applied by spray, mop or cloth and removed by mopping or wiping with a clean, dry cloth. It is used to remove airport film, dust, and small amounts of dirt and soil.

dry-bulb temperature — Temperature of the air without the effect of water evaporation.

dry-charge battery — The common way of shipping a lead-acid battery. The battery is fully charged, drained, and the cells are washed and dried. The battery is sealed until it is ready to be put into service. Electrolyte is added and the battery given a freshening charge, making it ready for service.

dry-sump engine — An engine in which most of the lubricating oil is carried in an external tank and is fed to the pressure pump by gravity. After it has lubricated the engine, it is pumped back into the tank by an engine-driven scavenger pump.

dual controls — Two sets of flight controls for an aircraft which allow the airplane to be flown from either of two positions.

dual indicator — An aircraft instrument which provides two sets of indications on one dial. For example, the oil pressure of both engines may be shown on one indicator using one dial but two pointers.

dual magneto — A single magneto housing which holds one rotating permanent magnet and one cam with two sets of breaker points, two condensers, two coils, and two distributors. For all practical purposes, this constitutes two ignition systems.

dual rotor system — The rotor system of a helicopter in which there are two separate main rotors spinning in such a direction that they tend to cancel the torque of each other.

duckbill pliers — Flat-nosed pliers used extensively in safety wiring.

duct — A hollow tube used to transmit and direct the flow of air through an aircraft.

duct support systems — Methods and apparatus used to support cabin air supply ducts.

ductility — The property which allows metal to be drawn into thinner sections without breaking.

ductwork — The channels or tubing through which the cabin air supply is distributed.

dump valve — The valve which allows the fuel in a tank to be dumped in flight to decrease the landing weight of the aircraft.

duo-servo brakes — Brakes that use the momentum of the aircraft to wedge the lining against the drum and assist in braking when the aircraft is rolling either forward or backward.

duplex bearing — A matched pair of bearings with a surface ground on each to make contact with the other matched surface. When three bearings are used they are called triplex bearings. When four are used they are called quadplex, etc. These are usually ball bearings.

duplexer — A circuit which makes it possible to use the same antenna for transmitting and receiving without allowing excessive power to flow to the receiver.

durability — The ability to last despite hard wear.

Duralumin — Original name of the aluminum alloy now known as 2017. First produced in Germany and used in their Zeppelin fleet of WWI.

Dutchman shears — Common name for compound-action aviation shears. Normally, they come in three forms: straight cut, those which cut to the left, and those which cut to the right.

dye penetrant inspection — An inspection method for surface cracks in which a penetrating dye is allowed to enter any cracks present and is pulled out of the crack by an absorbent developer. A crack appears as a line on the surface of the developer.

dynafocal engine mount — A mount which attaches an aircraft engine into the airframe in which the extended center line of all of the mounting bolts would cross at the center of gravity of the engine and propeller combination.

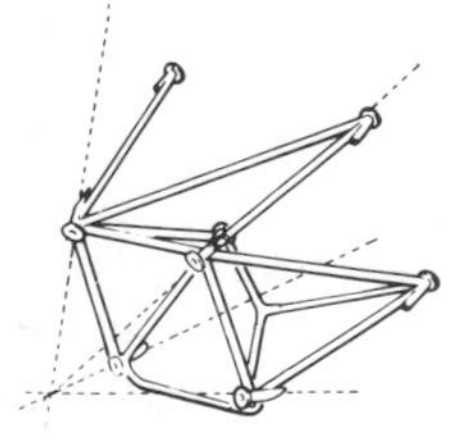

dynamic balance — The condition that exists in a rotating body in which all of the rotating forces are balanced within themselves and no vibration is produced by the body in motion.

dynamic damper — A counterweight on the crankshaft of an aircraft engine. It is attached in such a way that it can rock back and forth while the shaft is spinning and absorb dynamic vibrations. It, in essence, changes the resonant frequency of the engine/propeller combination.

dynamic factor — The ratio between the load carried by any part of an aircraft when accelerating and the corresponding basic load.

dynamic pressure — The product $1/2\ p\ V^2$, where p is the density of the air and V is the relative speed of the air.

dynamic stability — The stability of an aircraft which causes it, when once disturbed from straight and level flight, to return to a condition of straight and level flight in progressively diminishing oscillations.

dynamometer — An instrument used to measure torque force or power.

dynamotor — An electrical device containing two rotating armatures on the same shaft to convert low-voltage DC into high-voltage DC.

dynatron effect — The area of operation in a tetrode electron tube where plate current decreases as plate voltage increases. This effect is caused by secondary electrons to the screen grid.

dynode — The elements in a multiplier tube which emit secondary electrons.

Dzus fastener — A patented form of cowling fastener in which a slotted stud is forced over a spring steel wire and rotated to lock the wire in a cam.

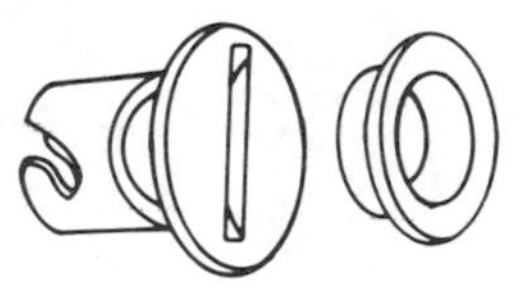

E

earplug – A rubber, wax, or soft plastic that is worn in the canal of the ear to keep loud noises from damaging the ear and to prevent hearing loss.

earth connection – "Earth" is the term that is used in the United Kingdom for the electrical term "ground" used in the United States.

earth induction compass – A form of direction indicator which derives its signal from the lines of flux of the earth cutting across the windings of the flux valve mounted in the airplane.

earth's magnetic field – Magnetic lines of flux which surround the earth. These lines enter and leave the earth at the magnetic north and south poles, different from, but near, the geographic poles.

Easy-out – A type of screw extractor used to remove broken screws or studs from their holes. It is made of hard steel and has a point with a tapered, left-hand spiral-like thread. A hole is drilled in the shank of the broken screw that is to be extracted and the Easy-out is screwed into it by turning it to the left to unscrew it from its hole.

eccentric – Two or more circular objects turning on the same shaft but having different centers.

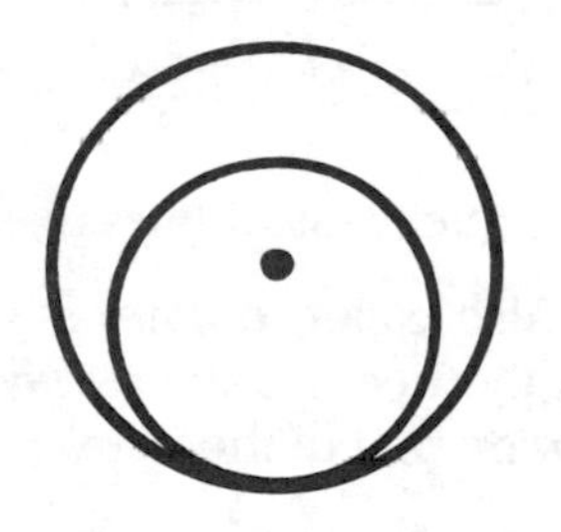

eccentric cam – A circular cam that has a center that is different from the center of the shaft that drives it. As the shaft rotates about its axis, the outside of the eccentric rises and falls, changing rotary motion into linear motion.

economizer system – Sometimes called power compensator and power enrichment system in a carburetor or fuel injection system that adds additional fuel, thereby increasing the mixture and richness of fuel during high power engine operations. The economizer is closed during cruising speeds.

E-core – The laminated core of an electric transformer, cut in the shape of the letter "E", on which the coil windings are mounted.

eddy current – Current induced into the core of a coil, transformer, or the armature core of a motor or generator by current flowing in the winding. Eddy currents cause power loss and are minimized by laminating the iron cores.

eddy current inspection – A form of nondestructive inspection used to locate surface or subsurface defects in a metal part. This is a comparison-type inspection, based on the difference in conductivity of a sound and a defective part.

eddy current losses – Electrical losses in the core of a transformer or other electrical machine. The induction of eddy currents into the core robs the machine of some of its power.

edge distance – The distance from the center of a bolt or rivet hole to the edge of the material.

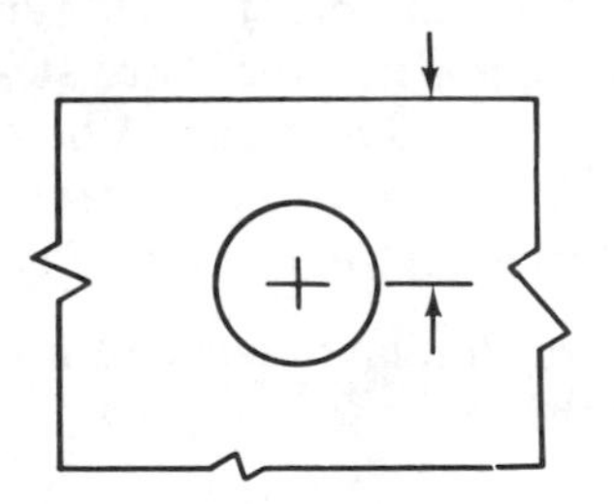

edge thickness – The thickness of the edge of a material.

edge-grain wood – Also called quarter-sawed wood. Wood that has been sawed from the tree in such a way that the edges of its grain are visible in the wide part of the plank.

Edison effect – The discovery of Thomas A. Edison in 1883 that a heated filament placed in an evacuated tube with another electrode will emit electrons.

eductor – A configuration of jet pump that is used in some aircraft fuel systems to remove fuel from a vent-drain tank and return it to the main tank.

effective pitch – The actual distance a propeller moves through the air in one revolution. It is the difference between the geometric pitch of the propeller and the prop slip.

effective thread – The effective (or useful) thread includes the complete thread and that portion of the incomplete thread having crests not fully formed.

effective value – Root mean square (rms) value of sine wave AC. It corresponds to the DC value which will produce the same amount of heat and is equivalent to 0.707 times the peak value.

efficiency – A measure of the effectiveness of a system or a mechanism found by dividing the output of the mechanism by its input. It is usually expressed as a percentage.

E-gap angle – The number of degrees of magnet rotation beyond its neutral position at which the primary breaker points, in a magneto, open. It is at this point that the primary current flow is the greatest, and therefore, the rate of collapse of the primary field will induce the greatest voltage into the secondary winding.

ejection seat – An emergency escape seat used in military aircraft which may be propelled or shot from the aircraft in the event of an emergency.

ejector pump – A pump that produces a low pressure as fluid flows through it. The low pressure is used to eject, or move, fluid from one place to another. *See also eductor.*

elastic limit – Maximum load in PSI a metal will maintain without causing a permanent set, or below which, it will return to its original dimensions when the load is removed, no matter how many times the load is applied.

Elastic Stop Nut – A self-locking nut using a collar of elastic material with an inside diameter slightly smaller than the outside diameter of the bolt or stud it fits. Forcing the threads through the collar fills the bolt and prevents the nut backing off inadvertently.

elasticity – The capability of an object or material to be stretched and to recover its size and shape after its deformation.

elastomeric bearing – A metal and rubber composite bearing used to carry oscillating loads where complete rotation is not needed. The bearing is made of alternate layers of an elastomer and metal bonded together. Elastomeric bearings can be designed to take radial, axial, and torsional loads.

elastomers – A rubber or synthetic rubber used in the layers between the metal in an elastomeric bearing. Elastomers can be stretched to twice their original length, and can return to their original size and shape when released.

E-layer — A layer of ionized air that exists between 60 and 75 mi. (100 to 120 km) above the earth.

elbow — A fluid line fitting that is used to join two pieces of tubing at an angle of 90°.

Elco connector — Also called Elcon connector. A special terminal for aircraft battery installation, similar to the Cannon connector. This is a slip-on-type connector, held in place with pressure from a hand screw.

electret — A permanently polarized dielectric material that is produced by heating barium titanate ceramic or carnauba wax and holding it in a strong electric field as it cools.

electret microphone — A device that changes sound pressure into an electrical signal. It consists of a diaphragm made of a thin foil of electret placed next to a metal coated plate. Sound pressure that is picked up by the microphone vibrates the diaphragm causing a voltage. The waveform of this voltage is a copy of the waveform of the sound that vibrated the diaphragm to produce it.

electric discharge machining — A process of machining complex metal shapes by a controlled electric arc which erodes the metal.

electric drill motor — An electric motor, usually of the universal type, which is geared down to provide additional torque and is equipped with a chuck to hold the twist drill.

electric inertia starter — A starter for aircraft reciprocating engines which uses an electric motor to spin a small flywheel to a high speed. The energy in the flywheel, when coupled to the engine crankshaft, spins it to start the engine.

electric strain gage — A device that is used to measure the amount of physical strain that is in a piece of material. A strain gage is made of a piece of very fine wire that is mounted between two pieces of tissue paper about the size of a postage stamp. The strain gage is bonded to the material in which the strain is to be measured, and the two ends of the wire are connected into a sensitive, resistance-measuring bridge circuit. When the material on which the strain gage is mounted is strained, the wire in the strain gage is stretched. It becomes longer and thinner. When its length and its cross sectional area changes, its resistance changes. The change in resistance of the strain gage is proportional to the amount of strain in the material, and the amount of strain is proportional to the amount of stress that caused it.

electric wave — One of the components of a radio wave produced along the length of the antenna.

electrical bonding – The connecting of metal structural parts together with electrical conductors in order to keep them at the same electrical potential. Bonding eliminates static electricity build-up which causes radio interference.

electrical bus – An electrical distribution point to which many circuits may be connected.

electrical charge – Electrical energy that is produced by a generator.

electrical diagram – Diagram or drawing showing the relationship of certain electrical components within a system.

electrical energy – Energy possessed by a substance or device because of a difference in electrical pressure. This can exist because of electromagnetic or electrostatic forces.

electrical equipment – Any electrical unit or combination of units that make up the electrical system.

electrical filter – An arrangement of a choke coil and condenser which is used in an electrical circuit to smooth out current flow.

electrical generator – A motor that converts mechanical energy into electrical energy.

electrical insulator – Material whose outer electrons are so forcibly held to the nucleus that they cannot be dislodged to flow in a circuit. Glass is an example of a good insulating material.

electrical lines – Wiring used on aircraft for powering equipment.

electrical resistance welding – The fusion of metals by clamping them together and passing a low-voltage electrical current through the joint. The resulting heat melts the metal, and the pressure causes the two pieces to fuse together. Spot and seam welding are forms of electrical resistance welding.

electrical shield – A housing that is made of a conductive material that encloses an electrical circuit. The shield picks up any electrical energy that is radiated from the circuit and carries it to ground so it cannot interfere with any other electrical or electronic equipment.

electrical short – An unintentional electrical system fault connection that provides a low-resistance path across an electrical circuit. Electrons can flow through the short to ground without passing through the load.

electrical steel – A low-carbon steel alloy that contains up to 5% silicon and is used in the form of thin laminations for the cores of transformers and the armatures of electrical motors and generators.

electrical strength — The maximum amount of voltage that can be placed on an insulator before the insulator breaks down and allows electrons to flow through it.

electrical symbols — Graphic symbols used in aircraft drawings to represent electrical wiring or components.

electrical zero — A reference position for meshing gears and for installing an indicating pointer.

electrically detonated squib — An explosive charge, usually installed in fire extinguisher systems, which is ignited by electrical methods.

electrically suspended gyroscope — A gyroscope rotor which is supported in an electromagnetic field. The gyroscope can spin with an absolute minimum of friction while it is supported in either of these fields.

electricity — Composed of electrons and protons. Can be observed in the attracting and repelling of objects electrified by friction and in natural phenomena (lightning and aurora borealis). Usually employed in the form of electrical currents generated by a mechanical motor or generator device which forces the flow of electrons from an area having an excess of electrons to an area with a shortage of electrons. In the process, heat is produced, and a magnetic field surrounds the conductor.

electroacoustic transducer — A device that changes variations in sound pressure into variations of voltage, or vice versa.

electrochemical action — The corrosive results of the electrode potential difference of two metals in contact with each other in the presence of an electrolyte.

electrochemistry — That branch of chemistry which deals with the electrical voltages existing within a substance because of its chemical composition.

electrode — A terminal element in an electric device or circuit. Typical electrodes are the plates in a storage battery, the elements in an electron tube, and the carbon rods in an arc light.

electrode potential — A voltage that exists between different metals and alloys because of their chemical composition. It will cause an electrical current to flow between these materials when a conductive path is provided.

electrodynamic damping — The diminishing of oscillations of the pointer of an electrical meter by the generation of electromagnetic fields in the frame of the moving coil.

electrogalvanizing — The process of coating metal with zinc by electroplating.

electro-hydraulic control — Hydraulic control which is electrically actuated.

electrolysis — A process by which a substance is reduced to its elements by the action on electricity.

electrolyte — A chemical liquid or gas which will conduct electrical current by releasing ions to unite with ions on the electrodes.

electrolytic — The action of conducting electrical current through a nonmetallic conductor by the movement of ions.

electrolytic capacitor — A capacitor that uses metal foil for the electrodes and a thin film of metallic oxide as the dielectric. The sheets of metal foil are separated by a piece of porous paper that is impregnated with an electrolyte. The capacity is affected by the thickness of the dielectric. For example, the thinner the dielectric, the greater the capacity of the capacitor.

electromagnet — A magnet produced by electrical current flowing through a coil of wire. The core is usually made of soft iron which concentrates the lines of flux and intensifies the magnetic field.

electromagnetic emission — The radiation of electromagnetic energy that is produced when electricity flows through a conductor.

electromagnetic induction — Transfer of electrical energy from one conductor to another by means of a moving electromagnetic field. A voltage is produced in a conductor as the magnetic lines of force cut or link with the conductor. The value of the voltage produced by electromagnetic induction is proportional to the number of lines of force cut per second. An emf of 1 volt will be induced when 100,000,000 lines of force are cut per second.

electromagnetic radiation — Electrical energy of extremely high frequency and short wavelength that will penetrate solid objects and expose photographic film after passing through the object.

electromagnetic vibrator — A device that interrupts the flow of DC through a set of contacts and changes it into pulsating DC. The contacts will vibrate between open and closed as long as the vibrator is connected to a source of DC electricity.

electromagnetic waves – Resonance of electric and magnetic fields that move at the speed of light.

electromagnetism – The magnetic field emanating around a conductor that is carrying electrical current. Its strength is determined by the amount of current that is flowing in the conductor.

electromechanical frequency meter – An instrument that uses the resonant frequency of a vibrating metal reed to measure the frequency of alternating current. An electromechanical frequency meter is also called a vibrating-reed frequency meter.

electromotive force – Abbrev.: emf. The force causing electrons to move through a conductor ("Electron moving force"). The symbol E is used in calculations until the actual number of volts is determined.

electron –
[1]The most basic particle of negative electricity. It spins around the nucleus of an atom, and under certain conditions can be caused to move from one atom to another. Electrons that travel in this manner are called free electrons.
[2]The negatively charged part of an atom of which all matter is made. Electrons circle around the nucleus of an atom in orbits or shells.

electron beam welding – A process of welding metal by the heat that is produced when a high-speed stream of electrons strike the metal.

electron current flow – The flow of electrons from negative to positive in a circuit outside of the source.

electron drift – The relatively slow natural movement of individual electrons that move from atom to atom within a conductor.

electron gun – The combination of an electron-emitting cathode together with accelerating anodes and beam-forming electrodes to produce the electron beam in a crt.

electron shell – A grouping of electron paths or energy levels. Beginning with the shell closest to the nucleus, the shells are labeled K, L, M, N, etc.

electron spin – The rotation of an electron about its own axis.

electron tube –
[1]A device consisting of an evacuated or gas-filled envelope containing electrodes for the purpose of controlling electron flow. The electrodes are usually a cathode "electron emitter", a plate "anode", and one or more grids.
[2]Also goes by the name of vacuum tube. It consists of a cathode and its heater, the grids, and plate which is usually housed in a glass envelope. Oxygen is removed leaving the tube in a vacuum, or it may be filled with an inert gas.

electron-flow – The current-flow in a circuit which is actually the flow of electrons. Electrons flow from negative to positive in the external circuit.

electronic emission – The freeing of electrons from the surface of a material usually produced by heat.

electronic leak detector – An electronic oscillator device which emits an audible tone if any refrigerant gas is picked up in its sensor tube. When a refrigerant leak is detected, the tone changes.

electronic moisture indicator – A device for checking moisture in a material. It operates on the principle of measuring the conductivity of the material.

electronic oscillator – An electronic device which emits an audible tone. This device is used in a leak detector. When a leak is detected, the tone changes.

electronic oscillator – An electronic circuit that converts DC into AC electricity.

electronic voltmeter – A special type of electronic instrument that is used to measure voltage.

electronics –
[1]The branch of science that deals with electron flow and its control.
[2]In physics the study and use of the movement and effects of free electrons and with electronic devices.

electroplating – An electrochemical method of depositing a film of metal on some object. The object to be plated is the cathode, the metal which will be deposited, the anode; and the electrolyte is some material which will form ions of the plating metal.

electrostatic charge – An electrical charge on an object caused by an accumulation of electrons or by a lack of electrons.

electrostatic deflection control — A method of controlling the position of a beam of electrons on the face of a cathode ray tube. The beam of electrons that forms the trace, or picture, on a cathode ray tube may be deflected to the correct position on the screen by electrostatic charges on plates that are placed above and below, and on each side of the beam.

electrostatic energy — In a capacitor, it is energy stored by an electric stress in the dielectric of a capacitor when two opposing electrical charges act across the dielectric.

electrostatic field — A field of force that exists around a charged body, sometimes called a dielectric field.

electrostatic stress — The electrical force which tends to puncture an insulator. It is caused by an accumulation of electrical charges on a body.

electrostatics — Branch of physics which deals with the attraction and repulsion of static electrical charges.

electrovalent or ionic bond — The bond formed by two atoms when one atom gives up one or more valence electrons to the other. The bond is based on the attraction between the positive and negative ions thus formed.

element — One of the basic known chemical substances that cannot be divided into simpler substances by chemical means.

elevator — A horizontal, movable control surface on the tail of an airplane. It is used to rotate the airplane about its lateral axis.

elevator angle — The angular displacement of the elevator from its neutral position. It is positive when the trailing edge of the elevator is below the neutral position.

elevator control tab — A metal tab located on the elevator that helps the pilot to control the elevator.

elevon — A control surface which combines the functions of both ailerons and elevators. Movement of the control wheel to the right or left causes the elevons to move differentially (the left elevon moves up and the right elevon moves down). The differential movement of the elevons causes the airplane to rotate about its longitudinal (roll) axis.

elliptical — Of, or having, the form of an ellipse. An elongated oval shape.

elongate — To stretch out or lengthen.

embarkation — The loading of passengers and cargo onto the airplane.

embossing – The process of raising a boss or protuberance on the surface.

emergency air pressure – Compressed air stored in high-strength steel cylinders used to provide emergency landing gear extension and emergency braking in the event of the failure of the main power system.

emergency locator transmitter – Abbrev.: ELT. A small, self-contained radio transmitter that will automatically, upon the impact of a crash, transmit an emergency signal on 121.5 and 243.0 MHz.

emery paper – A semi-abrasive paper composed of pulverized corundum or aluminum oxide, bonded by a layer of emery dust to one side of a sheet of flexible paper.

emery wheel – A semi-abrasive material composed of pulverized corundum or aluminum oxide and a suitable binder are combined into the form of a wheel.

emitter – The electrode of a transistor that corresponds to the cathode of a vacuum tube. Conventional current enters a transistor through the emitter.

empennage – The rear portion, or the tail section, of an airplane.

empty weight – The weight of the structure of an aircraft, its powerplant and all of the fixed equipment. It includes only unusable fuel and undrainable oil.

empty weight center of gravity – Abbrev.: EWCG. The center of gravity of an airplane which includes all fixed equipment, but only the unusable fuel and undrainable oil.

empty weight center of gravity range – The most forward and rearward empty weight center of gravity locations that will not allow the legally loaded center of gravity to fall outside of the loaded center of gravity range.

empty weight moment – The moment of an aircraft at its empty weight.

emulsion-type cleaner – A chemical cleaner which mixes with water or petroleum solvent to form an emulsion. It is used to loosen dirt, soot, or oxide films from the surface of an aircraft.

enamel – A material whose pigments are dispersed in a varnish base. The finish cures by chemical changes within the base.

encapsulate – To completely surround or cover something.

encased – Enclosed in a housing.

enclosed relay – An electrical relay in which both the coil and the contacts are enclosed in a protective metal housing.

encoding altimeter – Form of pneumatic altimeter which codes the transponder differently for each 100 ft. of altitude.

end spanner – Form of socket wrench which has a series of raised lugs around its end rather than splines broached inside.

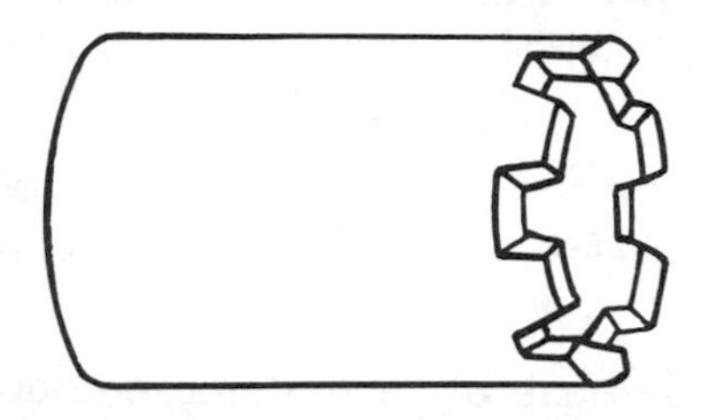

end voltage – The voltage across a chemical cell when the cell should be discarded or recharged. The end voltage of a particular type of cell is usually defined by the manufacturer. Also called end-of-life voltage.

endurance – The length of time that an aircraft can remain in the air. The power produced by the engines and the flight conditions can be regulated to give the aircraft the greatest speed, the greatest range, or the greatest endurance.

energy – The capacity for doing work.

energy level – A particular path that an electron may follow while revolving around an atomic nucleus. Each energy level may contain only up to a maximum number of electrons.

engaging solenoid – A solenoid used to engage an inertia starter with the engine.

engine – A machine that uses heat energy to develop mechanical power.

engine analyzer – An electronic instrument using a cathode ray oscilloscope as an indicator to analyze the condition of the ignition system and to visually examine the vibrations in the engine.

engine breather – The vent for the crankcase of a reciprocating engine. It allows fumes to escape from the crankcase and prevents a pressure build-up inside the engine.

engine compartment – The area of an aircraft in which the engine and its components and accessories are located.

engine compressor – The section of a turbine engine in which the air is compressed before it enters the burner section.

engine conditioning — An integrated system of engine checks and tests whereby engines can be brought up to or kept in top operating condition. Two of the most important checks of engine conditioning are the compression test and the cold cylinder check.

engine controls — All of the controls that are required for the proper operation of an aircraft engine. They include the throttle, mixture control, propeller pitch control, carburetor heat, engine cowl flap control, and the ignition switch.

engine cycle —
[1]The cycle of events which must be accomplished in the transformation of chemical energy into mechanical energy. The two most common cycles of events are the Otto cycle, which describes the events of the reciprocating engine, and the Brayton cycle, which describes the transformation taking place in a turbine engine.
[2]For ease of record keeping with airlines, cycles are recorded as one takeoff and landing. Also described as one start and one stop of the engine or sometimes as one full advance and retard of the throttle. The latter two situations require special recording procedures for maintenance runs if cycle times are needed.

engine gage unit — A 3-in-1 instrument used to show the operating condition of an engine. It houses a fuel pressure, oil pressure, and oil temperature gage in one case.

engine history recorder — An electronic data collection device on some newer engines which records the number of times certain normal operating parameters such as speed and temperature are reached.

engine logbook — A record book of the time in service, maintenance performed, inspections, etc. on an aircraft engine.

engine mount — The structure used to attach an engine in the airframe. It normally includes shock mounts.

engine mounting pads — The shock absorbing units connected between the engine and the engine mount.

engine nacelle — The streamlined, enclosed housing on a wing or fuselage in which the engine is mounted.

engine performance — The relationship between power, RPM, fuel consumption, and manifold pressure of an engine.

engine pressure ratio — Abbrev.: EPR. The ratio of turbine discharge pressure divided by compressor inlet pressure which is used as an indication of the amount of thrust being developed by a turbine engine.

engine ratings — Engine power ratings as type certificated by the FAA. These ratings list thrust or shaft horsepower at takeoff, cruise, etc.

engine ring cowl — Ring-shaped covering over the cylinders of a radial engine for the purpose of streamlining and improving the airflow through the engine.

engine seizure — The locking-up or stopping of an engine because of some internal malfunction.

engine stations — Numerical designations for specific locations along the length of the engine from inlet station one to jet nozzle station ten for ease of describing pressures and temperatures. Numbering differs between engine types according to MIL-E-5007.

engine sump — The lowest point in the engine from which the oil may be drained.

engine trimming — *See power trim.*

engine-driven air pump — An air pump, or vacuum pump as it is often called, that is driven from an accessory drive on the engine.

engineer — In the United States, it is a person who practices the profession of engineering and is the designer of aircraft. In the United Kingdom, it is an aircraft maintenance technician.

enrich — To make a fuel-air mixture ratio richer. When the amount of fuel metered into the engine is increased without increasing the amount of air, the mixture is enriched.

entrained air — Foam or bubbles in the scavenged oil caused by heat and the centrifugal action of the oil-wetted parts. Oil with large quantities of entrained air is a poor lubricant, this is the air that has to be removed. *See also deaerator.*

entrained water — Water held in suspension in aircraft fuel. It is in such tiny droplets that it passes through filters and will do no damage until the temperature of the fuel drops to the point that these tiny particles accumulate or coalesce to form free water in the tank.

envelope — A pre-sewn cover made of aircraft fabric which is slipped over the structure and attached.

envelope method of recovering — A method of recovering an aircraft structure in which a pre-sewn fabric envelope is slipped over the structure and attached. The opening is closed either by cementing the fabric to the structure or by hand sewing.

environmental – Referring to the conditions surrounding some object, in this case, pertaining to the properties of air around an aircraft.

environmental control systems – The control systems include supplemental oxygen systems, air conditioning systems, heaters, and pressurization systems in an aircraft that make it possible for an occupant to function at high altitude.

Eonnex – A fabric woven from polyester fibers.

epicyclic gear train – An arrangement of gears in which one or more gears travel around the circumference of another gear.

epoxy – A flexible, thermosetting resin made by the polymerization of an epoxide. It is noted for its durability and chemical resistance.

epoxy primer – A two-part catalyzed paint material used to provide a good bond between a surface and the topcoat.

EPR-rated gas turbine – A method of expressing the thrust of a gas turbine engine in terms of engine pressure ratio.

equalization – The process of restoring all of the cells of a nickel-cadmium battery to a condition of equal capacity. All of the cells are discharged, shorted out, and allowed to "rest". The battery is then said to be equalized and ready to receive a fresh charge.

equalizer circuit – A circuit in a multiple-generator voltage-regulator system which tends to equalize the current output of the generators by controlling the field currents of the several generators.

equilibrium – A condition that exists within a body when the sum of the moments of all of the forces acting on the body is equal to zero.

equipment – Any item that is secured in a fixed location to the aircraft and is to be utilized in the aircraft.

equipment list – A comprehensive list of equipment installed on a particular aircraft. This includes the required and optional equipment.

equivalent airspeed – Calibrated airspeed, shown on the airspeed indicator, corrected for errors, that are caused by the compressibility of the air inside the pitot tube or by the installation of the instrument.

equivalent circuit – A circuit containing only one or two components which has the same properties as a more complex circuit for the purposes of analysis.

equivalent flat plate area – The area of a square flat plate, normal to the direction of motion, which offers the same amount of resistance to motion as the body or combination of bodies under consideration.

equivalent monoplane — A monoplane wing equivalent as to its lift and drag properties to any combination of two or more wings.

equivalent shaft horsepower — Abbrev.: ESHP. A unit of measured power output of turboprops and some turboshaft engines. Where ESHP = SHP + HP from jet thrust (HP from jet thrust = thrust + 2.6 approximately).

equivalent specific fuel consumption — Abbrev.: ESFC. A means of comparison for turboprops and some turboshaft where ESFC = Wf – ESHP.

erosion — Gradual removal of materials by wear.

escape velocity — The speed an aircraft or missile must reach in order for it to escape from the gravitational field of the earth.

etch — To chemically remove part of a material. Clad aluminum alloy sheets are etched before painting to microscopically roughen them so that the primer can bond tightly to the surface.

etching — A process of detecting defects in aluminum alloy by use of a caustic soda and nitric acid solution.

ethylene dibromide — A chemical compound of bromine that is added to aviation gasoline used to convert the lead deposits from the tetraethyl lead into lead bromides which are volatile enough to vaporize and pass out the exhaust rather than foul the spark plugs.

ethylene glycol — A viscous form of liquid alcohol (C_2HSO_2) used as a coolant for high-powered liquid cooled engine.

eutectic metal — A metal alloy whose melting point, due to the proportion of its components, is the lower after combining than would be possible before the mixture of the same components. Solder is an example of an eutectic metal.

evacuated bellows — A set of bellows from which all of the air has been removed and the bellows sealed. They serve as the sensitive element in an aneroid barometer for measuring atmospheric pressure.

evaporation — The process of changing a solid or liquid into a vapor.

evaporator — The unit in a vapor-cycle air conditioning system in which liquid refrigerant absorbs heat from the cabin to change the refrigerant into a vapor. Air blown over the evaporator loses its heat and is cooled.

excitation — The application of electric current to the field windings of a generator to produce a magnetic field. Also, the input signal to an electron tube.

exciter — An electrical device that produces the magnetic field for a generator, alternator, or high energy ignition system.

exclusive OR gate — A logic device having two inputs on which a voltage on either input — but not both — will produce a voltage at the output.

exerciser jack — Hydraulic jack placed under the oleo shock absorber of a landing gear to exercise it up and down while it is being filled with hydraulic fluid. The object is to work all of the air out of the fluid.

exfoliation corrosion — A severe form of intergranular corrosion that attacks extruded metals along their layer-like grain structure.

exhaust back pressure — The pressure produced in the exhaust system of a reciprocating engine due to the diameter and length of the exhaust pipe. The pressure opposes the flow of exhaust gases as they are forced out of the cylinders when the exhaust valve is open.

exhaust cone — A separate outer casing housing the tail cone and located between the turbine case and the tail pipe on some engines. Also called the exhaust connector.

exhaust duct — A duct bolted to the rear of the basic turbine engine to configure the engine to a particular airplane. Same as tailpipe.

exhaust gas analyzer — A device that is used to analyze the chemical composition of the exhaust gas by measuring the conductivity of the exhaust gas and indicates the ratio of fuel and air in the mixture that produced it.

exhaust gas temperature — Abbrev.: EGT. The temperature of the exhaust gases as they leave the cylinders of a reciprocating engine or the turbine section of a turbine engine.

exhaust gas temperature indicating system: —
[1]Reciprocating engine: A system for measuring the EGT of a reciprocating engine to give the pilot an indication of the efficiency with which the fuel-air mixture inside the cylinder of the engine is burning.
[2]Gas turbine engine: An indicating system used that measures the temperature of the exhaust gas as it enters the tailpipe after passing through the turbine.

exhaust heater — Thin sheet metal shrouds around an exhaust muffler or exhaust stacks. It is used to heat the air that is taken into the cabin or into the carburetor.

exhaust manifold – The collector arrangement for exhaust gases of reciprocating engines. The manifold attaches to the individual exhaust ports and carries the exhaust gases overboard through a common discharge.

exhaust nozzle – The rear opening of a turbine engine exhaust duct. The nozzle acts as an orifice, the size of which determines the density and velocity of the gases as they emerge from the engine.

exhaust port – The hole in the cylinder of a two-cycle engine through which the exhaust gases leave.

exhaust stacks – Short, individual pipes attached to the exhaust ports of the cylinders of reciprocating engines, through which the exhaust gases are discharged overboard.

exhaust stroke – The stroke of the Otto cycle where the exhaust gases are forced out of the cylinder as the piston is moving away from the crankshaft and the exhaust valve is open.

exhaust valve – The valve in an aircraft engine cylinder through which the burned gases leave the combustion chamber.

exit guide vanes – Fixed airfoils at the discharge end of an axial flow compressor that straighten out the swirling air caused by the rotating rotors so that the air leaves the engine in an axial direction.

expand – To enlarge in dimensions.

expanded plastic – The increase in the volume of plastic resin generated when the materials that make up the plastic are mixed. The volume is increased by bubbles of gas.

expander-tube brake – A form of nonservo brake in which composition blocks are forced out against a rotating drum by hydraulic fluid, expanding a synthetic rubber tube on which they rest.

expansion boots – Another name for the inflatable deicer boots.

expansion coefficient – A number which describes the change in linear dimensions of a material with a specified change in its temperature.

expansion reamers – Precision cutting tools used to enlarge and smooth the inside circumference of a drilled hole. Diameter of the reamer may be changed by an adjustable wedge inside the blades.

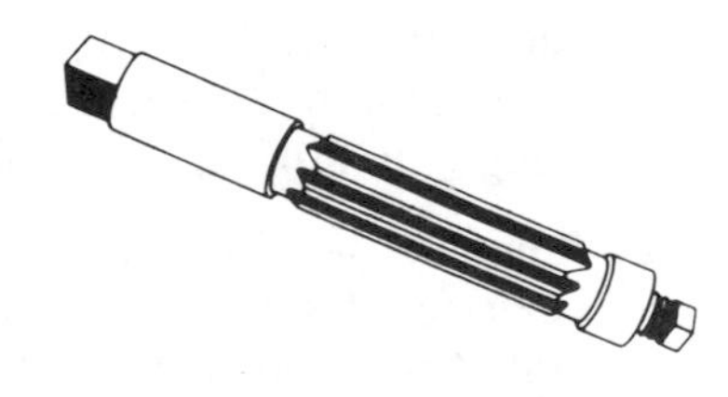

expansion turbine — A turbine wheel in an air-cycle air conditioning system used to extract some of the energy from the compressed air by causing it to do work in driving the turbine and then expanding it. The expansion turbine drives the air compressor.

expansion wave — The change in velocity and density of the air as it passes over the thickest part of an airfoil moving through the air at speeds greater than the speed of sound.

expel — To eliminate or get rid of.

expendable weight — Weight which is decreased in flight. The fuel on board is an expendable weight as it decreases in flight.

explosive charge — A unit incorporating an explosive to break a seal and discharge a substance.

explosive rivet — A patented blind rivet manufactured by the DuPont Company. Its hollow end is filled with an explosive and sealed with a cap. When the rivet is heated, it explodes, swelling its end and clamping the metal together.

exponential — A math term used with reference to a graph indicating that the gradient is not a straight line, but rather a curve.

extension lines — Lines on an aircraft drawing which extend from a view for the purpose of identifying a dimension.

extent of damage — An amount of damage sustained by an airplane or its components.

exterior angle — The angle between one of the sides of a polygon and an extension to an adjacent side.

exterior view — A view of an object showing only its outer or visible surfaces.

external combustion engine — A form of heat engine in which the chemical energy in the fuel is converted into heat energy released to the outside of the engine. Heating water to produce steam which, in turn, is put to mechanical use is a form of external combustion engine.

external inspection — Any inspection which is done externally to the airframe, engine, or unit component without having inspected the internal mechanism by disassembly, nondestructive inspection methods, etc. Usually, it is considered as part of the visual inspection.

external load — A load that is carried, or extends, outside of an aircraft fuselage.

external-control surface locks — Locks which are applied on the exterior of the control surfaces of a parked aircraft to prevent their moving in gusty wind.

external tooth lock washer — A thin, spring steel, shake-proof lock washer with twisted teeth around its outside circumference that holds pressure between the head of a screw or bolt and the metal surface to prevent the fastener loosening.

extinguishing agent — The agent used in a fire extinguishing system to either cool the fuel below its kindling point or to exclude oxygen from the surface of the fire.

extrude — To form by forcing through a die of the desired shape.

extrusion — A strip of metal, usually of aluminum or magnesium, which has been forced, in its plastic state through a die. This can produce complex cross-sectional shapes required for modern aircraft construction.

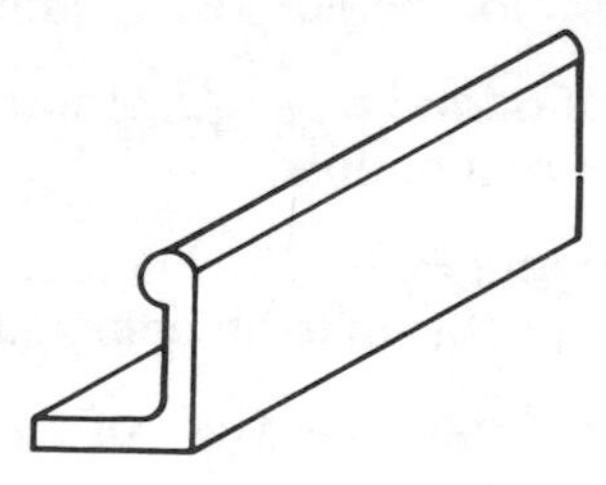

eye — The hole of an eyebolt.

eyebolt — A bolt that has a flattened head with a hole in it. An eyebolt is used to attach a cable to a structure.

eyebrow lights — Small shielded lights positioned over the two top corners of an instrument of an aircraft instrument panel. The lights illuminate the dial of the instrument, but they do not shine in the eyes of the pilot.

eyelet — A flanged tubular fastener designed for securing by curling or splaying the end.

F

FAA Form 337 – Major Repair and Alteration Form.

FAA-approved data – Data which may be used as authorization for the techniques or procedures for making a repair or an alteration to a certificated aircraft. Approved data may consist of such documents as Advisory Circular 43.13-1A and -2A, Manufacturers Service Bulletins, Manufacturer's kit instruction, Airworthiness Directives, or specific details of a repair issued by the engineering department of the manufacturer.

FAA-PMA – The identifying letters required on an aircraft part or component to signify it as being manufactured under a Federal Aviation Administration Parts Manufacturing Approval.

fabric material – Cloth used to cover truss-type aircraft structures. The basic fabric is grade-A long staple cotton, but Irish linen may be used interchangeably. Synthetic fabrics made of polyester resins and specially treated glass fibers may be used in place of cotton.

fabric punch tester – A hand tester used to give an indication of the relative strength of aircraft fabric by measuring the force required to press a specially shaped pointed plunger a specific distance into the fabric.

fabric repair – The repair to a fabric covered structure that produces the same strength and tautness in the fabric as it originally had.

fabricate – To construct or build something.

fabrication – The assembling of parts to make a complete unit or a structure.

face of a propeller – The flat side of the propeller blade which strikes the air first.

face of the drawing – The surface of the object as seen from the front view.

face shield – A transparent protective guard which covers the entire face to protect against flying objects or liquid spray.

face-end spanner – A type of semicircular, open-end wrench having short pins extending from its face and used to turn special circular-type nuts.

faceplate, lathe — Used for turning metal. A heavy, steel disk with a smooth face that is mounted on the headstock of the lathe and is turned by it.

face-to-face bearings — Bearing sets installed in such a way that one set carries thrust loads in one direction while the other bearing set carries thrust loads in the opposite direction.

facing — Facing is a machining operation on the end, flat face or shoulder of a part.

factor of safety — The ratio of the ultimate strength of a member to the probable maximum load. This ratio is larger than one.

fading —
[1]A decrease in the friction applied by a drum-type brake when it is hot. As the drum is heated, it expands in a bellmouth fashion, and part of it moves away from the lining. This decreases the friction area and causes a loss of breaking action.
[2]A decrease in strength of a received radio signal.

Fahnstock clip — An electrical type of spring clip connector that is used to temporarily connect a wire to an electrical circuit.

Fahrenheit — Abbrev.: F. A temperature scale that measures 32° at which pure water freezes, and 212° at which pure water boils under standard atmospheric pressure conditions and 180 equal degrees between them.

fail hardover — A failure of an automatic flight control system in which a steady signal is produced that drives the controls to the extreme end of their travel and holds them there.

fail-safe — A design feature which transmits the loads into a secondary portion of the structure in the event of the failure of the primary structure.

fail-safe control — A type of control that automatically puts the controlled device in a safe condition if the control system should fail.

faired curve — A smoothly curved object.

fairing — A smooth covering over a joint or a junction in an aircraft structure to provide a smooth surface for the airflow. Its primary purpose is to reduce drag.

fairlead — Wood or plastic guides for aircraft control cable, used to hold the cable away from the structure in straight runs.

false rib – Short, semi-rib extending from the leading edge of the wing back beyond the front spar. It is used to give rigidity and better shape to the leading edge of fabric covered wings.

false spar – A partial spar in an aircraft wing to which the aileron hinges attach.

false start – A condition in starting a turbine engine in which ignition occurs but the RPM will not increase. This condition is usually caused by the engine not being turned fast enough by the starter when ignition occurs. *See also hung start; hot start.*

fan – Constructed as an extension of some of the turbine blades rather than as an extension of the compressor blades. This fan pulls large volumes of air around the outside of the gas generator portion of the engine.

fan air – The portion of airflow through a turbofan engine which is acted upon by the fan stages of the compressor.

fan marker – An aircraft directional radio signal that is transmitted vertically upward from a transmitter that is located along a navigational radio range and is heard only when the aircraft is directly over the transmitter.

farad – Abbrev.: f. Basic unit of capacitance. A capacitor of one farad will hold one coulomb (6.28 × 1018 electrons) under a pressure of one volt.

fastener – A device such as a rivet or bolt used to fasten two objects together.

fatigue – Weakening and eventual failure of a metal due to continued reversal, or repeated stresses, beyond the fatigue limit.

fatigue crack – A crack in a structural member caused by flexing or vibration.

fatigue failure – The failure of a material due to flexing or vibration.

fatigue limit – The amount of flexure of vibration a body will stand before fatigue failure occurs.

fatigue resistance – The property which enables a metal to withstand repeated loads or reversals of loads and vibrations.

fatigue strength – The ability of a material to withstand vibration and flexing.

fault – A defect in an electrical circuit.

faying strip – The strip along the edge of a sheet metal skin where a lap joint is formed. This inaccessible area is highly susceptible to the formation of corrosion.

faying surface – The overlapping area of adjoining surfaces.

feather edge – A very thin, sharp edge that is subject to bending or breaking.

feathering propeller – A controllable pitch propeller with a pitch range sufficient to allow the blades to be turned parallel to the line of flight to reduce drag and prevent further damage to an engine that has been shut down after a malfunction.

feathering solenoid – A locking, electrical solenoid used with a Hamilton Standard hydromatic propeller. The solenoid keeps the feathering pump running after the feathering button has been momentarily depressed. The solenoid is deenergized when the propeller becomes fully feathered.

Federal Aviation Administration – Abbrev.: FAA. A part of the Department of Transportation. The FAA establishes the rules and regulations as well as enforces those ruling. The purpose of the FAA is to set the standards of civil aircraft for the public safety.

Federal Aviation Regulations – Abbrev.: FARs. Rules, regulations, and guidelines that govern the operation of aircraft, airways, and airmen established by the FAA for the safety and operation of civil aircraft.

Federal Communications Commission – Abbrev.: FCC. A government board, made up of seven commissioners, responsible for regulating all interstate electrical communications and all foreign electrical communications that originate in the United States.

feedback –

[1]A portion of the output signal of a circuit which is returned to the input.

[2]Positive feedback occurs when the feedback signal is in phase with the input signal. Negative feedback occurs when the feedback signal is 180° out of phase with the input signal.

feedthrough capacitor – A capacitor that serves a dual purpose. In some aircraft, a feedthrough capacitor is used in magnetos to serve as the normal capacitor to minimize arcing of the points, and to decrease the amount of radio interference that is caused by electrical energy being radiated from the ignition switch lead.

feedthrough connector – A connector that is used to carry a group of conductors through a bulkhead.

feel — Any sensation, especially to a flight control or brake system, that is power-controlled. This "feel" allows the pilot to sense how much pressure is needed for the operation of the system.

feeler gage — A measuring tool consisting of a series of precision-ground steel blades of various thicknesses. It is used to determine the clearance or separation between parts.

fence — A fixed vane that extends chordwise across the wing of an airplane. Fences prevent air from flowing along the span of the wing.

Fenwal spot-type fire detection system — A fire detection system utilizing bimetallic thermal switches between two insulated conductors. A fire or overheat condition closes the circuit between the two conductors and signals the presence of the fire.

ferrite — Particles of iron carbide scattered throughout an iron matrix and existing at ordinary temperatures.

ferritic stainless steel — Straight chromium carbon and low-alloy steels that are strongly magnetic.

ferromagnetic materials — Magnetic materials composed largely of iron.

ferrous metal — Iron or any alloy containing iron.

ferrule resistor — A group of resistors that have metal bands (ferrules) around each end so they can be mounted in standard fuse clips.

ferrule terminals — The terminals that are on each end of a tubular fuse. Used for making connections with the circuit.

ferry — The movement of an aircraft from one location to another.

ferry permit — A flight permit issued by the FAA allowing an unlicensed aircraft to be flown from its one location to another location.

fiber locknut — A type of self-locking fastener with a fiber insert which puts a pressure on the threads to lock the nut in place, thereby preventing the nut from turning when installed in areas subject to vibration.

fiber optics — An optical inspection procedure which uses a tool composed of tiny glass rods which conduct light and vision. The flexibility of a bundle of fiber rods makes inspection around corners practicable.

fiberglass — Extremely thin fibers of glass. May be woven into a cloth or lightly packed into a mat. Used to reinforce epoxy or polyester resin for aircraft structure.

fidelity — Degree of similarity between the input and output waveforms of an electronic circuit.

field — A space in which magnetic or electric lines of force exist.

field coil — Coil or winding used to produce a magnetic field.

field effect transistor — Abbrev.: FET. A special form of semiconductor device with a high input impedance. Electron flow between its source and drain is controlled by a voltage applied to the gate.

field excitation — DC supplied to the field of an alternator or a generator to produce magnetic flux which is cut by the conductors in the armature or stator. *See also excitation.*

field frame — The main structure of a generator or motor within which are mounted the field poles and windings.

field maintenance — Usually refers to maintenance performed on aircraft remotely or semi-remotely from the home station, where there are few tools and equipment to implement normal maintenance procedures.

field strength — The intensity of the magnetic strength of a magnet or electromagnet.

field strength meter — An electrical instrument that measures the strength of an electromagnetic field that is radiated from a radio transmitting antenna.

filament — The heated element in a light bulb or electron tube. The heat speeds up the molecular movement in the cathode which emits electrons.

file — A hand-operated cutting tool made of high-carbon steel and fitted with rows of very shallow teeth extending diagonally across the width of the tool. *See also files, classification.*

files, classification — The teeth on a file vary from very fine to coarse in the following sequence:

1. Dead-smooth cut.
2. Smooth cut.
3. Second cut.
4. Bastard cut.
5. Coarse cut.

Files with a single set of cutting teeth are called single-cut files. A file that has a second set of cutting teeth that cross the first set is called a double-cut.

filiform corrosion – A thread- or filament-like corrosion which forms on aluminum skins beneath polyurethane enamel.

fill – The direction across the width of fabric.

fill threads – Threads running across the width of a piece of fabric from one selvage edge to the other. Also called woof threads.

filler – A material added to a resin to increase its bulk.

filler material – Any material that is mixed or added to give body to the base material.

filler metal – Metal which is used to increase the area of a weld. Normally supplied in the form of the welding rod or the electrode used in arc welding.

filler neck – Usually, a cylinder-shaped neck or tube leading into a reservoir for replenishing fluids.

filler plug – The plug installed in a sheet metal or wood structural repair to make the surface of the repair coincide with the original skin contour. The filler plug is used only to make the surface aerodynamically smooth. The strength of the repair is in the doubler that is inside the structure.

filler rod – Thin metal rod or wire used in welding to provide the necessary filler metal, and is used to provide additional strength to the weld.

filler valve – A readily accessible valve which provides a means of servicing an installed oxygen, air, or fluid system.

fillet – A rounded-out part at the intersection of two plane surfaces to produce a smooth junction where the two surfaces meet. Fillets produce a smooth aerodynamic junction between the wing and the fuselage of an airplane.

fillister-head screw – A machine screw whose shape consists of a rounded top surface, cylindrical sides, and a flat bearing surface.

film resistor – A resistor formed by coating a ceramic, glass, or other insulating cylinder with a metal oxide or other thin resistive film.

film strength – A lubricants ability to maintain a continuous lubricating film under mechanical pressure without breaking down.

filter –

[1]A device for straining out unwanted solid particles in a fluid.

[2]A circuit arranged to pass certain frequencies while blocking all others. A high pass filter passes high frequencies and blocks low frequencies.

filtering – The separation of unwanted components from either a fluid flow or an electrical flow.

fin –
[1]The vertical stabilizer of an airplane to which the rudder is hinged. It produces directional stability.
[2]A form of key under the head of a fastener which serves to keep the fastener from turning during assembly and use.

final approach – That part of an airplane's flight in which it is being readied for landing at an airport to terminate the flight.

fineness ratio – The ratio of the length to the maximum diameter of a streamline body, such as an airship hull.

fine-wire spark plug – A spark plug using platinum or iridium electrodes. The small electrodes allow the firing end cavity to be open to provide better scavenging of the lead oxides from the plug. The heat transfer characteristics of the fine wires prevent their overheating.

finger brake – *See box brake.*

finger patch – A form of welded patch to go over a cluster in a steel tube fuselage. Fingers extend along all of the tubes in the cluster.

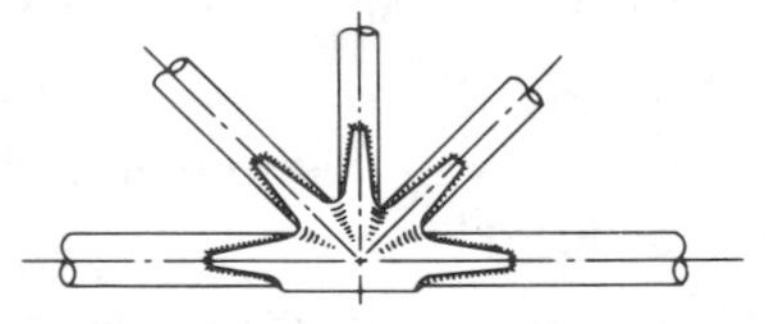

finger screen – Finger-shaped filter screen used on fuel tank standpipes used to screen or filter out large pieces of contamination.

finish turning – A final smoothing process in the machining of a metal part in which the part is turned to its correct dimension.

finite life – A limited part life. The part must be scrapped at the end of its predetermined operating life.

fire alarm relay – A relay actuated by the fire detection system which provides energy for the fire warning bell and the fire warning lights in the cockpit.

fire detection system – An installed system in an aircraft that informs the pilot of a fire on board the aircraft.

fire extinguisher – Any device containing an extinguishing agent used to either cool a material below its kindling point or to exclude oxygen from its surface.

fire extinguishing agent: – Any approved chemical that is used to extinguish a fire by either reducing the temperature of the fuel to a temperature that is below its kindling point or by excluding oxygen from the fire.

fire point – The temperature at which the vapors given off by a substance will ignite and continue to burn when a flame is passed above the surface of the material.

fire valve – A valve which automatically shuts off the supply of combustion air to a combustion-type cabin heater in the event of a fire or overheat condition.

fire zone – An area or region of an aircraft designated by the manufacturer to require fire detection and/or fire extinguishing equipment and a high degree of inherent fire resistance.

fireproof – The capacity to withstand the heat associated with fire without being destroyed.

fire-resistant structure – A structure that is able to resist fire or exposure to high temperature for a specified period of time without being destroyed or structurally damaged.

firewall – A fire-resistant bulkhead which must be installed between an engine compartment and the rest of the aircraft structure.

firewall shutoff valve – A valve located on the airframe side of a firewall that will completely shut off the flow of fuel, oil, or hydraulic fluid to the engine during an engine fire.

firing order – The order, or sequence, in which the cylinders of an internal combustion engine fire in a normal cycle of operation.

firing position – The position of the piston in the cylinder of a reciprocating engine at which time ignition should occur. Igniting the mixture at this position allows the peak cylinder pressure to occur shortly after the piston passes top center.

firmer chisel – A chisel, used in woodworking, that has a thin, flat blade.

first-class lever – A lever in which the fulcrum, or pivot point, is positioned between two forces that act in opposite directions.

fisheyes – Isolated areas on a surface which have rejected the finish because of wax or silicone contamination.

fishmouth splice — A welded splice in steel tube structure in which one tube telescopes over the other. The outside tube is cut into a "V", resembling an open fishmouth, to provide additional area for the weld.

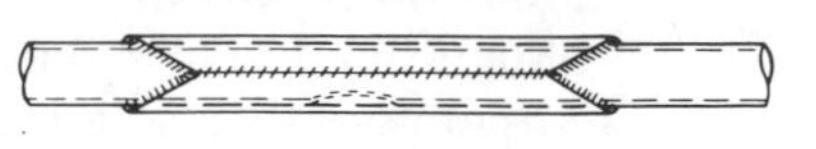

fishtail — A method no longer used to decrease the speed of an airplane on its approach for landing. Fishtailing consist of alternately skidding the airplane to the left and then to the right by using the rudder while keeping the wings level with the control stick.

fissure — A scratch or crack.

fit — Fit is the general term used to signify the range of tightness which may result from the application of a specific combination of allowances and tolerances in the design of mating parts.

fitting — A part used to join or attach assemblies together.

five-hour rating — The ampere-hour rating of a battery which will discharge the battery in five hours. This is the most commonly used rating for aircraft batteries.

five-minute rating — A rating of the ampere-hour capacity of batteries, normally used to indicate the capacity of a battery for high current drains such as starting current. The five-minute rating of a battery is an indication of the way the battery will function under the severe loads put on it by the engine starter.

fix — A geographic location.

fixed — The state of a permanently installed system, in contrast to any type of portable equipment system.

fixed equipment — Nonmovable, attached accessories.

fixed landing gear — A landing gear that is not retractable.

fixed tail surfaces — Often called stabilizers, mounted rigidly to the fuselage. Provide stabilizing effect to aircraft during takeoff, flights, and landing. Fixed tail surfaces provide anchorage for the rudder and elevators.

fixed-pitch propeller — A propeller whose blade angle is fixed.

fixed-wing aircraft — An airplane with rigidly attached wings, as distinguished from a helicopter or autogiro.

fixture — A small jig or device for holding parts in the proper position for assembly.

flameout — A condition in the operation of a gas turbine engine in which the fire in the engine goes out due to either too much or too little fuel sprayed into the combustors.

flammable — Any material that will burn or will support combustion.

flammable liquid — Any liquid that gives off combustible vapors that are easily ignited.

flange — Any design of a machine, motor, or other mechanism that has a ridge that sticks out from the device. It is generally used for attaching something to the device or for connecting two or more devices together.

flaperon — A special type of control used on swept-wing airplanes that serves as both aileron and wing flap.

flapper valve — A type of check valve that allows fluid to flow through it in the direction in which the flow forces the valve off of its seat, but it does not allow fluid to flow in the opposite direction as this causes the flapper valve to close.

flaps — Auxiliary controls that are built into the wings of an airplane used to increase both the lift and the drag of the wing. Extending the flaps increases the drag so the airplane can descend at a steep angle, increases lift for takeoff, and increases the lift so the airplane can fly at a slower speed.

flare —

[1]A flight maneuver made by pulling back on the control wheel just before touchdown to reduce speed and to settle the airplane onto the runway with the least amount of vertical speed.

[2]A signal device that was at one time carried in most airplanes in the event of a crash landing. The flare is usually fired into the air from a specially designed gun signaling the general location of downed aircraft position.

[3]A 37° cone-shaped expansion at the end of a piece of tubing to which a sleeve and nut are slipped over the tubing prior to flaring by a special cone-shaped flaring tool.

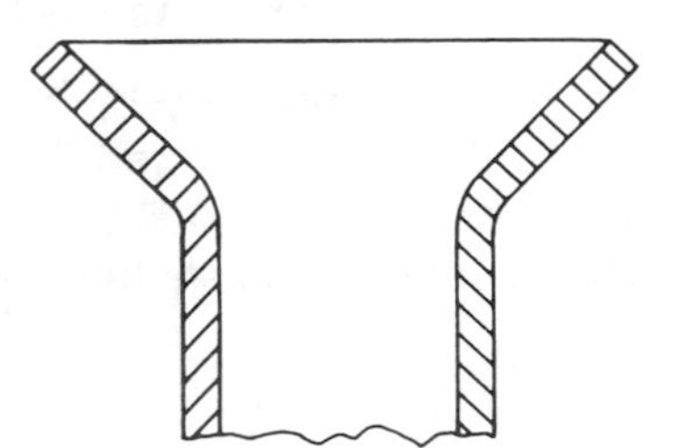

flareless fitting – A form of fluid line fitting used on some hydraulic lines. Instead of using a formed flare on the end of the tube, a compression sleeve is forced into the tube when tightened onto a recess in the attachment fitting to form the fluid-tight sealing surface.

flaring – An operation used to expand the end of a length of tubing in order to produce a tight seal when coupled to another unit.

flaring block – A split clamp, usually made of hardened steel, used to hold the tubing while it is being flared.

flaring tool – A split block with chamfered holes to clamp the various sizes of tubing while a hardened and polished cone is forced into the end of the tubing to form it against the chamfer resulting in a flared seal area.

flash – The thin fin of metal along the sides or around the edges of a forged or upset section. It is caused when metal flows out between the edges of the forging dies.

flash line – A raised line along the boundary of a cast part.

flash plating – A very thin deposit of metal sufficient to give a solid color.

flash point – The temperature to which a fluid must be raised before it will momentarily flash, but not sustain combustion when a small flame is passed above its surface.

flashback – A malfunction in an oxyacetylene torch in which the gases burn inside the mixing head. Flashback is very dangerous and can cause an explosion unless the gases are immediately shut off at the regulator in order to stop the fire inside the torch from burning back through the hoses to the supply tanks.

flasher mechanism – An automatic electrical switching device used for the flashing operation of lights.

flashing off – The drying process of a finish to which solvents have been added for proper spray paint viscosity. Although the surface feels dry to the touch, the film is not completely dry and hard throughout until the proper cure time has been established.

flashing the field – A procedure in which a battery is momentarily connected to the field coil of an aircraft DC generator so that current can flow through it for a few seconds to make a permanent magnet of the field frame in order to restore the residual magnetism

flashover — A condition inside the distributor of a high-tension magneto in which the spark jumps the air gap to the wrong electrode. This may be caused by moisture inside the distributor or by a dirty distributor block.

flash-resistant — Not susceptible to burning violently when ignited.

flat file — A file that is slightly tapered toward the point in both width and thickness. Cuts on all sides. Double-cut on both sides and single-cut on both edges.

flat lacquer — Any lacquer that dries with a non-glossy or flat finish.

flat machine tip — Compressor or turbine blade tips which have a constant cross section, as opposed to a squeeler-tip or a shrouded-tip configuration.

flat rating — A current means of referring to "rated thrust" at a specific temperature above Standard Day value.

flat spin — A dangerous flight condition or flight maneuver in which one wing is stalled while the other wing is flying, and the aircraft is autorotating in a downward spin.

flat washer — Same as plain washer. Used under the head of a bolt or nut in order to protect the connecting material from damage.

flat-compounded generator — A generator that has both a series and a parallel winding. The series field is adjusted by a regulator to keep the output voltage of the generator constant from a no-load condition to the maximum load the generator can produce.

flathead pin — This is often called a clevis pin and is a high-strength steel pin with a flat head on one end and a hole for a cotter pin on the other end, and is used as a hinge for control surfaces or for attaching a cable to a control horn. Clevis pins are designed to take shear loads only.

flathead rivet — An AN442 rivet used for internal structure where the head of the rivet will not be exposed to the airstream, and usually where it is driven with an automatic riveting machine.

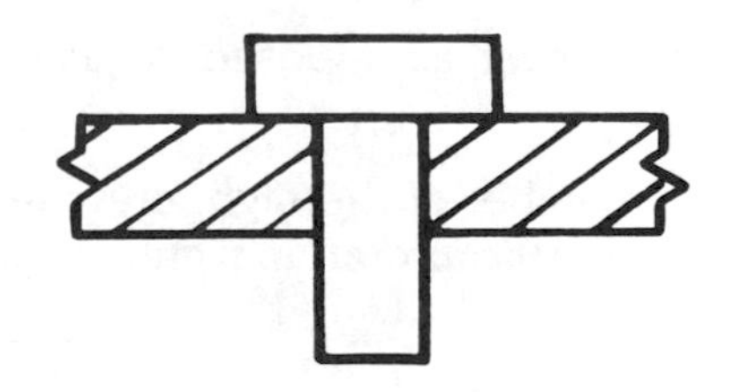

flatnose pliers — Pliers with deep, square jaws and a firm hinge, used to produce a sharp, neat bend in sheet metal and to make flanges along the edge of a part.

fleet weight — The average weight of several aircraft of the same model and with the same equipment. This weight may be used by FAR Part 121 and 135 operators.

flex hose — The colloquial term for flexible tubing used in an aircraft plumbing system to allow relative movement between the two ends of the hose.

flexibility — A material characteristic that allows it to be repeatedly bent, stretched, or twisted within its elastic limits and still return to its original condition each time the bending, stretching, or twisting force is removed.

flexible control cable — Steel aircraft control cable which consists of seven strands of steel wire, each strand having seven separate wires. This is called seven-by-seven cable.

flexture assembly — The flexible unit in Bell Helicopters' nodal vibration dampening system.

flight — Travel through the air from the time an airplane departs from the ground until it lands.

flight control surfaces — The movable airfoils used to change the altitude of the aircraft in flight.

flight controller — Command unit of an autopilot system. It is manually operated to generate signals which cause the aircraft to climb, dive, or perform coordinated turns.

flight deck — The area in an aircraft which houses all of the occupants who fly the aircraft, along with all of the controls used in flight. It includes the flight stations for the pilot, copilot, flight engineer, navigator, and radio operator.

flight director system — A form of automatic flight control in which all of the information is displayed to the pilot rather than being used to actuate control servos.

flight engine — A prime mover, as opposed to gas turbine engines, in an APU or GPU.

flight engineer — The member of the flight crew responsible for the mechanical operation of the aircraft in flight.

flight idle – Engine speed usually in the 70-80% range for minimum flight thrust.

flight level – A level of constant atmospheric pressure related to a reference datum of 29.92 inches of mercury. Each is stated in three digits that represent hundreds of feet. For example, flight level 250 represents a pressure altitude of 25,000 ft.

flight line – The area of an airfield where airplanes are parked. Also called a ramp,or tarmac.

flight manual – Approved information which must be carried in an airplane. This pertains to the speed, engine operating limits, and any other information that is vital to the pilot.

flight path – The flight path of the center of gravity of an aircraft with reference to the earth.

flight recorder – A recording device installed in transport aircraft which produces a record of the aircraft altitude, airspeed, heading, and of the voices in the cockpit. Housed in a crash-proof container, it is used to determine the probable cause of any accident the aircraft should be involved in.

flight simulator – A training mock-up that duplicates the flight characteristics of the aircraft controls and simulates actual flight conditions including instruments, controls, etc., which allows flight training and emergency procedures to be simulated on the ground at a vastly reduced cost.

flight time – The time from the moment the aircraft first moves under its own power for the purpose of flight until the moment it comes to rest at the next point of landing.

flight-path angle – The angle between the flight path of the aircraft and the horizontal.

flint lighter – A flint and steel friction lighter used to ignite an oxyacetylene torch for welding.

float charging potential – A charging potential which may be left connected across the poles of a chemical cell on standby service without damage or destructive overcharge.

floated battery – A permanently installed storage battery that is positioned across the output of a generator. The generator carries the normal electrical load, and it keeps the battery fully charged at all times.

float-type carburetor — A fuel metering device which uses a float-operated needle valve to maintain a constant fuel metering pressure or head.

flock — Pulverized wool or cotton fibers attached to screen wire used as an air filter. The flock-covered screen is lightly oiled, and it holds dirt and dust, preventing its entering the engine.

flood valve — A control valve used to direct the flow of extinguishing agent in a CO_2 fire extinguisher system.

flow control valve — A flow control valve that controls the direction or amount [illegible]luid flow.

flow indicator — A device in an oxygen system to provide users with a positive indication that they are getting sufficient oxygen.

flow meter — An autosyn electrical transmitter which provides a signal to a cockpit instrument. This indicator shows pounds per hour of fuel flow being consumed by an operating engine.

flow reverser — *See thrust reverser.*

flowchart — A diagram that uses special symbols connected by lines to indicate the sequence of the steps that must be followed in order to achieve a desired end result.

fluctuate —
[1]The swing or oscillation of a dial from low to high and/or high to low (back and forth).
[2]To continually change or vary in an irregular way.

fluctuating arc — A malfunction in an inert-gas arc welding system caused by improper grounding.

fluid —
[1]A substance, either gaseous or liquid, that will conform to the shape of the container that holds it. A gaseous fluid will expand to fill the entire container, while a liquid fluid will fill only the lower part of the container.
[2]Any material whose molecules are able to flow past one another without destroying itself. Gases and liquids are both fluids.

fluid mechanics — The science and technology of forces produced by fluids.

fluid ounce — A liquid volume which is equal to 1/16 liquid pint, or 1.8 cu. in.

fluid power – The transmission of force by the movement of a fluid. The best examples are hydraulics and pneumatics.

fluidics – The branch of science that studies the various shapes of ducts to sense, measure and control physical conditions.

fluidity – The ability of a liquid to flow easily and smoothly.

fluorescent – A substance is said to be fluorescent when it will glow or fluoresce when excited. Some types of dye penetrant material use fluorescent dyes which are pulled from the cracks by a developer and observed under ultraviolet light.

fluorescent finish – An aircraft finish which is highly light-reflecting.

fluorescent penetrant inspection – A form of nondestructive inspection in which a part is thoroughly cleaned and immersed in a vat of penetrating oil. When the part has soaked for a sufficient time, it is removed. The oil is washed from the surface, and the part is dried. It is then covered with a developer that will draw the oil from any crack into which it may have seeped. The part is inspected under ultraviolet light which will cause the crack to appear as a vivid green line.

fluorescent pigment – A paint pigment which can absorb visible or non-visible electromagnetic radiation and release it as energy in a wavelength.

fluoroscope – An instrument used for nondestructive inspection.

flush – To clean, wash, or empty out with a sudden flow of solvent or other cleaning agent.

flush patch – A type of sheet metal repair that leaves a smooth surface maintaining the skins' original contour. The repair is reinforced on the inside and the damaged area is filled with a plug patch.

flush repairs – Metal repairs to the aircraft skin designed to maintain the original contour.

flush rivet – A countersunk rivet in which the manufactured head is flush with the surface of the metal when it is properly driven.

flush riveting – Riveting using countersunk-head rivets which produces a perfectly smooth outside skin.

flute – A groove that is cut into a cylindrical object.

flutter – The rapid and uncontrolled oscillation of a flight control resulting from an unbalanced surface. Flutter normally leads to a catastrophic failure of the structure.

flux —
[1]A material used to cover the pool of molten metal in the process of welding, and which prepares the metal to accept the solder. The flux excludes oxygen from the weld by forming oxides which remain on the surface.
[2]Magnetic lines of force.

flux density — The number of lines of magnetic force per unit area.

flux valve — A special transformer that develops a signal whose characteristics are determined by the unit's position in relation to the earth's magnetic field. It is part of an earth inductor compass system.

flux valve spider — The framework around which the three pick-up coils of a flux valve are wound. The highly permeable material of which the spider is made accepts the lines of flux from the earth's magnetic field.

fly — To travel through the air as pilot or passenger in an aircraft.

fly cutter — A cutting tool used to cut round holes in sheet metal. It is turned by a drill press and the cutting is done by a tool bit held in an adjustable arm.

flying boat — Form of seaplane whose fuselage serves as the boat hull.

flyweights — L-shaped speed sensing units pivoted on a disc which rotates. When rotational speed is high enough, centrifugal force moves them to an angular position. This motion is utilized for various applications such as in propeller governors, mechanical tachometers, etc.

flywheel — A heavy wheel or weight that is used to smooth out the pulsations in a drive system.

flywheel effect — Characteristic of a parallel LC circuit which permits a continuing flow of current even though only small pulses of energy are applied to the circuit.

foam rubber — A form of rubber that has millions of tiny air bubbles which have been beaten into the latex before it was vulcanized

foamed plastic — Synthetic resin that is filled with millions of tiny bubbles. It's main characteristic is its light weight and resiliency. Foamed plastic is sometimes called expanded plastic.

foaming — An undesirable condition in a lubrication system in which oil passing through the engine picks up air which causes tiny air bubbles to form in the oil. Oil foaming reduces the ability of the oil to lubricate and to pick up heat as it should.

fogger oil jet – An air and oil spray mist device used on some engines for lubricating main bearings, as opposed to a fluid stream-type oil jet.

foil – A form of metal, such as that used in common household aluminum foil, that has been rolled out into very thin sheets.

folded fell seam – A type of machine-sewn seam recommended for use in sewing aircraft fabric.

folding – To make sharp, angular bends in sheets of material.

foot-pound – Abbrev.: ft.-lb.

[1]A unit of work: One pound of force moved through a distance of one foot.

[2]A unit of torque: The amount of torque that is produced when a force of one pound is applied one foot from the pivot point of rotation.

force – Energy, when applied to an object, that tries to cause the object to change its direction, speed, or motion.

forced exhaust mixer – A turbofan, long duct design which causes fan air and hot exhaust streams to mix. Used for sound attenuation primarily. Same as mixed exhaust.

forced landing – Any landing that is necessitated by a malfunction of the aircraft or engine, or by improper flight planning.

forceps – A small tool that is used to grasp or hold things.

foreflap – The first flap in a triple slotted segmented flap.

foreign object damage – Abbrev.: FOD. Internal gas turbine engine damage which occurs from the injection of foreign objects into the engine path, e.g. ground debris or objects in the air such as birds or flying debris.

foreign particle – Material or particle which enters a fluid system that may cause serious damage or contamination to the system.

forge – A method of forming metal parts by heating the metal to a plastic state (nearly, but not quite melted) and hammering it to shape.

forge welding – The joining of metal by forging. *See also forge, forging.*

forging – The process of forming a product by hammering or pressing. When the material is forged below the recrystallization temperature, it is said to be cold forged. When worked above the recrystallization temperature, it is said to be hot forged.

fork lift — A machine that has two long steel fingers that can be slid under a pallet on which a heavy load is placed for lifting and/or repositioning.

form factor — The ratio of the length of a wire coil to its diameter.

form of thread — The profile of a thread in an axial plane for a length of one pitch.

former — A frame of light wood or metal which attaches to the truss of the fuselage or wing in order to provide the required aerodynamic shape.

forming — Process of shaping a part.

forming block — A block, usually made of hardwood, around which metal parts are formed.

forming machine — Hand-operated or power-driven machine used to shape sheet metal.

forward bias — The polarity relationship between a power supply and a semiconductor that allows conduction.

forward center of gravity limit — The most forward location allowed for the center of gravity of an aircraft in its loaded condition.

forward fan — Turbofan with the fan located at the front of the compressor. It can be a part of the compressor or a separate rotor as in the three-spool engine.

fouled spark plug — A term describing the condition of the spark plug electrodes when they are contaminated with foreign matter. This condition provides a conductive path for the high voltage to leak off to ground, rather than building up high enough to jump the electrode gap.

four-cycle engine — The most common event cycle for aircraft engines. The four-stroke, five-event cycle consists of an intake stroke, in which the piston moves inward with the intake valve open, and a compression stroke in which the piston moves outward with both valves closed. Ignition occurs near the top of the compression stroke. The power stroke is an inward stroke of the piston with both valves closed, and the exhaust stroke occurs when the piston moves outward with the exhaust valve open. At this point, the cycle begins again.

Fowler flap — Wing flaps which are lowered by sliding out of the trailing edge of the wing on a track. These modify the shape of the airfoil and increase the area of the wing.

fractional distillation — A process of oil refining in which the crude oil is heated to a specified temperature and all of the products that boil off at this temperature are condensed. The temperature is then raised, and the next level of products are boiled off and condensed.

fractions — The various components of a hydrocarbon fuel which are separated by boiling off at different temperatures.

frame — A former ring which provides shape and rigidity to a semi-monocoque or monocoque structure.

free balloon — A lighter-than-air, helium-filled device that is used in weather observations to find the height of the base of the lower layer of clouds.

free electrons — Those electrons so loosely bound in the outer shells of some atoms that they are able to move from atom to atom when an emf is applied to the material.

free fit — A loose fit between moving parts such as a nut that turns easily on the threads of a screw or bolt.

free turbine — A turbine wheel which, rather than driving a compressor rotor, drives a propeller or helicopter transmission through a reduction gearbox. *See also power turbine.*

free water — Liquid water that has condensed out and is no longer entrained in a turbine engine fuel.

freeze — The process in which a liquid is changing into a solid by the removal of heat energy from the liquid.

freezing point — The temperature at which a liquid will change into a solid.

French fell seam — A type of machine-sewn seam. Recommended for sewing together sheets of aircraft fabric.

Freon — A fluorinated hydrocarbon compound used as a fire extinguishing agent and a refrigerant for vapor-cycle air conditioning systems made by E.I. DuPont de Nemours & Company.

frequency — The number of complete cycles of alternating current in one second.

frequency converter — A circuit device that changes the frequency of an alternating current.

frequency modulation — Abbrev.: FM. Method by which information is transmitted by radio waves by varying the frequency of the carrier with the audio.

frequency multiplier — Circuit designed to double, triple, or quadruple the frequency of a signal by harmonic conversion.

frequency synthesizer — An electronic circuit that is used to produce AC with an accurately controlled frequency.

fresh annual inspection — An annual inspection which has just been performed on an airplane. Sometimes this is used as a selling point for an airplane. It is actually meaningless unless you know the integrity of the person performing the inspection.

freshening charge — The charge given a dry-charged battery to bring it up to its rated capacity.

fretting — A condition of a surface erosion caused by a slight movement between two parts that are fastened together with considerable pressure.

fretting corrosion — Corrosion damage between close fitting parts which are allowed to rub together. The rubbing prevents the formation of protective oxide films and allows the metals to corrode.

friction — Relative motion or rubbing of one object against another.

friction brake — Any of a number of different mechanisms used with a rotating wheel or shaft in which friction is used to slow its rotation.

friction clutch — A mechanism that is used to connect a motor to a mechanical load.

friction damper — Rubber insert used to limit excessive movement in a pedestal-type dynafocal engine mount.

friction error, instrument — The error that is caused by friction in an instrument mechanism.

friction horsepower — The amount of horsepower required to turn the engine against the friction of the moving parts and to compress the charges in the cylinders.

friction loss — The loss of mechanical energy in a device that is caused by friction changing mechanical energy into heat.

friction mean effective pressure — Abbrev.: FMEP. Average working pressure within an engine used to overcome friction. IMEP – BMEP = FMEP.

friction tape — A type of cloth electrical insulating tape that is impregnated with a black tar-like material.

friction welding – A method of joining materials by vigorously rubbing the mating surfaces together while forcing them together with a large amount of pressure.

friction-lock Cherry rivet – A patented blind rivet made by the Cherry Rivet, division of Townsend, Inc., in which the stem locks in the hollow shank by friction.

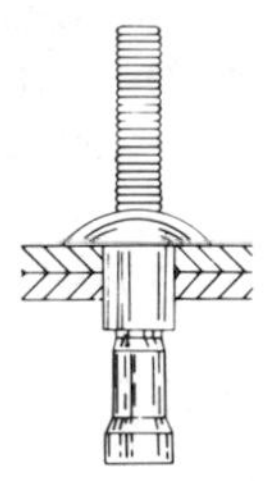

Frise-type ailerons – An aileron having the nose portion projecting ahead of the hinge line. When the trailing edge of the aileron moves up, the nose projects below the wing's lower surface and produces some parasite drag, decreasing the amount of adverse yaw.

front spar – The foremost spar of a multispar wing.

frustrum – The portion of a cone from which the top has been removed.

fuel – A substance that, when combined with oxygen, will burn and produce heat.

fuel boost pump – An auxiliary electrically operated pump located within a fuel tank to force the fuel from the tank to the engine. Usually a centrifugal type pump. Provides vapor free fuel with a slight head pressure to the main pump.

fuel cell – The compartment in an aircraft in which fuel is contained.

fuel consumption – Actual amount of fuel consumed by an engine under a specified set of conditions. It may be expressed in either pounds per hour or gallons per hour.

fuel control unit – The fuel-metering device used on a turbine engine that meters the proper quantity of fuel to be fed into the burners of the engine. It integrates the parameters of inlet air temperature, compressor speed, compressor discharge pressure, and exhaust gas temperature with the position of the cockpit power control lever.

fuel dump system — A portion of the fuel system of large jet transport aircraft that allows fuel to be dumped in flight to reduce the weight of the aircraft to the allowable landing weight.

fuel enrichment system — A control system on some turbine engines to enrich the fuel mixture for cold weather starting.

fuel evaporation ice — Ice formed due to the cooling effect of the fuel evaporating after it is sprayed into the induction system air stream of reciprocating engines. This evaporation process causes carburetor and induction system parts to become very cold causing moisture in the air to condense, collect and freeze on them. This type of ice is most troublesome in float-type carburetors.

fuel flow — Rate at which fuel is consumed by the engine in pounds per hour (PPH).

fuel flowmeter — A cockpit instrument used to indicate the rate of the fuel consumed by the engine during flight.

fuel grade — A classification of aviation gasoline according to its anti-detonation characteristics.

fuel heater — A radiator- like device which has fuel passing through the core. A heat exchange occurs to keep the fuel temperature above the freezing point of water so that entrained water does not form ice crystals which could block fuel flow.

fuel injection manifold valve — A valve used in a fuel injection system that distributes fuel from the fuel control unit to the various injection nozzles. The valve provides a metering force for conditions of low fuel flow and provides a positive fuel shut off when the engine is shut down.

fuel injection system — A fuel metering system used on some aircraft reciprocating engines in which a constant flow of fuel is fed to injection nozzles in the heads of all cylinders just outside of the intake valve. It differs from diesel fuel injection in which a timed charge of high-pressure fuel is sprayed directly into the combustion chamber of the cylinder.

fuel load — That part of the useful load of an aircraft which consists of the usable fuel on board.

fuel manifold — A fuel distribution manifold to which the fuel nozzle is attached. The manifold contains one single fuel line when used with single-line duplex nozzles and two lines for dual-line duplex nozzles.

fuel metering device — Apparatus such as carburetor, fuel injection, or fuel control unit which mixes fuel with intake air in the correct proportions and delivers the mixture to the engine.

fuel nozzle — A spray device which atomizes and directs fuel into a combustion chamber for best flame propagation.

fuel nozzle ferrule — Receptacle in combustion liner in which the fuel nozzle tip is inserted.

fuel pressure gage — A gage which indicates the pressure at which fuel is delivered to the carburetor.

fuel pump — An engine-driven pump used to provide a positive volume of fuel under pressure to the engine.

fuel shut-off valve — A valve in an aircraft fuel system which shuts off all of the flow of fuel to the engine.

fuel system — The system that stores fuel and delivers the proper amount of clean fuel at the right pressure to meet the demands of the engine.

fuel tank vent — A vent in a fuel tank which allows the air pressure above the fuel to be the same as that of the surrounding air.

fuel-air combustion starter — A fuel engine starting accessory which utilizes a combustion section similar to an engine combustor. Combustion products are exhausted through a turbine connected to a reduction gearbox to create starting torque.

fuel-air mixture ratio — The weight ratio in pounds of the fuel and air mixed together to be burned in an aircraft engine.

fuel-oil cooler — A heat exchange device which heats the fuel and cools the oil. It is a radiator-like unit in which fuel passes through the cores and oil passes around the cores. The oil flow is controlled by a thermostatic valve which routes the oil through the cooler only when a certain oil temperature is reached. On some engines, no fuel heater is required due to the exchange rate of this oil cooler.

fuel-oil heat exchanger — A heat exchanger device that is used on turbine engines to take heat from the engine oil and put it into the fuel. It is a radiator-like unit in which fuel passes through the tubes that pass through the hot engine oil. Heat from the oil raises the fuel temperature, and at the same time, lowers the temperature of the oil.

fulcrum — A point on which a lever is supported, balanced, or about which it turns.

full annealing – A process used to produce a fine grained, soft, ductile metal without internal stresses or strains. To obtain full anneal the temperature of the metal is raised to its critical temperature followed by controlled cooling.

full oil – Quantity of oil shown as oil capacity in aircraft specification.

full rudder – The movement of the rudder to its extreme travel when the rudder pedal has been pushed by the pilot.

full throw – The full range of control surface travel.

full-register position – The position of the rotating magnet in a magneto when the poles are fully aligned with the pole shoes of the magneto frame. At this point, the maximum number of lines of flux flow in the frame.

full-rich – That position of the mixture control which allows the maximum amount of fuel to flow to the engine for a given metering force.

full-scale drawing – A drawing of a part to its full size.

full-shroud turbofan – Same as a long duct turbofan.

full-wave rectifier – A form of rectifier which inverts one-half of the input AC signal and provides a pulsating DC output having twice the frequency of the input alternating current.

fully articulated rotor – *See articulated rotor.*

fumes – Vaporized liquids.

functional test – Testing a system through its normal operating range to determine whether or not it functions (works) as it should.

fungicidal paste – Paste which is mixed with clear dope to apply as a first coat on cotton. The fungicidal agent soaks into the fibers and prevents the formation of mold or fungus.

fungus spores – The seed of certain fungi which can enter organic materials such as cotton or linen and cause the material to rot.

furrow – A deep groove.

fuse – An aircraft electrical circuit protection device that consists of a link of low melting point metal which will melt and open the circuit when an excessive amount of current flows through it.

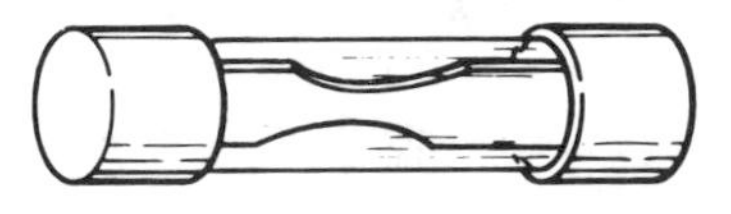

fuse holder — A device that mounts to an electrical fuse panel which holds tubular fuses and makes connections to both ends.

fuse link — A strip of low-melting-point metal used in an electrical circuit fuse device to protect a circuit. When an excessive current flows through the circuit the fuse link will melt and open the circuit.

fuselage — The body of an airplane. That part to which the wing, tail, and landing gear attach, and which, in a single-engine airplane, usually carries the engine.

fuselage stations — Distances measured perpendicular to the longitudinal axis of an airplane. Representing distances from the datum in inches.

fusible alloy — A filler material that melts at approximately 160°F. Used as a method of tube bending. It is heated in hot water, and poured inside the tubing. Once removed from the hot water and cooled, the tube can be bent by hand around a forming block or bender. After the tube is bent, it may be reheated to remove the fusible alloy for reuse.

fusible plug — A hollow plug in an aircraft wheel filled with a material having a specific melting point. If the melting point is reached from brake heat, the filler will melt out and deflate the tire rather than allowing the pressure to increase enough to burst the tire.

fusion — The melting together of metal parts.

G

gage – Any of a variety of measuring instruments used to indicate the amount of pressure that is being measured, the depth of a hole or a groove and its bottom, or the thickness or the clearance between close-fitting parts of a machine, etc. Also spelled gauge.

gage pressure – Pressure that is measured from the existing atmospheric pressure. Oil pressure in an engine, and hydraulic pressure are normally measured as gage pressure. If gage pressure is measured in pounds per square inch, it is spoken of as so many psig. Two other types of pressure that are often used are differential pressure (PSID), which is the difference between two pressures, and absolute pressure (PSIA). Absolute pressure is measured from zero pressure.

gain – The increase in signal power through a circuit.

galling – Fretting or chafing of a mating surface by sliding contact with another surface or body. The heat friction causes the material from one surface to be welded or deposited onto the other surface ultimately destroying the surface area.

gallon –
[1]Imperial: A unit of liquid measurement that is used outside of the U.S. One Imperial gallon weighs 10 lbs. and is equal to 277.4 cu. in., 4.55 liters, or 1.201 U.S. gal.
[2]U.S.: A unit of liquid measurement that is used in the U.S. One U.S. gallon weighs 8 lbs. and is equal to 231 cu. in., 128 fl. oz., or 3.785 liters.

galvanic action – Electrical pressure within a substance which causes electron flow because of the difference of electrode potential within the material.

galvanic corrosion – Corrosion due to the presence of dissimilar metals in contact with each other in the presence of an electrolyte such as water.

galvanic couple – Dissimilar metals that will produce an electrical voltage when they are both in contact with the same electrolyte. When dissimilar metals are in this condition, one metal forms the anode and the other the cathode, thereby producing a current between the two metals. *See also galvanic corrosion.*

galvanic electricity – Electricity produced by chemical action such as that produced in a dry-cell battery or storage battery.

galvanic grouping – An arrangement of metals in a series according to their electrode potential difference.

galvanic metal electrical series – The hierarchical arrangement of metals in the order of their chemical activity. The following list of metals indicates their hierarchical chemical activity, and act as the anode in any electrolytic action to those that follow (least active):

1. Zinc
2. Cadmium
3. Iron and steel
4. Cast iron
5. Chromium iron
6. Lead-tin solder
7. Lead
8. Tin
9. Nickel
10. Brass
11. Copper
12. Bronze
13. Copper-nickel alloys
14. Monel
15. Silver
16. Graphite
17. Gold
18. Platinum

galvanizing – A method of protecting steel parts from corrosion by dipping them in a vat of molten zinc or by electroplating. The protection actually comes from sacrificial corrosion of the zinc.

galvanometer – A type of electrical measuring instrument in which electrical current is measured by the reaction of its electromagnetic field with the field of a permanent magnet.

gamma rays – Electromagnetic radiation that results from nuclear fission.

ganged tuning – A mechanical arrangement to permit the simultaneous tuning of two or more circuits.

gap – Distance between two objects.

garnet paper – An abrasive polishing paper consisting of a sheet of flexible paper with a layer of finely crushed garnet.

gas —
[1]A fluid which will assume the shape of the container it is placed in,and will expand to fill all of the container.
[2]The physical condition of matter in which a material takes the shape of its container and expands to fill the entire container. Oxygen and nitrogen are two chemical elements that are gases at normal room temperature and pressure. The air we breathe is a physical mixture of gases, primarily nitrogen and oxygen.

gas generator — The basic power producing portion of the engine and excluding such sections as the inlet duct, the fan section, free power turbines, and tailpipe. Each manufacturer designates these exclusions but they generally consist of the compressor, diffuser, combustor, and turbine.

gas generator turbine — High pressure turbine wheel(s) which drive the compressor on a turboshaft engine.

gas path — The airflow or open portion of the engine front to back where air is compressed, combusted, and exhausted.

gas path analysis — Abbrev.: GPA. A computer analysis of engine parameters on some airliners. It is designed to assist the modular maintenance and on-condition maintenance concepts by giving continuous on-condition data or used for predicting engine component airworthiness.

gas storage cylinders — Long bottles of high-strength steel which are used to store compressed gases.

gas turbine engine — A form of heat engine in which burning fuel adds energy to compressed air and accelerates the air through the remainder of the engine. Some of the energy is extracted to turn the air compressor, and the remainder accelerates the air to produce thrust. Some of this energy can be converted into torque to drive a propeller or a system of rotors for a helicopter.

gas welding — The method of fusing metals together by a flame using gas as its fuel. The most common types of gas welding are oxyacetylene and oxyhydrogen.

gaseous — Having the nature of, or in the form of, gas.

gaseous breathing oxygen — A special type of oxygen containing practically no water vapor and at least 99.5% pure.

gaseous fuel — Any mixture of flammable gases used for fuel.

gas-filled tube – An electron tube with gas introduced into the envelope to produce certain desired operating characteristics.

gasket – Static, stationary seal between two flat surfaces.

gasoline – A volatile, highly flammable liquid mixture of hydrocarbons produced by the fractional distillation of petroleum and used as fuel in internal-combustion engines.

gasoline combustion heaters – Aircraft cabin heaters in which gasoline from the aircraft fuel tanks is burned to produce the required heat.

gassing, battery – The release of hydrogen and oxygen as a free gas at the end of the charging cycle of lead-acid storage batteries.

gate –
[1]A logic device having one or more inputs and/or outputs. The condition of the inputs determines whether or not a voltage is present at the outputs.
[2]The electrode of a silicon-controlled rectifier or a triac through which the trigger pulse is applied.

gate hold procedures – Standard operating procedures that ensure passenger confort whenever an aircraft is delayed more than five minutes.

gate-type check valve – A form of one-way flow valve using a swinging gate or flapper to isolate one of the vacuum pumps in a multi-engine aircraft from the rest of the system in the event of a failure of the pump.

gauge – *See gage.*

gauge pressure – *See gage pressure.*

gauss – The unit of magnetomotive force. It is equal to 1 maxwell per cm^2.

gear – A toothed wheel or disc which meshes with another toothed wheel or disc to transmit motion.

gear-driven supercharger – An internal engine-driven supercharger on a reciprocating engine that is driven from the crankshaft through a gear arrangement.

gear and pinion mechanism – A mechanical amplifying mechanism consisting of two gears: one a pinion, which is much smaller than the other. The mechanical advantage of the mechanism is determined by the ratio between the number of teeth on the pinion and the number of teeth on the large gear. Often the large gear is only a portion of the wheel and is called a sector gear.

gear backlash – The measured clearance between the teeth of meshed gears.

gear indicators – Indicators in the cockpit of an airplane having retractable landing gear to inform the pilot of the condition of the wheels. It will indicate whether they are down and locked, in transit, or up and locked.

gear preload – The pressure with which two gears mate or mesh together.

gear-type pump – A type of power-driven fluid pump, usually a constant displacement-type pump, that is driven from the engine accessory, used to pump fluid under pressure. The gear-type pump is made up of two meshed spur gears that are mounted in a close-fitting housing. Fluid is taken into the inlet side of the housing and fills the space between the teeth of both gears. As the fluid is carried around the housing of the rotating gears to the discharge side of the pump, the gear teeth mesh, and the fluid is forced out of the outlet side of the pump.

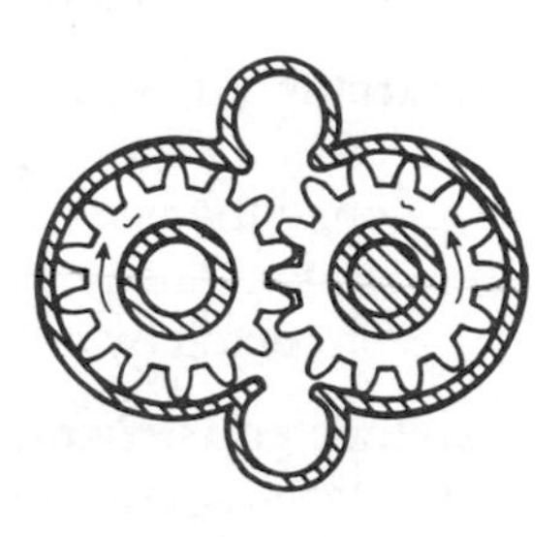

geared fan – A design which allows the fan to rotate at a different speed than the compressor rotor. The fan being geared down allows for higher tip speeds in the compressor.

geared fan gas turbine engine – A turbofan engine that uses a set of reduction gears between the first stage of the gas generator compressor and the fan.

geared propeller – A propeller driven from the crankshaft through a series of reduction gears. This allows the engine to operate at an efficient speed while holding the propeller RPM in its efficient range.

gel – A jellylike substance formed by the coagulation of a solution into a solid phase.

gelatinous – Having the consistency of gelatin or jelly.

gelled cell battery – A type of lead-acid battery that has a gelling agent added to the electrolyte to make it non-spillable and to retain a high level of electrolyte in the battery.

genemotor – Also called dynamotors. A type of electrically operated machine that changes low-voltage DC from a storage battery into high-voltage DC for operating certain types of communications equipment.

general aviation – That portion of the aviation industry that covers all of aviation with the exception of military aviation and the airlines.

General Aviation District Office – Abbrev.: GADO. Designated FAA Field Offices staffed to serve the general public and aviation industry on matters pertaining to the certification, maintenance, and operation of general aviation.

generator – A mechanical device consisting of a conductor being turned within a magnetic field used to produce electricity by electromagnetic induction.

generator current limiter – A special high-current fuse capable of carrying momentary current overloads. It will melt however, and open the circuit under current flows which might damage the generator. These are used in generator installations which are not protected by automatic current limiters.

geodetic construction – A form of aircraft construction in which the stress carrying portion of the skin is made up of a lattice work of thin metal or wood strips.

geographic poles – The poles of the axis about which the earth rotates. They form true north and true south.

geometric pitch – The distance a propeller should advance in one revolution if there were no slip.

German silver – Metals of the copper/nickel/zinc alloy family. Also called nickel silver.

germanium – A grayish-white metallic chemical insulating element having four valence electrons. It is often used in the manufacture of semiconductor devices such as diodes and transistors.

gerotor pump – A form of constant displacement pump using a spur gear driven by the engine and turning inside of an internal tooth gear having one more space than teeth on the drive gear. As the pump rotates, the volume at the inlet port increases, while the volume at the outlet decreases, moving fluid through the pump.

GIG – A military demerit.

giga – Billion.

gigacycle – Gigahertz.

gigahertz – One billion hertz.

gilbert – The unit of magnetomotive force. It is equal to approximately 0.7968 amp-turn.

gill-type cowl flap — A cowl flap used on the trailing edge of each cowling of a horizontally opposed engine. Its purpose is to regulate the flow of air through the engine for cooling.

gimbal — The frame in which a gyro spins. It is designed in such a way that it allow a gyroscope to remain in an upright condition while the base is tilted. Rate gyros use a single gimbal, while attitude gyros are universally mounted in a double gimbal.

gimlet point — A gimlet point is a threaded cone point usually having a point angle of 45-50°. It is used on thread forming screws such as Type A tapping screws, wood screws, lag bolts, etc.

glass cloth — A type of aircraft fabric. Made from fine spun glass filaments which are woven into a strong, tough fabric. These fabrics are used for reinforcing plastic resins to mold various types of products.

glass fiber — Filaments of fine spun glass which are woven into cloth or are packed together into a mat used for thermal and acoustical insulation.

glaze —
[1]The hard, smooth surface of a finishing system. This must normally be "broken" or roughened before another coat of material will adhere to it.
[2]A hard, glass-like surface that forms on the rotating disks of a multiple-disk brake. It forms when the surface is locally overheated without the sintered material heating uniformly. This slick surface does not produce uniform friction and will cause the brakes to chatter or squeal if it is not removed.

glide —
[1]A slow descent of an aircraft without the aid of the engine.
[2]To descend at a normal angle of attack with little or no engine power.

glide path — The path of an aircraft relataive to the descent profile while approaching a landing. *See also glideslope.*

glide ratio — The ratio of the forward distance the airplane travels to the vertical distance the aircraft descends when it is operating at low power or without power.

glide slope – Sometimes called the glide path. That part of an ILS which provides the pilot with a radio beam to follow in an aircraft descent from the outer marker to the point of touchdown. The radio beam is at angle of approximately 2½° from the approach end of the runway, and provides a vertical path for the aircraft to follow when it is making an ILS landing. The ILS indicator instrument shows the pilot the airplanes rate of descent along the glide slope.

glider – A heavier-than-air aircraft that is supported in flight by the dynamic reaction of the air against its lifting surfaces and whose free flight does not depend on an engine.

gliding angle – The angle between the flight path during a glide and a horizontal axis fixed relative to the air.

gliding ratio – The ratio of the horizontal distance an aircraft travels while gliding for every unit of vertical distance it descends.

glove valve – A hand-operated hydraulic valve which, when closed, will stop the flow of fluid.

glow coil igniter – A type of ignition igniter. Consists of a resistance wire wound into a coil around a pin extending from the body of the igniter. DC causes the coil to become red hot, igniting the fuel/air mixture until the heater is operating at a temperature sufficient to maintain the flame, after which the glow coil is automatically turned off.

glow discharge tube – A glass tube with a gas, such as neon, under low pressure. Two electrodes are embedded in opposite ends of the tube. When a sufficiently high potential difference is applied between the electrodes, the gas will ionize and glow.

glow plug igniter – A type of igniter that uses a coil of wire that is heated by high-voltage DC electricity. Air and fuel blowing through the coil ignites the fuel-air mixture inside the combustors.

glue – Any adhesive capable of holding materials together as it dries.

glue blocks – Wood blocks used as a backing support when making repairs to a wooden structure. They distribute the pressure applied by the clamps evenly over the area being glued.

glue joint – Glued wood joints. Any joint made between two pieces of wood by using a glue rather than using any type of mechanical fastener.

Glyptal – A registered trade name of an insulating varnish used in electrical machinery.

G_m – Transconductance.

go around – To abort a landing.

gold – A malleable, ductile, yellow corrosion resistant chemical element used on critical electrical contacts.

gold leaf – Pure gold that has been rolled into extremely thin sheets.

goniometer antenna – A fixed loop antenna used by automatic direction finding equipment, consisting of two coils oriented 90° to each other. It measures the angle between a reference direction, and the direction from which the radio signal is being received.

go/no-go gage – A type of measuring gage consisting of a part having two dimensions: the minimum size and the maximum size. An opening of the correct dimension will allow one side to go, or pass through, and the other dimension will not go.

gouge – A cut, groove or hole that is considered to be a defect.

gouging – A furrowing condition when metals are displaced. Usually caused by foreign material between close moving parts.

governor – A control which limits the maximum rotational speed of a device.

governor fulcrum – The roller on which the governor valve lever pivots.

grade-A cotton – Long-staple cotton fabric with 80 threads per inch in both the warp and fill directions. It is the standard material for covering aircraft structures.

gradient – The rate of regular ascent or descent of an aircraft or temperature, pressure or concentration in a specified direction.

gradient system – A device used to give "artificial feel" to hydraulic boosted flight controls.

grain boundary – The lines in metal that are formed by the surfaces of the grains in the metal.

grains – Grains are the individual crystals of the material.

gram – Abbrev.: g. The unit of weight or mass in the metric system. One gram equals about 0.035 on. or 1/1,000 kg.

granular – Containing or consisting of grains of granules.

graph – Pictorial presentation of data, equations, and formulas.

graphite – A soft, black form of carbon that usually has a greasy feel. Graphite is used as a dry lubricant and also as the "lead" in a pencil. Graphite is also known as black lead.

gravitational acceleration — The acceleration of a free-falling object that is caused by the earth's gravitational pull. The acceleration rate of a freely falling object is 32.2 ft./second, per second or 980.7 cm./second per second.

gravity — The force of attraction between the earth and any object on or near it. This force is proportional to the mass of the object.

green run — The first run of a new or freshly overhauled engine.

Greenwich mean time — Abbrev.: GMT. Also kkown as Zulu time. 0° meridian, the meridian at the Royal Observatory, London, England.

greig Dacron — A synthetic polyester fiber fabric in its unshrunk condition as it comes from the loom.

grid —
[1]The electrode of a vacuum tube to which the signal is applied.
[2]Framework of a plate in a lead-acid battery cell. It is made of lead and antimony to which the actual material (spongy lead or lead peroxide) is attached.
[3]The electrodes in an electron tube between the cathode and the anode. It is used to control the amount, shape, and velocity of the electron stream between the cathode and the anode.

grind — To remove metal form a part with an abrasive stone or wheel.

grinder — A machine having an abrasive wheel which removes excess material while producing a suitable surface.

grinding wheel — An abrasive wheel used on grinders to remove excess material.

grip length — The length of the unthreaded shank of a bolt or the length of a blind rivet between the manufactured head and the maximum extent of the pulled head. It is the maximum thickness of material that can be joined by a fastener.

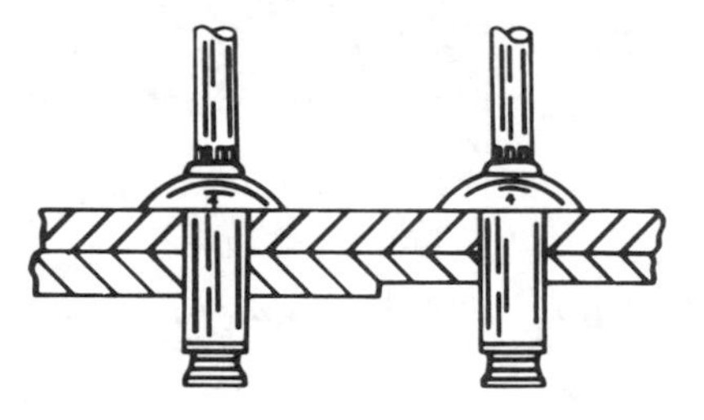

grip range — The difference between the maximum and minimum thickness of material that may be joined by a fastener.

grit blasting –
[1]*See abrasive blasting.*
[2]A process for cleaning metal in which an abrasive material such as sand, rice, baked wheat, plaster pellets, glass beads or crushed walnut shells are forcefully blown onto the part's surface.

grommet –
[1]A metal or plastic eyelet used for reinforcing holes in aircraft fabric.
[2]A small ring of metal, rubber, or plastic used as a fairlead and protector for tubing or wire going though a hole in a metal structure.

grooved surface – Shallow channel wider than a scratch and usually smooth resulting from wear effected by concentrated contact stress. It is caused by abnormal relative movement between contact surfaces or by foreign material on contact surfaces. The parts usually affected include cylinder barrels, valve faces and oil seal outer sleeves.

gross thrust – The thrust developed by an engine, not taking into consideration any pressure of initial air mass momentum. Also called static thrust (F_g).

gross weight – Total weight of a fully loaded aircraft including the fuel, oil, and the cargo it is carrying.

ground –
[1]A reference point for voltage measurement in an electrical circuit.
[2]To connect a part or component to the electrical ground that is normally the airframe.

ground crew – The people who maintain, service, and prepare the aircraft before and after flight.

ground effect – The condition of slightly increased air pressure below an airplane wing or helicopter rotor system that increases the amount of lift produced. It extends upward for approximately one-half wing span or one-half of the rotor diameter.

ground idle – Gas turbine engine speed usually in the 60-70% of the maximum RPM range, used as a minimum thrust setting for ground operations.

ground loop –
[1]The sudden reversal of direction of travel on the ground of an airplane having a tail wheel-type landing gear. The center of gravity swings around ahead of the wheels.
[2]An undesirable flow of electrical current through the braid around a shielded wire.

ground plane – The reflector used in a quarter-wave radio antenna, which serves as an additional quarter-wave element.

ground potential – The zero potential (no voltage difference) of electrical circuits.

ground power unit – A type of small gas turbine whose purpose is to provide either electrical power, and/or air pressure for starting aircraft engines. A ground unit is connected to the aircraft when needed. Similar to an aircraft installed auxiliary power unit.

ground resonance – A self-excited vibration of a helicopter occurring whenever the frequency of oscillation of the blades about the lead-lag axis of an articulated rotor becomes the same as the natural frequency of the fuselage.

ground return electrical circuit – An electrical circuit that uses the structure of the aircraft as one of the conductors in the circuit.

ground speed – The speed at which the aircraft is moving over the ground, taking in to consideration such factors as wind direction and velocity.

ground support equipment – Abbrev.: GSE. Equipment separate from the airplane, but in direct support to facilitate maintenance. GSE can include such items as engine hoist, auxiliary power units, testing equipment, compressed air units, etc.

ground wave – That portion of a radio wave which travels to the receiver along the surface of the earth. Compressed air.

ground-adjustable propeller – Propeller whose pitch may be adjusted and locked when the engine is not operating, but cannot be changed in flight.

ground-boost engine – An unsupercharged engine which strictly depends on atmospheric pressure for air density.

grounded – To declare an aircraft or airman unfit for flight.

grounded-base amplifier – *See common-base amplifier.*

grounded-collector amplifier – *See common-collector amplifier.*

ground-controlled landing approach – A directed approach to landing an aircraft by instructions provided by a ground controller The controller watches the aircraft on a radar scope, which shows the relative position of the aircraft to the glide slope and its horizontal position. The controller gives the pilot instructions necessary for keeping the aircraft on its intended path until it lands.

grounded-emitter amplifier – *See common-emitter amplifier.*

growler – Test equipment used to check generator and starter armature for shorts. The growler forms the primary of a transformer and the armature forms the secondary. Shorts show up as they cause vibration of a piece of metal–such as a hacksaw blade–held over the armature.

guarded switch – A switch that is protected against accidental movement by having a guard or type of shroud located directly over the switch. The guard must be raised before the switch can be actuated.

guide vanes – Stationary airfoil sections radially located in the engine which directs the flow of air on gases from one major part to another.

G-unit – The unit of acceleration as a measure of the force of gravity. One G-unit is the attraction of gravity for a body.

Gunk – A chemical-type degreaser used for loosening grease and soft carbon from the surface of metal parts.

gusset – A small reinforcing member used to support the corners of a structure.

gust – A sudden strong rush of air or wind. Also called wind gust.

gust lock – Gust locks are of two general types used to prevent the controls from being damaged by wind gusts while the airplane is parked on the ground. Internal control surface locks are set in the cockpit, and external locks are positioned between the movable surfaces and the fixed aircraft structure. External locks are usually painted red and will have a warning streamer to indicate their installation.

guttered surface – Severe erosion confined to narrow areas in the direction of the flow of gases. Causes may be the improper valve seating allowing escape of combustion gases through a narrow area. Areas effected are valve seats, cylinder heads, pistons, valve and spark plug inserts.

gyro horizon – An attitude gyroscopic instrument that indicates rotation about the pitch and roll axes.

gyrocompass – A form of navigational instrument that uses a gyroscope as a stable reference to keep the compass from oscillating.

gyrodyne – A rotorcraft whose rotors are normally engine-driven for takeoff, hovering, and landing, and forward flight through part of its speed range. It means of propulsion, consisting usually of conventional propellers, is independent of the rotor system.

gyroplane — A rotorcraft whose rotors are not engine-driven except for initial starting, but are made to rotate by action of air when the rotorcraft is moving. Its means of propulsion usually consists of a conventional propeller which is independent of the rotor system.

gyroscope — A device utilizing a rapidly spinning wheel with its weight concentrated around its rim. This wheel exhibits the characteristics of rigidity in space and precession.

gyroscopic precession — That characteristic of a gyroscope that causes it to react to an applied force at 90° to the point of application, in the direction of rotation.

gyroscopic rigidity — That characteristic of a gyroscope which causes it to remain rigid in space or not tilt its axis of rotation as the earth rotates.

gyroscopic turn indicator — A rate gyro used to measure the rate of yaw of an aircraft.

H

hacksaw — A hand-operated metal cutting saw with narrow, replaceable blades. The replaceable blade is held in the hacksaw frame under tension.

hailstones — A form of precipitation formed by drops of water carried by upward currents of wind inside a cumulonimbus cloud to a level where the temperature is low enough to freeze them into ice. When the hailstones are too heavy to be carried in the cloud, they fall to the ground.

hairline — A very thin line.

hairline crack — An imperceptible crack that is visible on the surface of a piece of material.

hairspring — A flat coiled spring used in aircraft instruments as either a calibrated restraint or as a preloading device for the gears.

halation — A form of distortion on a cathode ray tube that shows up as a blurred image caused by the reflection from the back of the fluorescent coating when the coating is too thick.

half hitch — A type of knot used for lacing wire bundles.

half life — One half-life is the measure of the rate at which a particular radioactive material decays, or loses one half of its radioactivity. In each half-life interval, the material loses one half more of its remaining life and so on.

half life inspection — A jet engine inspection required under conditions of warranty, completed at half the time between overhaul (TBO) interval. It includes primarily a hot section disassembly, inspection and repair as necessary.

half view — An aircraft drawing which shows only one-half of a symmetrical view. Center lines and break lines are used to show that there is more of the object than is shown in the drawing.

half-duplex communication — A type of communication in which signals can be sent in one direction at a time, but cannot be sent in both directions at the same time.

half-round file — A hand file that is flat on one side and curved on the other.

half-section — A view in which the cutting plane extends only halfway across the object, leaving the other half of the object as an exterior view.

half-wave radio antenna — An antenna with an electrical length that is approximately one half of the wavelength of the frequency for which the antenna is tuned.

half-wave rectifier — A form of rectifier that changes AC into pulsating DC using one diode and producing only one-half of the AC wave in its output.

Hall-effect generator — An electronic device used to measure the intensity of a magnetic field.

halo — A circle of light that appears around the moon when it seen through a thin layer of cirrostratus clouds. The halo appearance is caused by tiny ice crystals in the cloud that scatter the light that passes through it.

halogen — One of the five chemical elements (fluorine, chlorine, bromine, iodine, and astatine are the halogens) in Group VII of the periodic table of chemical elements used in some fire extinguishing systems.

hammer — A hand tool consisting of a heavy head and a handle. It is used for pounding, driving, or shaping.

hammer welding — A type of forge welding in which the edges of two pieces of metal are heated red-hot and then beat together with a hammer.

hand drill — A hand-operated, eggbeater-type tool used to turn a twist drill.

hand driving — A method of forming rivets in which the head is driven with a hand set and hammer, and the shank is bucked with a bucking bar.

hand file — *See file.*

hand forming — The process of shrinking, stretching, or forming sheet metal by using soft-faced mallets or hammers to force the metal down against suitable forming blocks or into dies.

hand inertia starter — An inertia starter for reciprocating engines in which the flywheel is brought up to speed by a hand-operated crank.

hand pump — A pump operated by hand to create a flow of fluid.

hand rivet set – A rivet set which may be clamped into a vise to hold the manufactured head of the rivet while the shank is upset with a hammer and a flat punch.

hand snips – *See aviation snips.*

hand tools – A general name for all of the hand-operated tools used by technicians in the performance of aircraft maintenance.

hand-bending tools – A hand-operated tube bending tool consisting of a clamp, a radius block, and a sliding bar. It is used to bend thin-wall aluminum alloy or copper tubing in such a way that it does not collapse the tube.

handbook – A manual that describes simple operations or a system of operations. A handbook normally does not contain specific detailed information on the maintenance of such systems.

hand-cranked inertia starter – A starter that uses a hand crank to store energy in a spinning flywheel. The crank is geared to the flywheel through a high-ratio gear system so that the flywheel can be spun at a high speed. The flywheel is coupled by a clutch to the crankshaft of the engine, thereby spinning enough to turn the engine fast enough for it to start.

hangar – A building that is used for the purpose of housing and maintaining aircraft.

hard – A condition of a material is considered hard when it is compact and solid and is difficult to bend or deform.

hard landing – An improper landing of an aircraft which has transmitted undue stresses into the structure. The degree of hardness of the landing will depend on the type of special inspection that will be performed to determine if there is structural damage to the aircraft.

hard X-rays – The degree of the penetration power of an X-ray as determined by the amount of voltage that is applied to the anode of the X-ray tube. The higher the voltage the greater its penetrating power.

hardboard – A wood composition material that is manufactured by bonding sawdust and chips of wood with an adhesive under heat and pressure.

hard-drawn copper wire – Copper wire that has been pulled through dies to reduce its diameter to a predetermined size. Pulling the wire also work hardens the wire and increases its tensile strength.

hardenability — In a ferrous alloy, the property of metal that determines the depth and distribution of hardness induced by heat treatment and quenching or by cold working.

hardened steel — Steel that has been hardened by a process of heating the steel above its critical temperature then quenching it in brine, water, or oil. Although the hardened steel is very strong, it is also brittle.

hardener — A chemical constituent of an adhesive that promotes its setting and hardening.

hardening —

1. A form of heat treatment of metal which increases its resistance to abrasion and its brittleness, but decreases its ductility and malleability.
2. Aluminum: The process of increasing the strength and hardness of aluminum after it has been solution heat treated. Age hardening takes several days at room temperature until the metal reaches its full hard state. *See also aging; age hardening; precipitation heat treating.*
3. Steel: A process whereby steel is made hard and brittle by heating it to a temperature above its critical temperature and immediately quenching it in water or oil.

hard-facing — A process of welding, plating, or spraying a hard material such as a carbide on the surface of a tool to increase its hardness and to keep the tool from wearing.

hardness — The property of a metal that enables it to resist penetration, wear, or cutting actions.

hardness test — An evaluation of the hardness of a material by measuring the depth of penetration of a specially shaped probe under a specified load. The surface hardness of aluminum alloy parts such as brake housings and wheels can be measured to determine whether or not their heat treatment has been affected by overheating.

hardware — The nuts, bolts, screws, rivets, etc. necessary for assembling parts.

hardwood — A type of wood with compact texture.

harmonic — A frequency of vibration which is an even multiple of the fundamental of another vibration frequency.

Hartley oscillator — A form of electronic oscillator which produces its feedback through a tapped inductor.

Hastalloy – A nickel-based alloy in the family of turbine super metals of high temperature strength.

hat channel – An extruded structural material that has the cross-sectional shape of a hat. Hat channels are used to stiffen and give rigidity to flat sheets of metal.

hatch – A covered opening of an aircraft located on fuselage and not a door.

haze – A cloudy appearance obstructing visibility caused by smoke, fiine dust or air pollution.

head –
[1]The upper portion of a bolt, or extremity or projecting part of an object.
[2]The enlarged shape preformed on one end of a headed fastener to provide a bearing surface.

head angle – The included angle of the bearing surface of the head of a countersunk rivet or screw.

head diameter – The diameter at the largest periphery of the head.

head eccentricity – The amount that the head of a fastener is eccentric with the fastener body or shank.

head pressure – Pressure exerted by a fluid by virtue of the height of the top of the fluid column.

head wind – A wind that is blowing in the opposite direction the aircraft is flying, thereby impeding its forward airspeed.

headed and threaded rod – A fastener similar to a machine screw except that it has a much greater length. It has a round, truss, or flat head and an end threaded for a nut.

headless fastener – A fastener, either threaded or unthreaded, of which neither end is enlarged.

headphone – Two small receivers installed in soft cups mounted on a band that is worn over the head covering each ear.

head-up display – Abbrev.: HUD. A type of flight instrument that increases the safety of flight during the transition from instrument flight to visual flight during a landing.

hearth furnace – A special type of furnace that is used to melt iron and steel.

heat — A form of energy associated with the motion of molecules within a material. The more heat energy there is in a material, the faster its molecules move.

heat dissipation — The loss of heat or the transfer of heat into another object or substance.

heat energy — Energy that is associated with the motion of the molecules within the substance.

heat engine — Any mechanical device that converts heat energy into mechanical energy. For example, reciprocating and turbine engines are heat engines.

heat exchanger — Any device that is used to transfer heat from one body to another.

heat lamp — An incandescent lamp that produces a maximum of infrared radiation with a minimum of visible light rays. Used for drying paint or for applying heat to glued parts to increase the curing time.

heat load — The amount of heat which the air conditioner is required to remove from an airplane cabin in order to maintain a constant cabin temperature.

heat of compression — Heat that is generated when a gas is compressed.

heat sink —
[1]Device on which semiconductors may be mounted to absorb the heat that would normally tend to damage them.
[2]A heavy plate of conductive material that will absorb or carry away heat. Especially useful in welding.

heat treatment of a plastic resin — The operation in which a cemented joint in a thermoplastic resin is held at an elevated temperature so the entrapped solvent can diffuse into a greater volume of the resin. This decreases its concentration and increases the strength of the joint.

heat treatment of metals — Any operation in which the physical characteristics of a metal are changed by heating. This includes annealing, hardening, tempering, and normalizing.

heat value — The heat energy available per unit volume of a fuel.

heater — Any device that produces controlled heating.

heating element — An electrical resistance wire that glows red-hot to produce heat.

heat-shrinkable – A quality in a synthetic fiber that allows a fabric to shrink when heat is applied.

heat-shrinkable fabric – A type of inorganic fabric that is used to cover light aircraft structures. The fabric is sewn and put on the aircraft structure so that it is taut but not tight. After securing the fabric to the structure with a special adhesive, it is shrunk to the correct tautness by ironing it with an ordinary household electric iron or by heating it with a high wattage hair dryer. The material is then giving a coating of non-tautening dope.

Heaviside atmospheric layer – A layer of ionized particles that surrounds the earth at a height of about 55-85 miles above the earth's surface.

heavy ends – The last parts of crude petroleum refining in the fractional distillation process, that have the highest boiling points.

hecto- – One hundred.

hedge-hopping – An aircraft flying very near the surface of the earth avoiding obstructions on the ground.

Heliarc – A welding process used extensively on aircraft parts. It is a gas shielded process to prevent oxidation of the base metal. The two types of Heliarc welding are tungsten inert-gas (TIG) welding and metal inert-gas (MIG) welding.

helical – A line or form that winds around a cylinder, or the line of threads on a bolt.

helical potentiometer – A potentiometer whose resistance element is made in the form of a spiral, and the wiper is moved over the element by turning a multi-turn screw.

helical spline – A spline that winds around a shaft. Helical splines are used to change the linear motion of the device that rides on the splines into rotary motion of the shaft on which the splines are cut.

helical spring – A spring wound in the form of a spiral.

Helicoil – A special helical steel insert that is screwed into specially cut threads to restore threads that have been stripped out or to provide durable threads in soft castings.

helicopter – A form of heavier-than-air, rotor-wing aircraft whose lift is produced by engine-driven rotors which are essentially long, narrow airfoils.

helipad – A pad, or location, where helicopters take off and land.

helium — An inert, gaseous, chemical element used to inflate lighter-than-air aircraft.

henry — The standard unit of inductance. It is the amount of inductance in which a current change of one ampere per second will induce a voltage of one volt. Named for Joseph Henry, an American physicist.

heptane — A liquid hydrocarbon material (C_7H_{17}) having a low critical pressure and temperature and whose detonation characteristics are used as the low point in determining the octane rating of aviation gasoline.

heptode — A vacuum tube that has seven active electrodes which include the anode, cathode, control grid, and four other special purpose grids such as screen grids, suppressor grids, and beam-forming grids.

hermetically sealed — A method of protecting an aircraft instrument by exhausting all of the air from its case and sealing it with solder so that no moisture can get into the it.

hermetically sealed integrating gyro — Abbrev.: HIG. A gyro mounted in a sealed case with a viscous damping medium. The output is therefore an indication of the total amount of angular displacement of the vehicle in which the gyro is installed, rather than the rate of angular displacement.

herringbone gear — A V-shaped gear whose teeth are cut across the face of the gear.

hertz — Abbrev.: Hz. The term named after the German physicist, Heinrich Hertz. The unit is used for measuring the frequency of vibrations or of AC electricity. It us used for the frequency of any type of repeating cycles of motion. One hertz is equal to one cycle per second.

Hertz antenna — A half-wave radio antenna.

heterodyne — To mix, or beat together, two frequencies to produce an intermediate frequency.

heterogeneous mixture — A mixture that is composed of dissimilar ingredients.

hexagon — A figure that has six sides.

hexagon head bolt — A bolt head shaped with six sides (a hexagon).

hexode — An electron tube having six active elements.

hidden surfaces — Any surface represented on an aircraft drawing which cannot be seen in a particular view, but is represented in outline form with hidden lines.

high blower — High impeller-to-crankshaft ratio of a single-stage, two-speed, internal supercharger system. Usually about 10:1 ratio.

high cycle fatigue — A condition seen as cracking or stretching caused by vibration stresses above the design limit of the engine.

high strength steel — Steel that has a tensile strength of between 50,000 and 100,000 PSI.

high-bypass turbofan — Usually engines with 4:1 fan-to-engine bypass ratio or higher. That is, four or more times as much air flows across the fan as across the core engine.

high-bypass turbofan engine — A turbofan engine in which the mass airflow in pounds per second that passes through the fan can be as much as four times greater than that which is moved by the gas generator or core of the engine.

high-carbon steel — Steel which contains more than 0.5% carbon.

high-level language — The language of computer instruction that the computer can understand.

high-lift device — Any lift-modifying device such as a slot, a slat, or any of the forms of flaps that which are used to allow an airfoil to achieve a higher angle of attack before airflow separation occurs.

high-pass filter — A type of electronic filter that allows AC, above a certain frequency, to pass with little or no opposition.

high-potential ignition lead test — A test that is performed on the spark plug electrical wires of an aircraft ignition system to see if there is a voltage leak to ground.

high-pressure compressor — The rear section of a dual-spool compressor, also called N_2 compressor or high speed compressor.

high-pressure oxygen system — Gaseous oxygen systems whose cylinders carry between 1,000 and 2,000 PSI pressure.

high-pressure system — In gaseous systems, this refers to 1,000-2,000 PSI maximum pressures. In liquid systems, it refers to approximately 300 PSI pressures.

high-pressure turbine — The forward most turbine wheels, also called the N_2 turbine or high-speed turbine that drives the high-speed compressor in a two-spool, axial flow gas turbine engine.

high-speed steel – Alloys of steel which maintain their strength when operating at red-hot temperatures. They are used for metal-cutting tools.

high-strength fastener – A fastener with high tensile and shear strengths attained through combinations of materials, work-hardening, and heat treatment.

high-tension magneto – A self-contained magneto ignition system that is used to provide a high potential voltage to the spark plugs. The magneto consists of a rotating magnet, cam, breaker points, capacitor, and a coil with a primary and a secondary winding. The output of the secondary winding goes to a distributor, then to the spark plugs.

high-voltage igniter plug – As opposed to low-voltage igniter plug. This plug utilizes an air gap between its center electrode and its outer casing. Used to start the engine combustion process.

high-voltage ignition system – A main system with a voltage output in the range of approximately 5,000-30,000 volts delivered to the igniter plug. *See also low-voltage ignition system.*

high-wing airplane – A monoplane with the single supporting surface mounted on top of the fuselage.

hinge – A form of fastener that allows one of the connected pieces to pivot with respect to the other.

hinge point – The pivot point about which a control surface or a door hinges.

Hipernik – A magnetic alloy that is made of 50% iron and 50% nickel.

Hi-Shear rivet – A special form of threadless bolt used for high-speed, high-strength, lightweight construction of an aircraft. A steel pin is held into the structure by an aluminum or mild-steel collar swaged into a groove around the end of the pin.

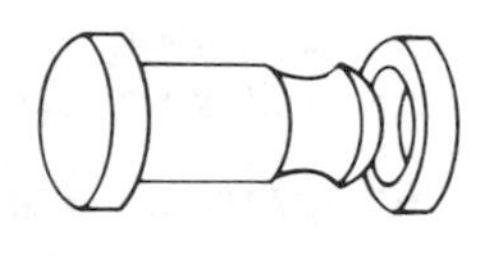

holding coil – An auxiliary coil in an electrical relay that keeps the relay energized after the current that caused the relay to close has stopped flowing through the main coil.

holding procedure – An aircraft maneuver in which a pilot is instructed by air traffic control to maintain a specified air space until given further instruction.

hold-out – The ability of a primer to hold the topcoats without their sinking into its surface.

hole –
[1]The vacancy in the valence structure of an element which will accept an electron from an outside source.
[2]The absence of an electron at a point where one might be expected. For most purposes, a hole may be treated as a positive charge.
[3]A serious discrepancy, flaw or weakness.
[4]An opening through which something is missing.

hole finder – A tool used in sheet metal work to determine the rivet hole locations to be drilled in a piece of sheet metal so they will match those in the piece of metal being covered.

hole punch – A hollow punch resembling a sharp-edged tube. It is used to punch holes in gasket material.

hollow drill – A drill with a hole through its center making it easier for the drill to be lubricated.

hollow-mill collar cutter – A tool used to remove collar material from pin rivets so the rivet can be tapped out of the work.

hone – A fine abrasive stone used to sharpen cutting tools.

honeycomb – A hexagonal cellular material made of thin metal, paper, or plastic, used as a core material for sandwich structure. Named after a bee's honeycomb because of its appearance.

honeycomb shroud ring – Honeycomb material which a rotating airfoil can cut into without damage and the shroud still provide its air sealing function. Usually in the hot section. Referred to as an abradable shroud.

honing – The process of removing a very small amount of material to produce a smooth finish on a surface or to produce a sharp edge on a cutting tool, such as a knife or a chisel.

hook rule – A steel scale with a hook or projection on one end so the rule can be used to measure to the edge of materials with edge radii.

hook spanner – A form of semicircular spanner wrench with the handle on one end and a hook on the other. The hook engages notches in the outside circumference of a ring-type nut.

Hooke's law – A law of physics that deals with the relationship between stress and strain in a material. It states that stress in a ductile material is directly proportional to the strain until the limit of elasticity of the material is reached.

hopper — A funnel-shaped container used for storing the abrasive in a sand blasting machine that has an opening in the top for loading and a smaller opening in the bottom for dumping.

hopper-type oil tank — A container within an oil tank used to hold oil diluted with gasoline for cold weather starting. The use of a hopper minimizes the amount of oil that must be diluted.

horizon — The line of sight boundary between the earth and the sky.

horizontal — Parallel to the earth's horizon or to the bas line of an object.

horizontal stabilizer — A fixed horizontal airfoil attached to the rear of the fuselage to provide stability in pitch.

horizontally opposed engine — A reciprocating engine with the cylinders arranged in two horizontal rows, one on either side of the crankshaft. The cylinders are slightly staggered, with the cylinders in one bank slightly ahead of those in the other bank. Staggering the cylinders allows each piston to be connected to a separate throw of the crankshaft.

horn — A lever or device fastened to, or connected to, a control surface to which an operating cable or rod is attached.

horsepower — Abbrev.: HP. The standard unit of power used for mechanical measurement. It is equal to 33,000 ft.-lbs. of work done in one minute, or 550 ft.-lbs. of work done in one second. Electrically, it is equal to 746 watts.

hose — A flexible plumbing line used in place of rigid tubing in areas subject to movement or vibration.

hose clamp — A metal clamp used to hold a rubber hose onto a piece of rigid tubing.

hot air muff — A jacket installed around a tail pipe. Air routed through the hot air muff picks up heat by convection through the tail pipe material. This heated air is then routed to an air-to-air heat exchanger, where its heat is given up to the air going to the cabin.

hot corrosion — Corrosion occurring in hot sections from a chemical reaction between sulfur in the fuel and salt in the airstream. This condition is more of a problem when operating near salt water.

hot dimpling — A type of coin dimpling, or countersinking, of metal for flush rivets or screws. A heating unit heats the metals to prevent cracking around the hole.

hot forming – Working operations such as bending and drawing sheet and plate, forging, pressing, and heading, performed on metal heated to temperatures above room temperature.

hot section – The portion of a turbine engine aft of the diffuser where combustion takes place.

hot section distress – Any of the metal deterioration conditions found in the hot section, such as warping, creeping, etc.

hot section inspection – An inspection of the hot section of a gas turbine engine.

hot shearing – A method of cutting heavy sheets of magnesium alloys in which the metal is cut while it is hot. This improves the smoothness of the cut.

hot spark plug – A spark plug with a long-nose insulator in which the heat transferring from the center electrode into the shell has a long path to travel. Hot spark plugs are used in engines which operate relatively cool, and they keep the center insulator hot enough to prevent the accumulation of lead oxides.

hot sparks – Localized areas in the cylinder of an internal combustion engine which are overheated to the point at which they become incandescent, or glow. They cause preignition.

hot spots – Localized discoloration on hot section parts indicating a breakdown of cooling air or harmful concentration of fuel at that point. This often is the result of a malfunctioning fuel nozzle.

hot stamping – A method of identifying or imprinting plastic materials, cloth, or paper by using heated metal dies.

hot start – A condition that develops when starting a turbine engine in which the exhaust gas temperature exceeds the allowable limits.

hot streak ignition – Afterburner ignition system in which a stream of raw fuel continues to burn through the turbines and finally provides ignition for afterburner fuel supply.

hot valve clearance – The clearance between the valve stem and the rocker arm when all of the engine parts have reached their operating temperature.

hot-tank lubrication system – A gas turbine engine lubrication system in which hot oil returns directly from the engine to the tank without being cooled because the oil cooler is in the pressure portion of the lubrication system.

hot-tank oil system – A lubrication system where the oil cooler is located in the pressure oil subsystem and the scavenge oil returns to the oil tank uncooled.

hot-wire ammeter – A form of radio frequency current measuring instruments which uses the heating effect of the rf current to heat a wire and change its length. As the wire lengthens, it moves a pointer across the dial to show the amount of current flowing through the wire.

hot-wire anemometer – A type of wind speed indicator that measures the wind speed that passes over a wire that is heated by an electrical current. The amount of heat the wind removes from the wire is proportional to the speed of the wind.

hourmeter – An odometer-type instrument used to measure hours of operating time. When incorporated into a mechanical tachometer, it is accurate only at a specified RPM.

housing – A frame, box, casing, etc. for containing some part, mechanism, etc.

hover – Action of a helicopter which allows it to sustain flight with no movement in relation to the ground.

hovering ceiling – A helicopters maximum altitude at which it can support itself without forward motion.

hub – That part of a propeller or rotor system which attaches to the main driving shaft and to which the blades are fastened.

Huck Lockbolt – A patented, threadless bolt used in the production of aircraft where high-strength, high-speed, lightweight fasteners are required.

hue – The graduation of colors. The characteristics of a color that differentiates between red, blue or yellow and any of the intermediate colors.

humidifiers – Devices used to increase the humidity of the air. They are primarily used in air-cycle air conditioning systems to increase the comfort within the cabin.

humidity – Amount of water vapor in the air.

hung start – A condition when starting a turbine engine in which ignition is achieved, but the engine refuses to accelerate to a self-sustaining RPM.

hunting —
[1]Self-induced and undesirable oscillation above and below a desired value in a control system.
[2]Oscillatory motion of the blades of an articulated rotor about the Alpha Hinge caused by coriolis forces.

hydraulic booster unit — A unit for moving the flight controls in a large, high-speed aircraft. It is actuated by the normal cockpit controls but greatly amplifies the force the pilot exerts.

hydraulic brake — An aircraft brake operated by the means of hydraulic fluid under pressure.

hydraulic filter — A unit which removes foreign particles from the hydraulic system.

hydraulic fluids — Liquids used to transmit and distribute forces to various units which are being actuated.

hydraulic fuse — A unit designed to stop the flow of hydraulic fluid if a leak occurs downstream of the fuse.

hydraulic lock — A condition which occurs in a reciprocating engine having cylinders below the crankcase. Oil leaks past the piston rings and fills the cylinder with an incompressible fluid. The engine cannot then be rotated without damage.

hydraulic motor — A motor which is driven by a flow of hydraulic fluid.

hydraulic pump — An engine, electric motor-driven, or hand-operated pump used to move hydraulic fluid through the system.

hydraulic reservoir — Container for the hydraulic fluid supply in an aircraft.

hydraulic system — The entire fluid power system of an aircraft including the reservoir, pump, control valves, actuators, and all of the associated plumbing.

hydraulic valve lifter — Hydraulic units in the valve train of a reciprocating aircraft engine used to automatically adjust for any changes in dimensions of the engine caused by expansion, and to keep the operating clearance in the valve mechanism to zero.

hydraulics — The branch of science that deals with the transmission of power by incompressible fluids under pressure.

hydrocarbon — An organic compound which contains only carbon and hydrogen. The vast majority of our fossil fuels such as gasoline and turbine fuel are hydrocarbons.

hydrodynamics — The study of forces that are produced by incompressible fluids in motion.

hydrofoil — An airfoil-shaped plate that is attached to the bottom of an airplane or boat that lifts the vehicle out of the water by hydrodynamic action when the vehicle is moved through the water at high speed.

hydrogen — One of the basic elements. In chemical formulas, free hydrogen appears as H_2 because there must be two atoms of hydrogen to form one molecule of free hydrogen gas.

hydrogen brazing — Braze welding in which hydrogen is used as the fuel gas.

hydrogen embrittlement — A brittle condition caused by the metal absorbing hydrogen while it is being electroplated.

hydrogen fuel — A proposed jet fuel of the future which could be stored as a gas or cryogenic liquid. The present high cost and storage problems prevent its current use.

hydromechanical fuel control — A type of fuel control which utilizes hydraulic and mechanical forces to operate its fuel scheduling mechanisms.

hydrometeor — Fog, rain, or hail formed as a product of the condensation of water in the atmosphere.

hydrometer — A device used to measure specific gravity of a liquid. It consists of a weighted float with a long stem in the enlarged glass tube of a syringe. Liquid is pulled up into the tube and the float rides vertically on the surface. The amount the float is submerged is a function of the density of a liquid. The number on the float's stem opposite the liquid level is the specific gravity.

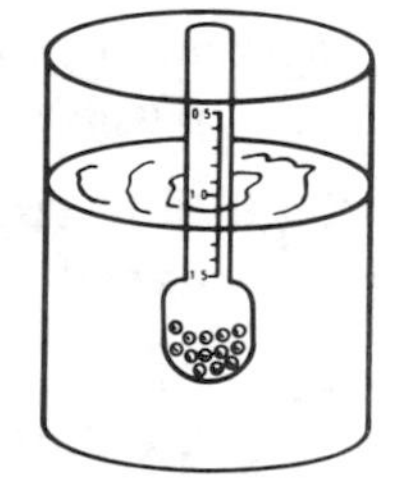

hydroplaning — A condition that can exist with some of the high-speed aircraft having small diameter, high pressure tires. When the tire is on a water covered runway and the brakes are applied, there is a possibility that the brake will lock up and the tire will ride on the surface of the water, much like a water skier. High-speed airplanes normally have relatively narrow, high-pressure tires that, when the brakes are applied, will lock up and skid on the surface of the water in much the same way a water ski rides on the surface of water. A tire that is hydroplaning can develop so much heat that it can be ruined. An effective anti-skid system can prevent hydroplaning.

hydropneumatic – Mechanical equipment that uses both hydraulic and pneumatic forces in order to accomplish its intended purpose.

hydro-ski – An hydrofoil that is mounted below the hull of a flying boat which hydrodynamically produces lift by the hydro-ski. As the flying boat begins to move through the water, the hydro-ski helps to lift the hulls out of the water.

hydrosonic – A regime of flight where speeds of Mach 5.00 are exceeded.

hydro-sorb – A hydraulic shock absorber used in a bungee shock cord landing gear to prevent rebound.

hydrostatic paradox – A condition that does not at first appear to be true, in which it can be observed that the pressure exerted by a column of liquid is dependent on its height and independent of its volume.

hydrostatic testing – Method of pressure-testing compressed gas cylinders with high-pressure water rather than a compressible fluid such as air. Water is used for reasons of safety.

hydroxide – A chemical compound that is made up of a metal or a non-metal (acid) base and a negative hydroxyl ion (OH).

hygrometer – An instrument used to determine the amount of moisture in the air.

hygroscopic material – A material (such as silica gel) that absorbs moisture from the air.

hypergolic – A self-igniting reaction upon contract of the components, without the presence of a spark.

hypersonic engine – *See variable cycle engine.*

hypersonic flight – Supersonic speeds at Mach 5 or above.

hypersonic flow – Flow at very high supersonic speeds. Mach 5 or above.

hyperventilation – Breathing at such an excess rate that the normal amount of carbon dioxide is depleted from the blood.

hypotenuse – The side of a 90° (right) triangle opposite the right angle.

hypoxia – Lack of sufficient oxygen reaching the body tissues.

hysteresis — The ability of a magnetic material to withstand changes in its magnetic state. When a magnetomotive force is applied to such a material, the magnetization lags the mmf because of a resistance to change in orientation of the particles involved.

hysteresis instrument indication error — An instrument error caused by the internal friction inside the instrument mechanism that occurs whether the instrument is indicating an increase or decrease.

I

I-beam — A structural beam made of extruded metal or built up of wood, whose cross section resembled the letter I.

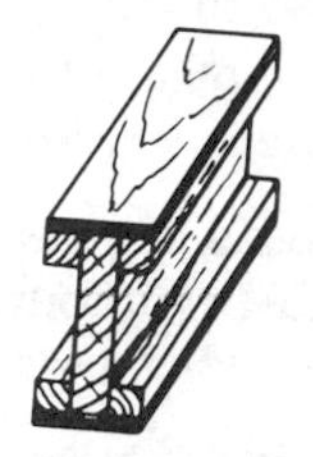

ice — The solid state or condition of water caused by a drop in temperature to 0°C or 32°F.

ice crystals — *See ice needles.*

ice fog — Tiny ice crystals, rather than the tiny droplets of water that make up ordinary fog.

ice light — A light mounted on an aircraft in such a way that it shines on the leading edge of the wing, allowing the pilot to see the build-up of ice on the wing at night.

ice needles — Slender ice particles that float in the air in clear cold weather. Also called ice crystals.

icebox rivet — Rivets made of 2024 or 2017 aluminum alloy, which are too hard to drive unless they are in a softened condition. These rivets must be heat-treated, quenched, and held in a subzero icebox until they are driven.

icing — To coat with ice. *See also ice-up.*

ice-up — A condition in flight in which ice forms on the aircraft structure.

ideal cycle — One in which no pressure loss occurs across the combustion section; but not practical for the gas turbine where a slight pressure loss is needed for correct cooling air.

identification symbol — A symbol used in an aircraft drawing to correlate a specific item with its description in the bill of materials or the revision block.

idiot light — A slang term for warning lights that are used instead of digital or analog instruments to indicate the condition or pressure of a system.

idle cut off – That position of the mixture control in which no fuel can flow from the metering system into the engine.

idle mixture – Fuel-air mixture used by an aircraft engine which provides proper operation at the idle RPM.

idle speed – The RPM of an aircraft engine when the throttle or power control lever is fully closed.

idle thrust – The jet thrust obtained with the engine power control lever set at the stop for the least thrust (idle stop) position.

idler gear – A gear that is used in a gear train to drive another gear in a reverse direction of rotation without changing the speed of rotation, and without adding to or taking away power from the gear train.

idler pulley – The idler pulley changes speed relationships between pulleys but does not change the direction of rotation of either pulley. An idler pulley is also used to adjust the tension on the belt that joins a drive pulley with a driven pulley.

idling current – A low output or operating current that flows in an electronic circuit when there is no input signal that requires the system to act by demands on the system.

igniter – The electrical device used to provide the spark for starting the fire in a turbine engine. Some igniters resemble spark plugs, while others, called glow plugs, have a coil of resistance wire which glows red hot when electrical current flows through the coil.

ignition – The process whereby the fuel-air mixture in either a turbine or reciprocating aircraft engine is ignited.

ignition harness – The complete set of wires that carry high-voltage current from the magneto to the spark plug.

ignition timing – Timing of the ignition of the fuel-air mixture in the cylinders of a reciprocating engine, so that the mixture will be burning before the piston reaches the top of its stroke and the maximum pressure will be produced in the cylinder as the piston starts downward.

I-head cylinder – A reciprocating engine cylinder whose intake and exhaust valves are mounted in the cylinder head. Also called in-head cylinders.

illumination – The lighted surface area given by a light source.

Illustrated Parts List – Abbrev.: IPL. An exploded-view drawing included in a service manual showing every part of a component, along with its proper name, part number, and number required for assembly. This is FAA-approved data and the use of parts not included in this list jeopardizes the approval of the component.

image frequency – The heterodyne action of an oscillator in a super heterodyne receiver. An image frequency is produced when an unwanted signal is of such a frequency that when mixed with the oscillator frequency, it produces a different frequency equal to the intermediate frequency of the receiver.

immersion heater – An electrical heater that is used to heat liquids by immersing the heater in the liquid to keep it warm in order to insure adequate temperature and flow during cold weather operations.

immersion-type oil heater – An electrical heater that is immersed in the engine oil reservoir to keep the oil warm when the engine is not operating, and to ensure an adequate flow of lubricant for starting in extremely cold weather.

immiscible – Liquids that do not mix with each other such as oil and water.

immunize – To remove small particles of iron or grit from the surface of stainless steel by pickling in an acid solution.

impact area – That portion of a damaged aircraft structure which has received the majority of the damage from a collision or other impact.

impact extrusion – Metal forming in which hard metal is forced through a die by striking it with a hard blow.

impact ice – Ice which forms on the wings and control surfaces or on the carburetor heat valve, the walls of the air scoop, or the carburetor units during flight. Impact ice collecting on the metering elements of the carburetor may upset fuel metering or stop carburetor fuel flow.

impact pressure – The pressure of the air as it strikes the stagnation point of a body.

impact test – A test to determine the energy absorbed in fracturing a test bar at high velocity. The test may be in tension or in bending, or it may properly be a notch test if a notch is present, creating multiaxial stresses.

impact wrench — A power wrench, usually air-driven, used to spin nuts onto bolts. Its torque forces are in a series of blows or impacts. Because of the uneven torque it produces, it should not be used for any threaded fastener where the amount of torque is critical.

impedance — Symbol: Z. The vector sum of the opposition to the flow of AC caused by the circuit resistance, and the capacitive and inductive reactance.

impedance matching — The matching of the impedance of a source of electrical power with the impedance of the load that uses the power for maximum transfer of power to occur.

impeller — A vaned disc which picks up and accelerates the air outwardly to increase the pressure in a supercharger for a reciprocating engine, or to provide the pressurized air for a centrifugal-type turbine engine.

impingement starting — A starting process requiring no engine-mounted starter; developed primarily for the Navy. Air from a ground source is directed onto the turbine wheel(s) to cause engine rotation for starting, then the air source is removed.

implode — To burst inward. The reverse of explosion.

impulse — A change in momentum caused by a surge or pulse of energy.

impulse coupling — A spring-loaded coupling between a magneto and its drive gear which causes the magneto to produce a hot and late spark for starting the engine. When the engine is being turned over slowly, the magnet is restrained by stops, and the spring is wound. At the proper time for the starting spark to occur, the spring is released and the magnet is spun, producing a hot, late spark. When the engine starts, centrifugal force holds the coupling engaged so that it acts as a solid unit.

impulse turbine — A stator vane and rotor blade arrangement whereby the vanes form convergent ducts and the blades form straight ducts. The rotor is then turned by impulse as gases impinge on the blades. Design common to turbine driven accessories such as air starters.

impulse-reaction turbine — A stator vane and rotor blade arrangement whereby the base area is an impulse design and the tip is a reaction design. This design is common to flight engines.

impurities —
[1]Harmful foreign objects in a fluid.
[2]A chemical element such as arsenic or phosphorus, that is added to silicon or germanium to give them the electrical characteristics they need.

in phase — A condition in an electrical circuit which the voltage and current rise and fall together. In an AC circuit, the two pass through 0° and 180° at the same time, going in the same direction.

inactive aircraft — An aircraft that is no longer operational.

inboard — Toward the center of the aircraft.

incandescent — Glowing because of intense heat.

incandescent lamp — An electric lamp that produces light by a white-hot filament enclosed in a glass bulb from which all of the air has been removed and replaced with an inert gas.

inches of mercury — Abbrev.: in. mg. A measurement of air pressure, normally used for pressures below atmospheric. 1 in. mg. is equal to approximately ½ PSI.

inches per second — Abbrev.: IPS. A vibration measurement used in electronic balancing.

incidence board — A device used to measure the angle of incidence of a wing.

inclined plane — A machine used to gain a mechanical advantage. It consists of a flat surface positioned at an angle with the horizon.

inclinometer — An instrument consisting of a curved glass tube, housing a glass ball, and damped with a fluid similar to kerosene. It may be used to indicate inclination, as a level, or, as used in the turn and slip indicator, to show the relationship between gravity and centrifugal force in a turn.

inclusions — Inclusions are particles of non-metallic impurities contained in material.

incompressible fluids — Any liquid, such as oil or water, etc., that cannot be compressed, but can be used is used in a regulated fluid power system, such as a hydraulic system, to gain a mechanical advantage.

Inconel — A chromium-iron alloy similar to stainless steel, but which cannot be hardened by heat treatment.

incrementally — Moving in steps rather than in continuous motion.

indicated airspeed — Airspeed as indicated on the airspeed indicator with no corrections applied.

indicated altitude — The altitude read on an altimeter when the barometric scale is set at the proper altimeter setting adjusted to standard sea level barometric pressure of 29.92 inches of mercury or 1013.2 millibars.

indicated horsepower — Abbrev.: IHP. The total horsepower developed in the engine. It is the sum of the brake horsepower delivered to the propeller shaft and the friction horsepower required to drive the engine.

indicated mean effective pressure — Abbrev.: IMEP. The average, actually measured, pressure inside the cylinder of an engine, expressed in pounds per square inch, during the power stroke.

indicating fuse — A small neon light installed in parallel with the fuse that shows when a fuse in an electrical circuit has blown.

indicating instrument — *See indicator.*

indicator — A device, as a gage, dial, or pointer, that measures or records, and visibly indicates. An apparatus that shows fluid pressures, temperatures, or quantities.

induced current — Electrical current that is generated in a conductor when it is crossed by magnetic lines of flux.

induced drag — That part of the total drag which is caused by the same dynamic factors that effect the production of lift. The shape and area of an airfoil, the angle of attack and the air density determines the amount of induced drag.

induced voltage — Voltage generated by a conductor when lines of magnetic flux cut across it.

inducer — The center inlet portion of a centrifugal impeller, sometimes made of a different, harder metal than the impeller for FOD protection.

inductance — That property of a conductor which causes an electromotive force, or voltage, to be generated when lines of magnetic force cut across it.

inductance coil — A coil designed to introduce inductance into a circuit.

induction compass — *See earth induction compass.*

induction furnace — An electric furnace which melts metal by the induction of high-frequency electromagnetic energy.

induction heating — A method of heating a conducting material by rapidly reversing current that flows in the material which causes it to get hot.

induction motor — An AC electric motor that has the AC line voltage connected across stationary windings in the motor housing. Current induced into the rotor causes a magnetic field that reacts with the field of the stator, and this reaction causes the rotor to turn.

induction period — The time period after catalyzed material is mixed in which the material is allowed to begin its cure before it is sprayed onto the surface.

induction system — The complete system of air passages into a reciprocating engine, beginning at the inlet to the air filter and ending with the intake valve of the cylinder.

induction system fire — A fire in the carburetor or air inlet system of a reciprocating engine usually caused by flooding and a backfire.

induction vibrator — A coil and set of contact points which produce pulsating DC from straight DC. Pulsating DC may be used in the primary winding of a magneto to produce a high voltage in the secondary winding.

induction welding — A method of welding in which the metal is melted by the induction of high-frequency electromagnetic energy.

inductive circuit — An AC circuit in which the capacitive reactance lags behind the inductive reactance.

inductive kick — A slang term used for inductive reactance in a coil of wire. It is a high voltage produced across a coil when current stops flowing through the coil. When current flows through a coil, a magnetic field is set up around each of the turns of wire in the coil. But when the current stops flowing, the current cuts across the coil producing many times the voltage that was originally in the coil. This high voltage is called an inductive kick.

inductive reactance — Opposition to the flow of AC caused by the generation of an induced voltage whose polarity is opposite to that of the voltage that created it.

inductive time constant — A measurement of the amount of time needed for the current flowing in an inductive circuit to reach 63.2% of its maximum value.

inductive tuning — A method of selecting or changing the resonance of a radio frequency circuit by changing the inductance. This can be accomplished by rotating a tuning coil which increases or decreases the inductance which in turn tunes the circuit.

inductor — A coil or other device used to introduce inductance into a circuit.

industrial diamond — A diamond that is used as a cutting tool.

inert agent — Fire extinguishing agent that extinguishes a fire by excluding the oxygen from its surface.

inert gas — A gas, such as argon and helium, which does not form other chemical compounds when combined with other elements.

inert gas arc welding — A process of arc welding in which the arc is submerged in an envelope of an inert gas, such as argon, to exclude the oxygen from the molten metal and prevent the formation of oxides.

inertia — The opposition which a body offers to a change of motion.

inertia anti-icer — A movable vane in the induction air system that, in the extended position, causes the velocity of the incoming air to increase, thereby discharging the heavier ice-laden air overboard while directing the lighter ice-free air into the engine plenum.

inertia force — A force due to inertia, or the resistance to acceleration or deceleration.

inertia starter — A starter for large reciprocating engines which uses the energy stored in a flywheel spinning at a high rate of speed to turn the engine for starting.

inertia switch — An electrical switch, that is built into an emergency locator transmitter, which is designed to close and start the ELT when there is a sudden change in its velocity.

inertia welding — Advanced technology process of welding by rubbing friction at high speeds. Developed to join super alloys that are difficult to weld with traditional methods.

inertial navigation — Navigation by means of a device which senses changes of direction or acceleration, and automatically corrects deviations in planned course.

inflammable – A word that has been replaced with "flammable" to avoid confusion.

infrared lamp – An incandescent lamp that produces white light energy in the infrared range.

ingest – To pull in something, such as air, or to ingest FOD in a gas turbine engine.

ingot – A large cast bar of metal, as poured with no working.

inherent stability – That built-in characteristic of an aircraft which causes it, when disturbed from straight and level flight, to return to straight and level flight.

inhibited sealer – A material used to exclude moisture and air from a honeycomb repair. In addition to sealing, it inhibits the formation of corrosion.

inhibitive film – A film of material on the surface of a metal which inhibits or retards the formation of corrosion. It does this by providing an ionized surface which will not allow the formation of corrosive salts on the metal.

inhibitor –
[1]An agent added to a resin to retard its curing and increase its shelf life.
[2]Any substance which slows or prevents a reaction.

initialization – To facilitate the start up of a program. In computers it is the start up of computer language instructions that the computer understands in order to operate as intended.

injection molding – A method of forming thermoplastics by forcing resin, under high pressure, in a mold and allowing it to harden.

injection pump – A high pressure fuel pump that is used in a reciprocating engine fuel injection system. Fuel is pumped, under high pressure, into the combustion chamber of the engine where it is atomized and ignites as it leaves the injector nozzle.

inlet buzz – An audible sound which sometimes occurs in inlets of supersonic aircraft when shock waves alternately move in and out. This condition appears when design speeds are exceeded.

inlet case – The front compressor supporting member, usually one single casting.

inlet duct – That portion of the structure of a turbine-powered aircraft that directs the air into the engine compressor.

inlet gearbox — An auxiliary gearbox driven from and located in front of the compressor in the engine inlet area. Not all engines are configured with this gearbox.

inlet guide vane — A sometimes stationary or sometimes variable vane set installed in front of the first stage rotor blades. The purpose of these vanes is to direct airflow at the optimum angle into the rotors and reduce aerodynamic drag.

inlet particle separator — An inlet device on some turbine powered rotorcraft which prevents sand and other FOD causing debris from entering the engine. On some, the trap has to be cleaned while, on others, the debris is directed overboard.

inlet pressure — Abbrev.: Pt_2. Pressure total taken in the engine inlet as a measure of air density, a parameter sent to the fuel control for fuel scheduling purposes.

inlet screen — An anti-FOD screen used on turbine powered rotorcraft and most stationary turbines. Not generally used on other aircraft installations due to icing and other aerodynamic problems that result.

inlet spike — Moveable inlet device used to control inlet geometry and shock waves. This inlet design diffuses supersonic airflow and reduces it to subsonic speed for entry into the engine.

inlet strut assembly — Spoke-like stationary airfoils which are part of the inlet case. They are used to support the front bearing housing and provide passageways for oil and air line routing from outside the engine to inside.

inlet temperature — Abbrev.: Tt_2. Temperature signal taken in the engine inlet to measure air density and sent to the fuel control unit as a fuel scheduling parameter.

in-line engine — An engine with all of the cylinders in a single line. The crankcase may be located either above or below the cylinders. If it is above, it is called an inverted in-line engine.

in-line reciprocating engine — An engine in which all of the cylinders are arranged in a straight line, with each cylinder piston connected to a separate throw of the crankshaft.

inner exhaust cone — *See tail cone.*

inner line — Refers to can-annular combustion liner; the innermost section.

inner tube — An air-tight rubber tube that has a stem for inflating it. Used inside a pneumatic tire to hold the air that inflates the tire.

inoperative — Not working.

inrush current — The high current that flows in an electrical machine or circuit when the switch is first closed.

inside caliper — A measuring instrument with two adjustable legs used to determine an inside measurement. Once the distance has been established, the actual measurement is made with a steel scale, a micrometer, or a vernier caliper.

inside diameter — Diameter measured from inside surface through center, to corresponding opposite inside surface.

inside micrometer — A form of micrometer caliper used to measure the inside diameter of a circular object such as a cylinder bore. It measures in increments of 1⁄1,000″ or less. It works on the same principle as an outside micrometer caliper.

inspect — The determination of the condition of something by sight, feel, measurement, or other means.

Inspection Authorization — Abbrev.: IA. An authorization issued by the FAA to experienced A&P technicians meeting certain requirements. This authorization allows them to return aircraft to service after annual inspections or certain major repairs.

inspection door — A small door, or hinged plate, on the surface of an aircraft structure that may be opened for inspecting the interior of the aircraft.

inspection hole — A hole in the skin of an aircraft, closed with a cover, and usually held in place with screws. The hole may be opened for inspection or repair inside the structure.

inspection plate — *See inspection hole; inspection door.*

instability — The characteristic of an aircraft that causes it, when disturbed from a condition of level flight, to depart further from this condition.

installation drawing — A drawing which shows all of the parts in their proper relationship for installation.

installation error — An error in pitot static instruments (the airspeed indicator, the altimeter and the rate of climb indicator) caused by a change in alignment of the static pressure port with the airflow as the angle of attack of the aircraft changes.

instantaneous rate of climb indicator — Abbrev.: IVSI. A form of vertical speed indicator which uses internal accelerometer-type air pumps to overcome the inherent lag of this type of instrument, and to provide an instantaneous indication of pitch attitude changes.

instrument — A device to show visually or aurally the attitude, altitude, or operation of an aircraft or aircraft part. It includes the electronic devices used for automatically controlling an aircraft in flight.

Instrument Flight Rules — Abbrev.: IFR. Flight by reference to instruments rather than outside visibility. These rules apply when the ceiling and/or visibility are below certain minimums set by the FAA.

Instrument Flight Rules conditions — Abbrev.: IFR conditions. Condition of weather unsafe for flight under visual flight rules.

Instrument Flight Rules flight — Abbrev.: IFR flight. Aircraft flight conducted entirely by reference to instruments and radio navigation.

instrument landing system — Abbrev.: ILS. A precision instrument approach system which normally consists of the following electronic components and visual aids: localizer, glide slope, outer marker, and approach lights.

instrument panel — A panel, typically located in front of the pilot, that holds all of the indicating instruments that show the condition of the aircraft flight and mechanical systems.

instrument shunt — An electrical shunt that is used with an ammeter to make it possible for it to measure current.

instrumentation — The installation or use of indicating instruments.

insulating electrical tape — A flexible, adhesive-backed tape made of a polyvinylchloride material that is used as an insulation over wire terminals and wire splices.

insulation — A heavy material that is used in an aircraft to prevent the conduction of heat into or out of any of its operating components.

insulation blanket —

[1]A layer of fireproof insulating material used to keep the heat of a jet engine tail pipe from radiating into the engine compartment.

[2]Any material such as fiberglass, aluminum, etc. that is used to insulate against sound, heat, or cold.

insulation grip — A plastic-covered, thin metal reinforcing sleeve on a preinsulated terminal lug that grips the insulation of the wire when the lug is crimped, adding strength and durability to the installation.

insulation resistance — The electrical resistance of an insulating material separating two conductors.

insulator — A material or device used to prevent passage of heat, electricity, or sound from one medium to another.

intake valve — A reciprocating engine valve, located in the head of a cylinder, which provides the passage of the fuel-air mixture into the combustion chamber.

integral fuel tank — A portion of the aircraft structure, usually a wing, which is sealed off and used as a fuel tank. When a wing is used as an integral fuel tank, it is called a "wet wing".

integrated circuit — Abbrev.: IC. A microminiature circuit incorporated on a very small chip of semiconductor material through solid state technology. A number of circuit elements such as transistors, diodes, resistors, and capacitors are build into the semiconductor chip by means of photography, etching, and diffusion.

integrated engine pressure ratio — Abbrev.: IEPR. Used on some turbofans to include fan discharge total pressure and compressor inlet total pressure.

integrating circuit — A network circuit whose output is proportional to the sum of its instantaneous inputs.

intensity control — A cathode ray tube control that calibrates the quantity (intensity) of electrons in the beam that strikes the phosphorescent screen inside the cathode ray tube.

intercom — A communication system within an airplane for the purpose of communicating between flight crew members or to passengers.

interconnector — A small tube connecting multiple burner cans together for the purpose of flame propagation during starting.

intercooler — A device used to remove heat from air or liquid.

intercostal — Longitudinal structure member similar to a stringer, but which is attached to a wing rib or fuselage frame and ends at an adjacent rib or frame. Intercostals are usually used to support access doors, equipment, etc.

intercylinder baffles — Sheet metal air deflectors installed between and around air-cooled cylinders to aid in uniform cooling.

interelectrode — The capacitive effect between two elements in an electron tube. At high frequencies, signals may be fed across the interelectrode capacitance between the plate and a grid.

interelectrode capacitance — The capacitance that exists between two electrodes in an electron tube.

interface — A surface that forms the common boundary between two parts of matter, i.e., water interfaces in jet fuel, and crystal interfaces in metals.

interference fit — A fit between two parts in which the part being put into a hole is larger than the hole itself. In order to fit them together, the hole is expanded by heating, and the part is shrunk by chilling, then put together. When the two parts reach the same temperature, they will not separate. The area around the hole is subject to tensile stress and thus vulnerable to stress corrosion.

intergranular corrosion — The formation of corrosion along the grain boundaries within a metal alloy.

interlock — A type of automatic control device that prevents an action until the device that is protected with the interlock is actuated.

intermediate case — Refers to the high pressure compressor outer case on a turbine engine.

intermediate frequency — A frequency generated in a superhetrodyne receiver equal to the difference between the received radio frequency signal and that produced by the local oscillator.

intermediate position — Position of some movable unit which lies approximately midway between the extreme positions of movement.

intermediate turbine temperature — Abbrev.: ITT. Temperature taken usually at a station between the high and low pressure turbine wheels.

intermittent fault — An intermittent condition in which a fault in a system does not occur with consistency.

intermittent load — A load that is not continually on the system.

internal air pressure — Air pressure within a vessel or container.

internal baffles — Deflector plates installed inside a tank or reservoir to prevent the fluid from sloshing or surging in flight.

internal combustion engine — An engine which obtains its power from heat produced by the combustion of a fuel air mixture within the cylinder of the engine.

internal control lock — A device which may be used to lock a control surface in place when the airplane is parked. It is actuated from a control in the cockpit.

internal damage – Damage that occurs inside a part, component, or mechanism which is not visible externally.

internal resistance – The resistance of the battery itself to the flow of current. It causes a voltage drop proportional to the amount of current flow.

internal supercharger – A gear-driven centrifugal blower in the accessory section of a reciprocating aircraft engine used to increase the pressure of the induction system air.

internal thread – A thread on the internal surface of a hollow cylinder or cone.

internal timing – The timing of the relationship of the E-gap position of the rotating magnet and the opening of the breaker points in a magneto.

internal wrenching bolt – A high-strength steel bolt with its head recessed to allow the insertion of an allen wrench.

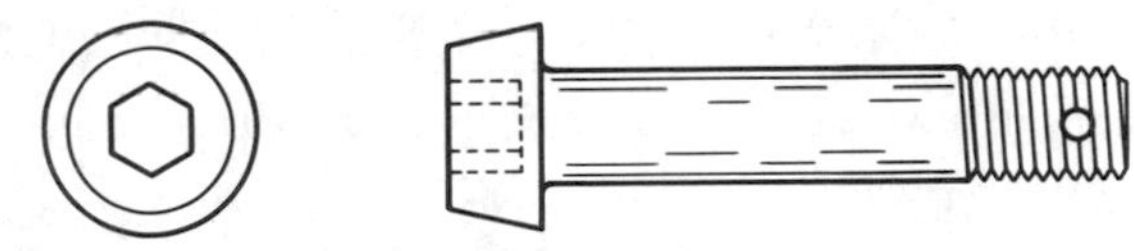

international airport – An airport where international air travel enters and departs. The international airport has facilities of immigration, customs, public health, etc.

International Civil Aviation Organization – Abbrev.: ICAO. A United Nations agency designed to promote and develop international civil air transport.

International Morse Code – Dots and dashes used in combination for transmitting messages. Each combination represents a letter of the alphabet.

international system of units – The system of metric units which includes: meter, kilogram, second, ampere, candela, degrees Kelvin, hertz, radian, newton, joule, watt, coulomb, volt, ohm, farad, tesla, and weber.

interphone system – A communication system normally carried out between in-flight crew members using microphones and earphones.

interplane struts – Struts which run vertically near the wing tips between the wings of a biplane.

interpolation — To compute intermediate values of a quantity between a series with of given values such as in a table of pressures.

interpole —
[1]Field poles placed between the regular generator fields. The windings around them are in series with the armature. Interpoles are used to prevent arcing at the brushes caused by armature reaction.
[2]A field pole in a compound-wound electrical generator that is used to correct for armature reaction. Armature reaction is the distortion of the generator field flux by the current flowing in the windings of the armature.

interrib bracing — Reinforcing tape of a fabric-covered wing which runs diagonally from the top of one wing rib to the bottom of the next throughout a truss-type wing to hold the ribs upright and in line until the rib-stitching is done.

interstage transformer — A type of transformer that is used to prevent the flow of DC from one stage of a multi stage transformer system to the other, and it provides the correct amount of impedance for the output of one stage and for the input of the following stage.

interstate air commerce — Aircraft that carry passengers/cargo for hire among the states of the United States.

Invar — A nickel-iron alloy with an extremely low temperature coefficient of expansion.

inversion — An increase in temperature with altitude. In meteorology, temperature normally decreases with altitude.

invert — To rotate an object vertically.

inverted engine — A reciprocating engine whose crankshaft and crankcase are above the cylinders.

inverted spin — A maneuver having the characteristics of a normal spin except that the airplane is in an inverted attitude.

inverter — An electrical device that changes DC to AC.

investment casting — Casting as in a vacuum furnace or spin chamber to produce a denser, better quality material. Used to produce some steels in turbine engines.

ion — An atom that has either gained or lost an electron. If an atom has a shortage of electrons, it is a positive ion. If it possesses an excess of electrons, it is a negative ion.

ion engine — A type of reaction engine that ejects a stream of ionized particles to produce a forward thrust.

ionize — To convert totally or partially into ions.

ionosphere — A series of atmospheric layers of ions that extends approximately 30 miles above the surface of the earth.

IR — I = current (intensity of current flow) R = resistance. A simple way of designating the voltage generated by the flow of current (I) through a resistor (R).

IR drop — The amount of voltage drop in a given conductor due to the resistance of the conductor. The IR drop is found by the formula: $I \times R = V$, or by multiplying the amount of current (I) in amps, by the amount of resistance (R) in ohms.

IRAN — An acronym used by the military services for a form of maintenance known as Inspect and Repair As Necessary.

iridium — An extremely hard and brittle metallic element of the platinum group, which is used for electrodes of fine-wire spark plugs which must operate in extreme lead fouling conditions.

iris exhaust nozzle — A design similar to a camera shutter. It can be a two position, partially-open/full-open, or a variable opening type. The widest opening is for afterburner mode. The variable opening type is controlled to continuously seek the optimum position for existing conditions.

Irish linen — A strong fabric made from flax that was used to cover many of the older aircraft. It is still popular in Europe, but is no longer readily available in the United States. It may be used as a direct replacement for grade-A cotton.

iron — A heavy, malleable, ductile, magnetic, silver-white metallic element. It is used in the production of steel and all ferrous metals.

iron-core coil — A type of inductor that consist of a soft laminated iron core around which wire is wound.

iron-vane movement — An AC electric measuring instrument which depends upon a soft-iron vane or movable core operating with a coil to produce an indication of AC current flow.

isobar — An irregular line across a meterological chart indicating areas on the earth's surface having equal barometric pressure.

isobaric metering valve — A metering valve in a cabin pressurization system which maintains a constant cabin altitude.

isobaric range — That range of cabin pressurization in which the cabin maintains a constant pressure, or cabin altitude, as the flight altitude changes.

isogonic lines — Irregular lines across an aeronautical chart, along which the magnetic north and true north are separated by a given angle. This angle is known as compass variation.

isolation mount — A rubber and metal composite used to prevent vibration transfer from one component to another.

isolation valve — A valve in an aircraft pneumatic system which may be shut off to isolate the components from the source of air pressure so maintenance may be performed without discharging the system.

isometric drawing — The representation of an object in isometric projection in which the lines parallel to the edges of the object are drawn to true length.

isometric projection — An anonometric projection in which the three faces of an object are equally inclined to the surface of the drawing, and all of the edges are equally foreshortened.

iso-octane — A hydrocarbon (C_8H_{18}) which has a very high critical pressure and temperature. It is used as a reference for measuring the anti-detonation characteristics of a fuel.

isopropyl alcohol — The fluid in the fluid anti-icing system for propeller blades which prevents the formation of ice on the blades during flight.

isostatic forging — Similar process to isothermal forging.

isotherm — A line on a meterological chart which denotes locations having the same temperature.

isothermal forging — A forging process used with super alloy production. One such process utilizes an alloy in powder form which when compressed results in closer near-net-shape than older methods. An advanced technology process for fabricating hot section parts with less waste during final machining.

J

jack —
[1]A piece of equipment that uses mechanical, hydraulic, or air pressure as a means of operation to lift an aircraft or other types of equipment off of the ground.
[2]An electrical socket and plug arrangement that can be pushed into each other to make an electrical connection. Some of the commonly used types of jacks include phone jacks, microphone jacks, and banana plug jacks.

jack pads — Structural locations capable of supporting the weight of an aircraft when it is jacked off the ground.

J-block — A precision block ground to an accuracy of approximately 0.00001″, used as a reference in precision machining operations.

jacket — A metal blanket or shroud used to insulate a portion of a turbine engine.

jacks — Hydraulic or mechanical devices used to lift an aircraft off of the ground for testing or servicing.

jackscrew — A threaded, hardened steel rod which may be rotated to lift an object or to apply a force.

jagged edge — An irregularly shaped edge on a piece of metal, wood, or plastic material.

jam acceleration — The rapid movement of the power control lever of a gas turbine engine, calling for maximum rate of engine speed increase.

jam nut — A thin check-nut screwed down against a regular nut to lock it in place.

jerry can — A specially designed five gallon container that is used for carrying fuel.

jet —
[1]A calibrated, restricted orifice in the fuel passage of a carburetor that is used to control the amount of fuel that can flow under a given pressure. The size of the hole (jet) determines the amount of flow through the jet.
[2]A forceful stream of fluid discharged from a small nozzle.
[3]An aircraft powered by a turbojet engine.

Jet A — A kerosene-type turbine engine fuel similar to the military JP-5. It has a very low vapor pressure and a relatively high flash point.

Jet A-1 — A kerosene-type turbine engine fuel similar to the military JP-8 fuel with additives to make it usable at very low temperatures of approximately –58°F.

Jet Assist Takeoff — Abbrev.: JATO. An auxiliary means of assisting a heavily loaded aircraft to takeoff particularly on a short runway. The JATO consists of small rockets attached to the aircraft and provides the required additional thrust needed for takeoff as the airplane rotates on takeoff.

Jet B — A wide-cut blend of hydrocarbon fuels for use in turbine engines. Used primarily in the military as JP-4 fuel.

jet efflux — Refers to gas flowing from the exhaust nozzle.

jet fuel control — Abbrev.: JFC. The fuel metering system for a turbine engine, which measures the operating conditions of the engine and meters into the burners the correct amount of fuel for the condition.

jet nozzle — A specially designed device shaped to produce a jet stream.

jet nozzle area — The area in square feet of the opening through which the engine exhaust gases pass to the atmosphere. The area will not include any of the exhausting area occupied by a tail cone.

jet propulsion — That form of propulsion produced when a relatively small mass of air is given a large amount of acceleration

jet pump — A pump which operates by producing a low pressure by moving fluid along with it or for removing vapors at a high velocity through a venturi. Seen in oil scavenge systems as oil pumps and fuel systems as vapor eliminators.

jet silencer — A device used to reduce and change the lower frequency sound waves emitting from the engine's exhaust nozzle, and thus reducing the noise factor.

jet stream — A strong narrow band of wind or winds in the upper troposphere or stratosphere, moving in a general direction from west to east and often reaching velocities of hundreds of miles per hour.

jet thrust — Thrust produced by a jet.

jet wake — The hot, high velocity gas stream issuing from the tail pipe of a gas turbine engine.

Jetcal analyzer — A trade name for an electronic test apparatus for checking the calibration of the EGT system, the RPM system, and the accuracy of their associated instruments.

jettison — To cast, or drop from an aircraft in flight.

jewel bearing — A cup-type bearing surface which rides on a hardened steel pivot that is used extensively in many types of indicating instruments.

jeweler's rouge — A very fine ferric oxide abrasive that is used for polishing hard metal surfaces.

jig — Framework or alignment structure used in the construction or repair of an aircraft to hold all the parts in proper alignment while they are fastened together.

jigsaw — A type of electric or pneumatically operated saw tool that uses a variety of narrow blades to cut small curves in wood, metal or plastic.

Jo-bolts — A particular type of internally threaded three-piece rivet.

joggle — A small offset in sheet metal formed to allow one part to overlap another.

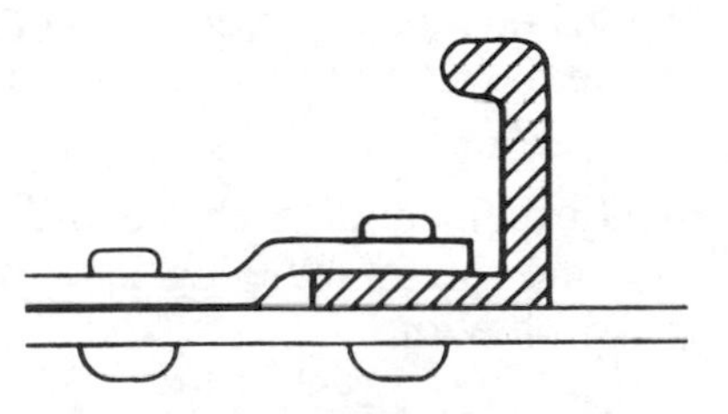

jointer — An electrically powered woodworking machine used to smooth the edges of a piece of wood.

joule — The international system unit of energy equal to the work done when a current of 1 ampere is passed through a resistance of 1 ohm for 1 second.

joule rating — Refers to the ignition system of a turbine engine.

journal — The polished surface of a crankshaft on which the bearings ride.

joystick — An slang term used for the control stick that controls an airplanes ailerons, and elevator. Movement of the joystick fore and aft moves the elevators, and side to side moves the ailerons.

JP-4 — Also known as Jet B. A gas turbine engine kerosine-type fuel that is made up of approximately 65% gasoline and 35% distillates.

JP-5 — Also known as Jet A. A highly refined gas turbine kerosine-type engine fuel.

jumbo jets — Name given to wide bodied airplanes such as the Boeing 747, 757, 767 McDonald Douglas DC-10, Lockheed L-1011, and the A-300 Airbus, etc.

jump seat — A compact portable seat positioned slightly behind the pilot's and copilot's seat in an airplane where a flight engineer sits to monitor certain engine-operating instruments and can operate some of the auxiliary controls.

jumper — A temporary electrical lead wire used to bypass a circuit for purposes of troubleshooting.

junction — The point at which two conductors or circuits join.

junction box — A metal or fiberglass box attached to an aircraft structure for holding the electrical terminal strips to which wire bundles are attached.

junction transistor — A transistor consisting of a single crystal of p- or n-type germanium between two electrodes of the opposite type. The center layer is the base and forms junctions with the emitter and collector.

jury strut — A small strut extending from approximately the mid-point of a wing strut to the spar of the wing. Its purpose is to stabilize the main strut against vibrations.

K

K – The abbreviated use for kilo or one thousand. In computer language, a computer that has a memory of 64K has a memory that can store 65,536 bits of information.

K monel – A high-strength, non-magnetic, heat-treatable, corrosion-resistant alloy made up of nickel, copper, and aluminum.

K-band – A radar wavelength.

K-chart – A sheet metal fabrication chart providing the multiplier to use when determining the setback for bends of other than 90°.

keel – A longitudinal member or ridge along the center bottom of a seaplane float or hull.

keeper – A soft iron bar or plate that is placed across the poles of the magnet when the magnet is not being used in order to keep the magnet from losing any of its magnetism.

Kelvin – Abbrev.: K. Absolute temperature scale with minus 273°C as absolute zero. Used in many engine performance calculations.

Kelvin bridge – A resistance measuring instrument used for accurate measurements of low resistances.

Kennelly-Heaviside layer – An ionized layer in the upper atmosphere which reflects radio waves to earth. Also called E-layer or ionosphere.

kerosine – Formerly spelled kerosene. A thin, colorless, flammable hydrocarbon material distilled from crude oil and used as a base for turbine engine fuel and as a solvent for cleaning parts.

kett saw – Metal cutting, hand-held power-operated tool. The head may be turned to any desired angle for cutting sheet stock aluminum. Uses various sizes of blades.

keyhole saw – A small U-shaped hand saw with a stiff, narrow blade used to cut a larger hole from a small drilled hole.

keying – The process of modulating a continuous carrier wave (cw) with a key circuit to provide interruptions in the carrier in the form of dots and dashes for code transmission.

keyway – A slot or groove machined into a hole or a shaft unto which a key if fitted.

kickback — The reverse rotation of a reciprocating engine due to premature ignition during starting.

kilo — Abbrev.: K. A metric term used to indicate one thousand. Kilo is also used as the prefix for kilogram, kilometer, and kilohertz.

kilogram — Abbrev.: kg. 1,000 grams.

kilohertz — A frequency of 1,000 cycles per second.

kilovolt — 1,000 volts.

kilovolt amperes reactive — Abbrev.: KVAR. A measure of reactive power.

kindling point — That temperature at which combustion can take place between a fuel and oxygen.

kindling temperature — The temperature required for a material to burn when combined with oxygen.

kinematic viscosity — The ratio of absolute viscosity and density, expressed in units of centistokes.

kinetic energy — Energy possessed by an object because of motion.

kink — A sharp twist of bend in a piece of wire, sheet metal, or a piece of tubing.

Kirchhoff's current law — "The algebraic sum of the currents entering and leaving any junction of conductors is equal to zero." Simply stated, this says that all the current which arrives at a point must leave that point.

Kirchhoff's voltage law — The law of electrical circuits which states that the sum of the voltage drops around a series circuit is equal to the applied voltage.

kirksite — An alloy of aluminum and zinc.

kite — A framework covered with paper, cloth, metal, or other material, intended to be flown at the end of a rope or cable, and having as its only support the force of the wind moving past its surfaces.

Klyston tube — A special electron tube for UHF circuits in which modulation is accomplished by varying the velocity of the electrons flowing through the tube.

knife edge — A sharp-edged piece of steel or other hard material used as a balance point or fulcrum for precision balance of a control surface or a propeller.

knife edge filter — A cylindrically shaped piece of metal whose surface is cut with incoming and outgoing sharp cut grooves. When inserted in an oil passageway, contaminants remain in the incoming groove and clean oil squeezes between the knife edge and the casing and through the outgoing groove to the oil jet.

knife edge tip — Thin metal rims on a shrouded-tip turbine blade. These sealing tips establish their own clearance to the shroud ring by contact loading and wear.

knock — A loud knocking or banging noise that is made inside a reciprocating engine cylinder during the compression stroke. The knock is an explosion rather than a smooth burning process, and is caused by The almost instantaneous release of heat energy from fuel in an aircraft engine caused by the fuel air mixture reaching its critical pressure and temperature.

knot — One nautical MPH, or 1.1508 statute MPH.

knot, wood board — A round, hard, section of a tree branch that is embedded in a board and which weakens its overall strength.

knuckle pin — The hardened steel pin that holds an articulating rod in the master rod of a radial engine.

knurl — A series of small ridges on the surface of a material to aid in gripping.

kraft paper — Strong brown paper such as grocery sacks are made of.

Kreuger flap — A type of leading edge wing flap hinged at the bottom side of the airfoil. When it is actuated, the leading edge bends downward, increasing the overall wing camber which allows the wing to develop additional lift at lower airspeeds.

L

labyrinth air seal – Thin sheet metal, either rotating or stationary, rims which control air leakage from the gas path to the inner portions of the engine. Same as knife edge air seal.

labyrinth oil seal – A main bearing oil seal. It is configured with thread-type grooves which allow gas path air to leak inward to the bearing sump and keep oil mist from escaping. Unlike the carbon seal which rides on a surface, the labyrinth has a small clearance between its sealing lands and the rotating shaft.

lacing cord – A strong cotton, linen, or synthetic fiber cord used to rib-stitch fabric covering to an aircraft structure. Also called rib-stitching cord.

lacquer – Pigments dissolved in a volatile base (solvents, plasticizers, and thinners) in preparation for spraying as a liquid, and the cure is effected by the evaporation of the solvents.

lag – A delay in time. Falling behind.

lagging current – An occurrence in an AC inductive reactance circuit whereby changes in the voltage occur before changes in the current. Current, therefore, is lagging the voltage.

lagging material – An insulating material that is wrapped around aircraft plumbing to prevent the unwanted transfer of heat loss to the outside air.

laminar – Layer-like; arranged in layers.

laminar flow – The flow of air over a surface in smooth layers without turbulence.

laminated – Composed of thin layers of material firmly bonded or united together.

laminated core – The core of a coil, transformer, or other electrical device consisting of a stack of thin soft iron sheets insulated from each other by a film oxide or varnish. Laminated cores are used to minimize eddy current.

laminated plastic material – Any type of reinforced plastic resin that is made up of layers of bonded material such as cloth, paper, or wood impregnated with plastic resin together to form complex shapes or to produce a material with high strength for its weight.

laminated structure — A structure of an aircraft made up of layers of material bonded together to form complex shapes or to produce a material with high strength for its weight.

laminated wood — Layers of wood glued together so that the grain in each layer runs in the same direction.

lampblack — A form of essentially pure carbon that is used for making generator brushes and is mixed with rubber for making tires.

land plane — An airplane that has been designed to operate from the surface of land using wheeled landing gear rather than water or snow.

lander space vehicle — A specially designed vehicle for landing on a celestial body.

landing direction indicator — Wind cones and wind sock devices used to indicate wind direction for takeoff and landings.

landing flaps — A secondary control surface built into the wing by which the overall wing area or lift-drag ratio can be increased. The increased wing area permits a lower landing speed. The increased drag reduces airspeeds on landing and shortens the afterlanding roll.

landing gear — The wheels, floats, skis, and all of the attachments which support the airplane when it is resting on the ground or water.

landing gear door warning system — A system used on aircraft having retractable landing gear which will warn the pilot of an unsafe landing gear door condition by the use of a horn, indicators, or red warning lights.

landing gear position indicating system — A system of lights or instruments that shows the pilot the position of the landing gear on aircraft that have a retractable-type landing gear.

landing gear warning system — A warning system incorporated on all retractabale gear aircraft to warn the pilot of an unsafe landing gear condition. The pilot is warned of an unsafe condition by a warning light and aural device. The horn blows and the light come on when one or more throttles are retarded and the landing gear is in any position other than down and locked.

landing lights — High-intensity lights located on the inboard portion of the wing or in the nose of the fuselage. Used to illuminate the runway for landing and taking off at night.

landing roll — The distance an aircraft travels on the ground after touchdown to the point it can be stopped or exits to the taxiway.

landing weight — The weight of the aircraft at touchdown. Often limited to less than takeoff weight by the manufacturer for structural reasons.

landing wires — Streamlined wires attached to the upper wing above the fuselage and extending to the lower wing near the outboard end in a biplane. These wires brace the wings against the forces opposite the normal direction of lift.

lap belt — A seat belt, or a safety belt, that crosses a person's lap while seated in an aircraft.

lap joint — A joint in a sheet metal structure where the edge of one sheet overlaps the other and are fastened together by welding, riveting, or bolting.

lap winding — A method of manufacturing the armature of a DC generator by connecting the ends of each coil that are wound on the armature to the next adjacent commutator segment with the coils lapping over each other.

lapping — To rub two surfaces together with a very fine abrasive between them producing an extremely close fit.

lapping compound — An abrasive paste used to polish surfaces.

laps — Surface defects caused by folding over fins or sharp corners into the surface of the material.

lapse rate — The rate of decrease of temperature with an increase in altitude.

large aircraft — Aircraft of more than 12,500 lbs. certificated takeoff weight.

large calorie — Abbrev.: Cal. A unit of heat energy. It is the amount of heat energy necessary to raise the temperature of 1 kg of water 1°C.

laser beam welding — Use of laser beam heat for welding engine parts. Currently used to weld titanium fan blades which formerly could not be welded by conventional methods.

laser memory — A method of storing billions of bits of digital information permanently on a disk that is similar to a phonograph record.

laser printer — A type of printer that prints a high quality paper copy of the information that is being processed by a computer.

last chance filter – A filter located just before the spray nozzles of a turbine lubrication system. Its purpose is to act as a final filter to prevent clogging of the spray nozzles by foreign matter.

latch – A fastening device used to hold a door closed.

latching relay – An electrically operated relay that, once it is energized, holds the contacts in the energized position by a mechanical latch and can only be released by some mechanical means.

latent heat – The amount of heat required to change the state of a material without changing its temperature.

latent heat of condensation – Amount of heat given off when a substance changes from a vapor to a liquid without changing its temperature.

latent heat of evaporation – The amount of heat absorbed by a substance when it changes from a liquid to a vapor without changing its temperature.

latent heat of fusion – The amount of heat which must be removed from a substance to change it from a liquid to a solid without changing its temperature.

latent heat of vaporization – The amount of heat absorbed by water before it is cha ged to water vapor.

lateral – Crosswise. The span of an airplane is the lateral dimension from wing tip to wing tip.

lateral axis – The pitch axis of an airplane. That axis which extends through the center of gravity of an airplane from wing tip to wing tip. Movement of the elevators rotates an airplane about its lateral axis.

lateral stability – The stability about the longitudinal axis of an aircraft. Rolling stability or the ability of an airplane to return to level flight due to a disturbance that causes one of the wings to drop.

lateral vibration – A vibration in a helicopter in which the movement is in a lateral, or side-to-side direction. The cause of lateral vibration is normally an unbalanced main rotor.

lathe – A metal-working tool in which the work is turned about its longitudinal axis, and cutting tools are fed into its outer circumference.

latitude — The location of an exact position, in degrees, minutes, and seconds, measured on the surface of the earth north and south of the equator.

launch — The releasing of an aircraft or aerospace vehicle for flight.

launching pad — A specially designed location from which rockets can be fired to launch them.

lay of a control cable — The twist of the strands of a wire cable.

layer — A single thickness.

layout — A drawing, pattern or format of a piece of sheet metal in which provision is made for all of the bends, and includes the locations of all of the drilled holes.

L-band radar — A type of airborne radar.

LC circuit — A circuit network containing inductance and capacitance.

LCR electrical circuit — An AC circuit that has inductance (L), capacitance (C), and resistance (R). The total opposition to the flow of current in an LCR circuit is the difference between the inductive and the capacitive reactances

lead — Symbol: Pb. A heavy, pliant, silvery metallic chemical element.

lead of screw thread — The distance the nut will move forward on the screw if it is turned one revolution.

lead-acid battery — A commonly used secondary cell having lead as its negative plate and lead peroxide as its positive plate. Sulfuric acid and water serve as the electrolyte.

leading current — An AC electrical circuit in which the current that flows in the circuit has more capacitive reactance than it has inductive reactance.

leading edge — The foremost edge of an airfoil section.

leading edge flap — A portion of the leading edge of an airplane wing which folds downward to increase the camber of the wing to increase both its lift and drag. Leading-edge flaps are extended for takeoffs and landings to increase the amount of aerodynamic lift that is produced at any given airspeed.

lead-lag hinge — A hinge at the root of a helicopter rotor blade with its axis perpendicular to the plane of rotation. This hinge is also known as the alpha, drag, or hunting hinge.

leaf brake — A bending tool used to form straight bends in sheet metal. The material is clamped in the brake, and a heavy leaf folds the metal back over a radius block to form the desired bend. Also called a cornice brake.

leaf spring assembly — A series of flat springs that are hinged at one end and arched in the center. The load is applied to the center of the arch and it is absorbed as the spring alternately straightens out and returns to its arched shape.

leakage —
[1]The breakdown of the dielectric strength of an insulator which allows current to pass through it.
[2]An amount or a liquid or gas that leaks in or out of something.

leakage current — The amount of current that flows from the battery terminals to the battery case through any moisture or contamination on top of the battery.

leakage flux — Magnetic flux that does not follow a direct path between the poles of a magnet, and which does not provide any useful work.

lean blow out — A condition in jet engine fuel combustion during which fuel supply is decreased to maintain or reduce engine speed. During this condition burning may be so slow that the flame is carried out of the combustion chamber and extinguished.

lean flame out — A condition of turbine engine operation in which the fire goes out in the engine because the fuel air mixture is too lean to support combustion.

lean mixture — A fuel-air mixture in which there is an excess of air over that required for the amount of fuel.

Leclanché cell battery — Another name for a common carbon-zinc cell flashlight battery.

left-hand rule –

[1]Left-hand generator rule: The fingers of the left hand are arranged in such a way that the thumb, first finger, and second finger point 90° to each other. If the thumb points in the direction of movement of the conductor, the first finger will point in the direction of the lines of flux (north to south), and the second finger will point in the direction of the induced voltage (back voltage, from positive to negative).

[2]Direction of magnetic flux: If the fingers of the left hand encircle a conductor in the direction of the lines of magnetic flux, the thumb will point in the direction of electron flow.

[3]Polarity of an electromagnet: If we grasp the coil of an electromagnet in such a way that our fingers encircle the coil in the same way the electrons are flowing, our thumb will point to the north pole of the electromagnet that is formed by the coil.

left-hand thread – A thread is a left-hand thread, if when viewed axially, it winds in a counterclockwise and receding direction. All left-hand threads are designated LH.

legs of a right triangle – Sides of a right triangle which are joined by the right angle (90°).

LEMAC – The leading edge of the mean aerodynamic chord which is often used as a location or reference for many aerodynamic measurements in aircraft operations and designs.

lenticular cloud – A cloud, shaped like a lens, that forms on the downwind side of a mountain which indicates severe air turbulence that should be avoided.

Lenz's law – The law of induced current that, simply stated, says the current induced in a conductor will produce a magnetic field which opposes the field producing the original current.

level –

[1]The horizontal condition of a body: A flat horizontal surface of an object is level when all points on the surface are perpendicular to the center of the earth.

[2]Spirit or bubble level: An indicating device that has a curved glass tube filled with a slightly colored liquid, except for a small air bubble. The tube is mounted in a housing so that when it is place in a perpendicular position to the ground, the bubble will rise to the center of the tube indicating a level condition.

leveling lugs – Points on the aircraft on which a level may be placed for leveling the aircraft.

leveling means – The way in which the aircraft may be checked for level flight attitude as specified by the aircraft manufacture. This may be longitudinal or lateral and sometimes both.

leveling scale – A scale built in the aircraft for checking the leveling of the aircraft in conjunction with a plumb bob.

lever –
[1]A device such as a bar used for prying.
[2]A flat bar turning about a fulcrum.

Leyden jar – A primitive capacitor. In effect, an apparatus for storing an electric charge on the inside foil lining of a glass jar.

L-filter – An inductor-input filter consisting of an inductor and a capacitor used to smooth the ripple from the output of a rectifier.

life-limited part – A part which has a specified number of operating hours and then must be removed for overhaul. Most turbined parts, however, are used until failure or are airworthy until the TBO of the engine.

life-support systems – Oxygen and pressurization systems in an aircraft that make it possible for the occupants of the aircraft to operate an aircraft at high altitudes.

lift – One of the aerodynamic forces acting on an aircraft in flight. It opposes gravity and is produced by downwashing a mass of air equal to the weight of the aircraft.

lift fan – Turbofan engine which can be swiveled up and down or an exhaust duct which can be turned to provide upward thrust for vertical or short takeoff. Used in VSTOL aircraft.

lift wires – Biplane wing support wires installed between the wings of a biplane to hold the wings in alignment against the forces of lift. Lift wires extend from the inboard end of the lower wing to the interplane struts on the upper wing.

lift-drag ratio – The efficiency of an airfoil section. It is the ratio of the coefficient of lift to the coefficient of drag for any given angle of attack.

lifting body – A wingless aircraft developed by NASA where lift is created by the shape of the craft itself.

light – Electromagnetic radiations of a frequency range that is visible to the human eye.

light aircraft – Aircraft having a total gross weight of 12,500 lbs. or less.

light emitting diode – Abbrev.: LED. A semiconductor diode which emits light when current flows through it.

light ends – The products of petroleum that boil off first in the process of fractional distillation.

light plane – Aircraft having a total gross weight of 12,500 lbs. or less. Light plane is also used as a colloquial term to mean a single-engine airplane that is used for private use or for other non-commercial flying use.

lightening hole – A hole cut in a structural part to decrease weight, and strength is maintained by flanging the area around the hole.

lighter-than-air aircraft – Aircraft that can rise and remain suspended by using contained gas weighing less than the air that is displaced by the gas.

lightning – An immense spark caused by the discharge of static electricity from a highly charged cloud. This cloud has become charged because of friction in turbulent air, and the lightning occurs between the air that is displaced by the gas.

light-up – The point at which combustion occurs in a turbined engine as indicated by an exhaust temperature rise on the cockpit indicator.

lime grease – A grease that is made up of oil and calcium carbonate.

limit switch – A switch designed to stop an actuator at the limit of its movement.

limiter – A stage in a frequency modulated receiver that limits the amplitude of the signal and thus removes static.

limits – The bounds of travel that are allowed.

limits of size – The limits of size are the applicable maximum and minimum sizes (commonly referred to as "limits").

Lindberg fire detection system – A continuous-element-type fire detector consisting of a stainless steel tube containing a discrete element which has been processed to absorb gas in proportion to the operating temperature. When the temperature rises to the operating temperature set point, it causes the pressure in the stainless steel tube to increase and close a switch, which actuates the warning light and bell.

line loss– The voltage loss in a conductor due to its length.

line maintenance – Inspection and repairs which are accomplished on the flight line as opposed to shop maintenance.

line of sight radio reception – A limitation of high-frequency radio signal transmissions that requires a clear path between the transmitting and the receiving antennas.

line voltage – The main power line voltage that operates a system.

line voltage regulator – Used to stabilize the line voltage by sensing and regulating the voltage demands supplied to a piece of electrical or electronic equipment.

linear actuator – An actuator that changes hydraulic or pneumatic pressure into linear motion.

linear movement – A type of movement or progression in which the output or result is directly proportional to the input.

linear operation – The type of operation in which the output of a device is directly proportional to its input. If the input increases by 10%, the output will also increase its value by 10%.

linear resistance curve – The characteristic illustrated by a load when any increase or decrease in the voltage across the load results in a proportional change in the current through the load.

linen – A type of fabric made from flax which, in the past, was a favorite covering material for truss-type airplanes. In the United States, it has been almost totally replaced by grade-A cotton or by one of the synthetic fibers.

lines of flux – Lines of magnetic force connecting the poles of a magnet.

link – A short connecting rod used for transmitting power and/or force.

link rod – *See articulated connecting rod.*

linseed oil – A drying oil used in some aircraft finishes. Also used to coat the inside of steel tubing to prevent its rusting. It is obtained from flax seed.

liquid – A fluid which will assume the shape of the container in which it is held, but will not expand to completely fill the container.

liquid air – A slightly bluish, transparent liquid that has been changed into a liquid by lowering its temperature to –312°F.

liquid lock – *See hydraulic lock.*

liquid nitrogen – Nitrogen that has been changed into its liquid state by lowering its temperature to –195°C (78° Kelvin) or lower.

liquid oxygen — Abbrev.: LOX. Oxygen that has been changed into its liquid state by lowering its temperature to –113°C (160° Kelvin) or lower.

liquid-cooled — An engine or mechanism that is cooled by a liquid.

liquid-cooled engine — A reciprocating engine that uses a mixture of water and ethylene glycol to remove excess heat. This is accomplished by circulating the liquid around the cylinders in jackets which then transmits the absorbed heat through a radiator where the heat is given up to the oncoming air.

liter — Metric unit of volume used for gaseous or liquid measurement, slightly more than a quart, measured at 1.0567 qt.

lithium — An alkaline-metal element; the lightest chemical element known.

lithium cell — One of a family of chemical cell types incorporating lithium in one pole piece.

lithium grease — A water-resistant, low operating temperature grease that is made of lithium salts and fatty acids.

lithosphere — The earth's most outer crust area that extends downward towards the center of the earth for approximately 50-60 mi.

litmus — A water-soluble powder that changes its color when it is acted on by an alkali.

litmus paper — An indicator paper which will change color when it comes in contact with an acid or an alkall. It turns red when wet with an acid, and blue with an alkali.

live center — A component on a lathe that has a sharp-pointed center that fits into the headstock of a lathe and turns with it.

load — An energy-absorbing or energy-using device of any sort connected to a current.

load bank — A heavy-duty resistor used to discharge a storage battery.

load cell — An electronic weighing system component, which contains the strain gages, that are placed between an aircraft jack pad and the jack. Used to measure the weight of the aircraft load.

load chart — A chart used for weight and balance purposes, specifying the location and distribution of weights. Aids the pilot in determining the loaded center of gravity condition.

load, electrical – Any apparatus that uses electrical power to perform some other function, such as operating a motor.

load factor – The ratio of the weight of an aircraft to the load imposed during a maneuver.

load manifest – An itemized list of weights and moments of a particular load taken on a specific flight. Used by FAR Part 121 and 135 operators.

loading graph – A method of computing the loaded weight and center of gravity of an aircraft.

loadmeter – A current measuring instrument which is calibrated in terms of the percentage of the total rated current of the power source.

lobes – The eccentric portions of a cam or camshaft.

LOC mode – The operating position of an automatic pilot when it is receiving its signals from the localizer portion of an instrument landing system.

local action – The formation of tiny chemical cells in one or both of the poles of a chemical cell due to impurities in the material. Local action may lead to exhausting of the service capacity of a cell or corrosion of the pole pieces.

local oscillator – The internal-oscillator section of a superheterodyne circuit.

local traffic – Any and all air traffic operating within the sight and control of an airport tower.

localizer – The electronic portion of an instrument landing system that indicates the center line of the runway to the pilot for the final approach in an instrument landing.

lock tabs – A special type of safetying device in which tabs made onto a special washer are bent up beside the nut to prevent its loosening.

lockclad cable – Conventional, flexible steel cable with aluminum tubing swaged over it, locking the cable inside the tubing.

lockout debooster – A hydraulic component which decreases the pressure applied to aircraft brakes. Its lockout function shuts off all flow of fluid to the brake in the event of a rupture of the brake line below the debooster.

lockring — A type of safety device consisting of a horseshoe-shaped ring which may be expanded by spreading its ends. It is then snapped into a groove in a shaft to lock it into a bearing.

lockstitch — The use of a modified seine knot to lock the stitches when hand sewing aircraft fabric. The baseball stitch is used for sewing, and it is locked with the seine knot every eight to ten stitches to prevent loosening.

locktab — A type of mechanical lock that is used to prevent a nut from coming loose. Its appearance is similar to a washer but with notches cut from the periphery. When a locktab is place under a nut on the shaft of a bolt, one or more of the external locking tabs are bent up against the flats of the nut to keep it from backing off and becoming loose.

lockwire — Another name for safetywire. It is a stainless steel, brass, or galvanized steel wire used to exert a pulling motion on the head of screws or bolts to prevent their becoming loosened.

lockwiring — A method of safetying two or more screws or bolts together by twisting lockwire between them in such a way that tension is held on the head of the fastener in the direction of tightening.

lodestone — Natural rock having magnetic characteristics.

log — Books containing a record of activities. Pilots keep a log of their flight time, and ground crews keep logs on the mechanical operating components of the aircraft, the airframe, engine, propeller and rotor to show the amount of time in service, and to record all of the maintenance that has been completed on each device.

logarithmic or audio taper potentiometer — A volume control potentiometer whose resistance increases logarithmically as the control shaft is rotated in a clockwise direction.

logbooks, mechanical — Books containing records of the total operating time, any repairs, alterations or inspections performed, and all AD notes complied with. One should be kept for the airframe, one for each engine, and one for each propeller.

logic — The science of dealing with the basic elements of truth and the use of truth tables.

logic circuit — A circuit designed to operate according to the fundamental laws of logic.

logic state — The state or condition (logic one or zero) of a digital electronic conductor.

logical flowchart — A type of flowchart that resembles a pert chart, which graphically shows the flow of information through a computer program, and the decisions that must be made at various points. Boxes show information, a diamond shaped figure is a logical decision point, and a parallelogram is a point where data is put into the program or where it is taken out.

logical one — Logical one represents a YES, or a TRUE, condition in digital electronics. It is produced by a closed switch or by the presence of a voltage.

logical zero — Logical zero represents a NO, or a FALSE, condition in digital electronics. It is produced by an open switch or by no voltage.

long duct turbofan — A design which ducts the cold stream to the rear of the engine and to the atmosphere. On some engines, the cold and hot streams mix; on others, they do not.

longeron — The main longitudinal strength-carrying member of an aircraft fuselage or engine nacelle.

longitudinal — Running lengthwise.

longitudinal axis — That axis of an airplane which extends through the fuselage from nose to tail, passing through the center of gravity. It is also called the roll axis which is controlled by the movement of the ailerons.

longitudinal dihedral angle — The difference between the angle of incidence of the wings and the angle of incidence of the horizontal stabilizer.

longitudinal stability — Pitch stability, or stability about the lateral axis of an airplane.

long-range navigation — Abbrev.: LORAN. A radio navigation system utilizing master and slave stations transmitting timed pulses. The time difference in reception of pulses from several stations establishes a hyperbolic line of position which may be identified on a loran chart. By utilizing signals from two pairs of stations, a fix in position is obtained.

long-wire antenna — An antenna for radio energy whose length is greater than one-half the wavelength of the frequency of the energy being transmitted or received.

loom – A type of braided, yet flexible insulating material used for wire protection.

loop –
[1]A control circuit consisting of a sensor, a controller, an actuator, a controller unit, and a follow-up or feedback to the sensor; also any closed electronic circuit including a feedback signal which is compared with the reference signal to maintain a desired condition.
[2]A flight maneuver executed in such a manner that the airplane follows a closed curve approximately in a vertical plane.

loop antenna – A highly direction sensitive antenna wound in the form of a coil, used to find the direction between the loop and the station that is transmitting the signal that the loop is receiving.

loopstick antenna – A loop antenna consists of a large number of turns of wire wound on a powdered iron (ferrite) rod to increase the amount of radio signal the coil picks up. Loopsticks are particularly useful in small portable radio receivers.

louver – A slotted or multi-slotted opening usually for the passage of air.

low blower – The lower ratio of blower to crankshaft speed of a two-speed internal supercharger.

low bypass turbofan – Usually engines with a one to one by-pass ratio. That is, approximately the same mass airflow flows across the fan as across the core engine.

low pitch, high RPM setting – The setting of a controllable-pitch propeller which allows the engine to produce its highest RPM with the propeller at its lowest pitch.

low-lead 100-octane aviation gasoline – 100-octane aviation gasoline which contains a maximum of 2 ml. of tetraethyl lead per gallon. Normal 100-octane avgas is allowed to contain 4.6 ml./gal.

low-pass filter – A filter circuit designed to pass low-frequency signals and attenuate high-frequency signals.

low-pressure compressor – The front section of a dual compressor gas turbine engine, also called the N_1 compressor or low speed compressor.

low-pressure compressor gas turbine engine – The front section of a dual compressor, also called the N_1 compressor or low speed compressor, which is driven by the last stages of the turbine.

low-pressure oxygen system – A gaseous oxygen system formerly used in military aircraft in which the oxygen is stored under pressures of approximately 450 PSI.

low-tension ignition system – *See low-tension magneto.*

low-tension magneto – A magneto system used for reciprocating engine airplanes that fly at high altitudes. It consists of a rotating magnet, a cam, breaker points, condenser, a coil with only the primary winding, and a carbon brush-type distributor. The primary current is directed through the distributor to a coil for each individual spark plug. These coils have a primary and a secondary winding which generates the high voltage at the spark plug.

low-voltage ignition system – A main ignition system used on turbine engines with a voltage output in the range of approximately 1,000 to 5,000 volts delivered to the ignitor plug.

low-wing airplane – An airplane having one main supporting surface flush with the bottom of the fuselage.

lubber line – The reference line of a magnetic compass or directional gyro. The line represents the nose of the airplane.

lubricant – A natural or artificial substance used to reduce friction between moving parts or to prevent corrosion on metallic surfaces.

lubricating – The application of a lubricant to machinery to reduce friction.

Lucite – A transparent acrylic thermoplastic resin made by E.I. DuPont de Nemours and Co. which is used for windshields and side windows of small aircraft.

lug – A projection from a structural member used as an attachment point.

luminescence – Illumination of light with little or no heat.

luminous paint – Paint used for marking aircraft instrument dials and pointers, which glows in the dark when it is excited with ultraviolet, or black, light.

lumped load –
[1]The theoretical load obtained by assuming that the resistance of a distributed load has been replaced by a single resistor or load.
[2]A load whose resistance is so much greater than the distributed resistance in the circuit, such as that of connecting wires, etc., that the circuit load may be considered concentrated at that point.

lye – An alkaline solution consisting of potassium hydroxide or sodium hydroxide.

M

Mach number – The ratio of the speed of an airplane to the speed of sound in the same atmospheric conditions.

machine bolt – The common name for a hex head bolt with uniform threads.

machine language – A language used in a computer system made up of zeros and ones, which uses a special program called a compiler that converts a programmed language, into machine language that can be used by the computer.

machine screw – A screw fastener with uniform threads that can be screwed into a tapped hole or into a nut. The head of a machine screw may be round, flat, truss, oval, or a fillister-type head.

machine-sewn fabric seams – Aircraft fabric seams that are machine sewn. The general types of machine-sewn fabric seams include the French fell, folded fell, and plain overlap.

machining – The process of forming the surface by cutting away material by turning, planing, shaping, and milling, with machine-operated tools such as lathes, milling machines, shapers, and planers.

machinist – A skilled person in the operation of such metal-working machine tools as lathes, shapers, planers, and milling machines.

machmeter – A direct-reading indicator installed in the instrument panel of high-speed aircraft which gives the pilot an indication of his flight Mach number. The internal mechanism of a Machmeter includes a bellows for measuring the difference between pitot pressure and static pressure, and includes an aneroid that modifies the output from this differential pressure bellows to correct for the changes in altitude.

mackerel sky – A meteorological condition of clouds that resemble the scales on a mackerel fish. The clouds consists of rows of altocumulus or cirrocumulus clouds.

magamp – A contraction of magnetic amplified. An amplifier system using saturable reactors to control an output to obtain amplification. *See also magnetic amplifier.*

magnesium – A silver-white, malleable, ductile metallic element used to produce light alloys for aircraft construction.

Magnesyn system – An AC remote indicating system in which a permanent magnet is used as the rotor. The stator is a saturable-core, toroidal-wound coil, excited with low-voltage AC. When the core is saturated, it will not accept lines of flux from the rotor.

magnet – A device or material that has the property of attracting or repelling other magnetic materials. Lines of magnetic flux link its external poles, and a conductor cutting across the flux will have a voltage inducted into it.

magnet keeper – A soft iron bar placed across the north and south poles of a U-shaped magnet. The iron bar produces a closed path through whcih magnetic lines of force pass.

magnet wire – A small-diameter, varnish-insulated copper wire used to wind coils for electromagnets, transformers, motors, and generators.

magnetic amplifier – An electronic control device which uses a saturable reactor. The condition of saturation is controlled by the input signal to modulate the flow of a much larger current in the output circuit. It is essentially a multi-coil transformer that controls the amount of load current that is allowed to flow in the load winding by a small amount of current in the magnetic core. Changing the amount of DC flowing in the magnetic core changes the permeability of the core. This, in turn, changes the amount of inductive reactance that opposes the AC flowing in the load winding.

magnetic brake – A type of friction brake that is spring-loaded to keep the armature from turning when the current stops flowing through the winding of a motor. When the current flows through the winding of the motor, the friction brake is released allowing the rotor to rotate.

magnetic bubble memory – Magnetic bubble memory, used in a digital computer, allows a very large amount of information to be stored in a very small area.

magnetic chuck – A metal machining tool that consists of a special work surface that uses electromagnetism to hold the material being machined tightly to the surface.

magnetic circuit – Any complete path of magnetic lines of flux that leaves the north magnet pole of an electrical machine, such as a motor or generator, and enters the magnetic south pole.

magnetic circuit breaker – A circuit breaker that opens a circuit whenever there is an excess of current flow in the circuit. It works on the principle of electromagnetism. When more current than the circuit breaker is rated to carry flows through it, the magnetic field develops enough strength to open a set of contacts and opens the circuit.

magnetic compass error – *See Acceleration error.*

magnetic course – The path of an airplane as measured from magnetic north.

magnetic deviation – A compass error caused by localized magnetic fields in the airplane attracting the floating magnets in the magnetic compass and deflecting it away from magnetic north.

magnetic drag cup – The aluminum or copper cup surrounding the rotating magnetic in a simple mechanical tachometer. Eddy currents are generated in this cup, causing magnetic fields which oppose those of the rotating magnet.

magnetic drain plug – Similar to a chip detector, except some types may not be powered to show contamination on the light in the cockpit. The drain plug consists of two small permanent magnets built into it to attract and hold any ferrous metal particles that may be in the lubricating oil system. Ferrous metal chips on the drain plug indicates the possibility of internal engine failure. Usually located in the lower portion of a sump in the scavenge oil subsystem.

magnetic field – The space around a magnet or conductor where magnetic flux is found.

magnetic flux – Invisible lines of magnetic force that are assumed to exist between the poles of a magnet, and which follow the path of least resistance. Traditionally, they are given the direction from north pole to south pole. When an electrical conductor cuts across the lines of magnetic flux, a voltage is produced in the conductor. One line of magnetic force is called one maxwell.

magnetic flux density – The unit of field intensity is the gauss. An indicidual line of force, called a maxwell, in an area of one aquare centimeter produces a field intensity of one gauss.

magnetic heading – *See magnetic course.*

magnetic north – True north direction corrected for variation error.

magnetic particle inspection — A form of nondestructive inspection for ferrous metal parts in which the part is magnetized, producing north and south poles across any discontinuity, either on the surface or subsurface. Iron oxide, mixed with a fluorescent dye, is attracted and held over the discontinuity. An ultraviolet light (a black light) flashed on the part shows the iron oxide as an incandescent line.

magnetic pickup RPM system — A newer fan speed indicating system which uses a magnetic pickup in the fan case. Blade motion produces an eddy current powering a cockpit indicator.

magnetic poles — A suspended magnet swinging freely will align itself with the earth's magnetic poles. One end is labeled "N" meaning north-seeking. The opposite end of the magnet marked "S" is the south-seeking end.

magnetic saturation — A magnet's saturated condition in which all of the magnetic domains are lined up in the same direction, and any increase in the magnetic field is not possible.

magnetic shunt — A piece of soft iron that is shunted across the air gap of a magnet used in an electrical measuring instrument. The position of the magnetic shunt can be changed to calibrate the instrument by varying the amount of magnetic flux that crosses the air gap.

magnetic variation — The angular difference between the geographic north pole and the magnetic north pole.

magnetic wave — The component of a radio wave that is perpendicular to the antenna.

magnetic yoke — The mechanical support which completes the magnetic circuit between the poles of a generator.

magnetic-drag tachometer — *See magnetic drag cup.*

magnetism — The ability to attract certain materials containing iron and to influence moving electrons.

magneto — A self-contained, permanent-magnet AC generator with a set of current interrupter contacts and a step-up transformer. It is used to supply the high voltage required for ignition in an aircraft reciprocating engine.

magneto safety check — An operational check on an aircraft reciprocating engine in which the magneto switch is placed in the OFF position with the engine idling to ascertain that the switch actually does ground out both magnetos.

magnetomotive force – Abbrev.: mmf. Magnetizing force in a magnetic field. Measured in gilberts or ampere-turns.

magnitude – A condition of size, quantity, or number.

magnitude of a force – *See magnitude.*

main bus – A common tie point for electrical circuits to obtain their voltage.

main fuel system – The fuel distribution system used for all normal engine operating conditions. *See also emergency or primer fuel systems.*

main rotor – The rotor that supplies the principal lift to a rotorcraft.

main wheels – The wheels of an aircraft landing gear that support the major part of the aircraft weight.

maintenance – Inspection, overhaul, repair, preservation, and the replacement of parts, but excludes preventive maintenance.

maintenance manual – A manual produced by the manufacturer of an aircraft, aircraft engine, or component which details the approved methods of maintenance.

maintenance release – A return to service approval of an aircraft by an authorized A&P technician or IA, and logged in the appropriate maintenance record.

major alteration – An alteration not listed in the aircraft, aircraft engine, or propeller specification; and one which might appreciably affect weight, balance, structural strength, performance, powerplant operation, flight characteristics, or other qualities affecting airworthiness, or that is not done according to accepted practices, or cannot be done by elementary operations.

major diameter – The diameter of the threads of a bolt or a screw.

major overhaul – The complete disassembly, cleaning, inspection, repair, and reassembly of an aircraft, engine or other component of an aircraft in accordance with the manufacturers specifications, and which will return the device to a serviceable condition.

major repair – A repair that, if improperly done, might appreciably affect weight, balance, structural strength, performance, powerplant operation, flight characteristics, or other qualities affecting airworthiness, or that is not done according to accepted practices, or cannot be done by elementary operations.

make-and-break ignition – One of the earliest forms of electrical ignitions for internal combustion engines. It produced an arc when two contacts carrying low-voltage current within the cylinder were separated.

male electrical connector – The pin contact that completes a circuit by sliding inside a socket (the female connector).

male fitting – A fitting designed to be placed, screwed, or bolted into another unit.

malfunction – The failure of a part of a component to function; or deviation in the operation of a unit from its intended purpose or design.

malleability – *See malleable.*

malleable – The characteristic of a material that allows it to be stretched or shaped by beating with a hammer or passing through rollers without breaking.

mallet – A hammer with a heavy wood, plastic, rubber, or leather head.

mandrel –

[1]Lathe: A tapered shaft, that fits into a hole used to support and center a device or piece of material so that it can be machined.

[2]Tube bending: A long steel rod with a rounded end inserted into a piece of metal tubing in order to keep the tubing from flattening while it is being bent.

maneuverability – The ability of an aircraft to be directed along a desired flight path and to withstand the stresses imposed upon it.

manganese – A non-magnetic chemical element of grayish white, used in the manufacture of iron, aluminum and copper.

manganese dioxide – A chemical compound that is used in carbon-zinc batteries to absorb the hydrogen gas that would otherwise insulate the carbon rod when electrons flow from the zinc can to the carbon.

Manganin – An alloy of copper, manganese, and nickel.

manifest – A list of passengers and cargo carried on any one flight.

manifold – A chamber having several outlets through which a liquid or gas is distributed or gathered.

manifold pressure – The absolute pressure, measured in inches of mercury, existing in the intake manifold of an engine. This is the pressure that forces the fuel-air mixture into the cylinder.

manometer – An instrument consisting of a glass tube filled with a liquid for measuring the pressure of gases or vapors either above or below atmospheric pressure.

manual depressurization valve – A back-up valve used to control cabin pressurization by manually controlling the outflow of air from the cabin if the automatic system malfunctions.

manufactured rivet head – The preformed head of an aircraft rivet when it is manufactured at the factory. *See also shop head rivet.*

manufacturer – A person or a company who manufactures aircraft, aircraft engines, or aircraft components.

manufacturing – The process of taking raw materials and changing them into a finished and usable product.

Marconi antenna – A non-directional, quarter-wave antenna, utilizing a ground plane, which serves as a quarter-wave reflector used for transmitting and receiving radio communications in the higher frequency bands.

marine grommet – A plastic or metal reinforcement ring designed with a special shield used to keep water spray caused by takeoff and landings from entering the structure. It is normally attached to the underside of wings and control surfaces of fabric covered aircraft, and are also used to reinforce drain holes that are cut into the fabric.

marker beacon – Abbrev.: MB. An electronic navigation system that indicates specific locations by a highly directional signal, receivable only directly above the transmitter. This signal illuminates as an indicator light and produces an audible tone when the aircraft is directly above it, and show the pilot his exact position.

married needles – A term used regarding an engine-rotor tachometer when the hands are superimposed. One hand indicates engine RPM and the other the RPM of the main rotor.

Marvel balancer – A universal balancer commonly used throughout the helicopter industry.

masking material – Aluminum foil or special paper used to cover areas of an aircraft surface which are not to be sprayed or on which a finish is not to be applied.

masking tape – Paper tape that has a sticky surface on one side, and generally comes in rolls of varying widths.

Masonite – A type of fiberboard.

mass – A measure of the amount of material or matter contained in a body. It is the property of a body which causes the force of gravity to give a body weight.

mass flow rate – Symbol: G. The result of a fluids density and its linear velocity.

mass production – The production of objects in very large quantities in a relatively short time period by the use of complex, and often computerized, equipment.

massive-electrode spark plug – Spark plugs using two, three, or four large nickel-alloy ground electrodes.

mass-type fuel flowmeter – A type of fuel-flow measurement system used with turbine engines that indicates the mass flow rather than the volume flow.

mast bump – The action of the rotor head striking the mast, occurring on underslung rotors only.

master cylinder – A combination cylinder, piston, and reservoir used in an aircraft brake system. Fluid is stored when the brakes are not applied, and is forced into the brake cylinder when braking is needed.

master rod – The only connecting rod in a radial engine whose big end passes around the crankshaft. All of the other rods connect to the master rod and oscillate back and forth rather than encircling the crankshaft.

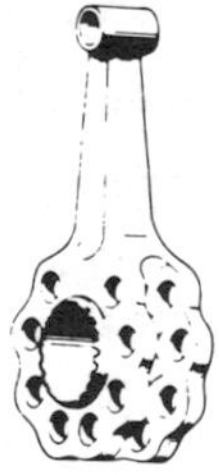

master switch – A single switch designed to control all electric power to all circuits in a system.

matched gears – Two gears that are used in a set and only replaced in a set.

mathematics – That branch of science dealing with numbers and their operation.

mating surfaces – Two surfaces that come together to form a seal.

matter – Any substance that has weight and occupies space.

maximum allowable zero-fuel weight – The maximum weight authorized for an aircraft excluding the fuel load.

maximum except takeoff power – Abbrev.: METO. The maximum continuous power an engine is allowed to develop without any time restrictions.

maximum landing weight – The maximum weight of the aircraft authorized for landing.

maximum range – The maximum distance a given aircraft can cover, under given conditions, by flying at the economical speed and altitude at all stages of the flight.

maximum speed – The maximum velocity obtainable in steady level flight regardless of altitude.

maximum takeoff power – The maximum power an engine is allowed to develop for a limited period of time; usually about one minute.

maximum takeoff weight – The maximum design weight of any aircraft on takeoff without exceeding its load factor.

maximum weight – Maximum allowable weight for an aircraft under any conditions.

maxwell – A unit of magnetic flux; one magnetic line of force.

Mayday – The international call for help used with voice radio transmitter when an aircraft is in serious danger.

mean – The mid-point; the average of a number of factors.

mean aerodynamic chord – Abbrev.: MAC. The chord of an imaginary rectangular airfoil having the same pitching moments throughout the flight range as that of the actual wing.

mean sea level – Abbrev.: MSL. The reference that is used for measuring altitude above sea level. Mean sea level is the average height of the surface of the sea.

mean solar day – The average time it takes the earth to rotate about its axis in one day.

measuring circuit – Any combination of resistors, batteries, and meters which make it possible to measure electrical values.

mechanic – A person, authorized by the FAA, who is skilled in repairing aircraft. Technician is the preferred term.

mechanical advantage — The increase in force or speed that is gained by using such devices as levers, pulleys, gears, or hydraulic cylinders.

mechanical blockage thrust reverser — A thrust reverser usually of the post exit-type (clamshell) used to reverse the hot exhaust stream of a gas turbine engine to help slow the airplane down during landings.

mechanical bond — The joining of two or more parts or pieces by mechanical methods such as bolts, rivets or pins.

mechanical efficiency — The ratio of the brake horsepower delivered to the output shaft of an engine to its indicated horsepower.

mechanical energy — Energy that expresses itself in mechanical movement or the physical production of work.

mechanical linkage — A direct connection between a control and a unit with no remote actuator.

mechanical mixture — A mixture where two or more elements or compounds are mixed, but can be identified by microscopic examination.

mechanical properties — Mechanical properties are those properties which involve a relationship between strain and stress.

medium bypass turbofan — Engines with 2:1 or 3:1 bypass ratios, which is the ratio of the amount of air the fan moves (or bypasses) in relation to the core engine. Sometimes referred to as "moderate bypass".

medium-frequency radio transmission — The band of frequencies of electromagnetic radiation that lies between 300 kHz and 3 MHz.

mega — 1,000,000.

megahertz — Abbrev.: MHz. 1,000,000 cycles per second.

Megger — A high-voltage, high-range ohmmeter that has a built-in, hand-turned generator for producing the voltage that is needed to measure insulation resistance and the resistance between a component and electrical ground.

megohmmeter — High-resistance measuring instrument incorporating a high-voltage DC generator in the instrument case. Not only does this measure high resistance, it does it with a high enough voltage to cause insulation breakdown if it has been weakened.

melt — A change in the physical state of a material when it goes from a solid to a liquid as a result of absorbing sufficient heat to produce the change.

melting point – A specific temperature at which a specified solid becomes a liquid.

member – Any portion of the aircraft structure that is essential to the whole.

memory effect – A reduction in the service capacity of nickel-cadmium cells which occurs when cells on standby service are regularly discharged to only a small fraction of their full service capacity.

meniscus – The curved upper surface of a column of liquid in a tube. If the liquid wets the tube, the curve will be concave; but if it does not wet the tube, it will be convex.

mensuration – The act or process of measuring.

mercerize – *See mercerizing.*

mercerizing – The process of dipping cotton yarn or fabric into a hot solution of caustic soda. It gives the material greater strength and luster, and is stronger and more pliable than untreated fabric.

mercury – A heavy, silver colored, toxic, liquid, metallic chemical element that remains in a liquid state under standard conditions of pressure and temperature. Mercury is approximately 13 times as heavy as water.

mercury barometer – A closed glass tube partially filled with mercury, used to determine the pressure exerted by the atmosphere. The standard atmospheric pressure at sea level will hold the mercury in the tube to a height of 760 mm or 29.92″.

mercury cell – A primary cell using zinc for the negative electrode, mercuric oxide for the positive electrode, and potassium hydroxide as the electrolyte.

mercury clutch – A centrifugal clutch in which mercury is used to engage the clutch.

mercury oxide cell – A chemical cell using powdered mercuric oxide and powdered zinc as its pole pieces. The electrolyte is a water solution of potassium hydroxide.

mercury thermometer – A liquid thermometer consisting of a glass tube with an extremely small inside diameter to which is attached a small reservoir at the bottom that contains mercury. A temperature scale marked alongside the tube is used to indicate the temperature when the mercury expands up the tube due to a rise in temperature.

mercury trap – A container in the pick-up tube of a vacuum cleaner used to retrieve spilled mercury. The mercury is sucked up by the cleaner and deposited in the bottle, which prevents it being sprayed out by the discharge of the cleaner.

mercury-vapor rectifier – A rectifier tube containing mercury which vaporizes during operation and increases the current-carrying capacity of the tube.

mesh – The engagement of the teeth of gears.

mesh rating – A U.S. sieve number and filtration rating common to fuel filters. Similar to a micron rating, e.g. a 74 micron filter carries an equivalent U.S. sieve number of 200 and has 200 meshes per linear inch.

mesosphere – A layer of the atmosphere between the top of the stratosphere or the ionosphere and the exosphere (about 250-600 mi. above the earth.)

metal – A chemical possessing most of these characteristics: usually rather heavy, with a bright and shining surface, malleable, ductile. and a good electrical conductor.

metal fatigue – Form of work hardening, or cold working, of a metal which results from flexing or vibration and which increases the brittleness of the material to its breaking point.

metal foil – A very thin sheet of metal, such as aluminum foil.

metal sheath – A closely fitting metal cover.

metal spinning – A process of metal forming in which sheet metal is clamped into a lathe along with the male die. A shaping tool is used to force the spinning metal against the die.

metal spraying – A method of covering or repairing a material with a coating of metal. The metal to be used for the coating is melted, and is sprayed out with hot, high-velocity compressed air.

metallic – Having the nature of metal or containing metal.

metallic ion concentration cell corrosion – Corrosion that results from a concentration of metallic ions in the electrolyte. The area of high concentration of metallic ions is the cathode.

metallic pigment – Extremely tiny flakes of metal suspended in paint to produce a sheen.

metallic ring test – A test for delaminations in a bonded structure in which a coin or similar object is used to tap on the surface. If the bond is good, a metallic ringing sound will be produced, but if it is delaminated, a dull thud will be heard.

metallizing –
[1]To replace the fabric covering on an aircraft structure with sheet metal.
[2]A welding process in which worn metal surfaces are restored to their original dimensions by spraying molten metal onto the worn area.

metallurgy – The science and technology dealing with metals and their use.

metalworking tools – Machines and tools used in the construction and repair of sheet metal aircraft.

metamerism index – A measurement used for scientific color matching. It indicates the way a pigment will look under varying light conditions.

metastable compound – A chemical compound that is inconsistent in the conditions in which it will remain stable.

meteor – A small particle of solar matter that is visible only as it burns due to the high temperature caused by friction as the meteor falls through the earth's atmosphere.

meteorology – The study of weather and atmospheric phenomena.

meter –
[1]Any device used to measure, indicate, or record.
[2]The basic unit of length measurement in the metric system equal to 39.37″.

metering device – A device used to measure or control the amount of fluid flow.

metering jet – The calibrated orifice in any fluid-flow system used to control the amount of flow for a given pressure drop across the jet.

metering pin – A flow control device, such as a tapered pin in an oleo shock absorber that is used to progressively restrict the passage of fluid from one chamber into the other, cushioning the landing impact. The shape, or contour, of the metering pin determines the amount of fluid that can flow with the pin in any position other than full in or full out.

metering valve – A valve used to control the flow of a fluid.

meter-kilogram — The amount of work that is produced when one kilogram of force acts through a distance of one meter.

methanol wood alcohol — A liquid alcohol that is produced by the distillation of wood pulp.

methyl bromide — A fire extinguishing agent (CH_3Br). More effective than CO_2 from a standpoint of weight, but more toxic than CO_2. It will seriously corrode aluminum alloy, zinc, and magnesium. Cannot be used in areas where harmful concentrations can enter personnel compartments.

methylene chloride — A liquid solvent (CH_2CL_2) used as the active agent in many paint strippers.

methyl-ethyl-ketone — Abbrev.: MEK. An important, yet harmful, low-cost solvent similar to acetone. Used as a cleaning agent to prepare a surface for painting and as a stripper for certain finishes.

mho — A unit of electrical conductance; the reciprocal of ohm.

mica — A transparent silicate mineral. It is used as an electrical insulator in capacitors, and as insulators for electric irons and heaters.

Micarta — A phenolic-type thermosetting resin impregnated cloth. It is used as an electrical insulator and for the manufacture of control pulleys.

mice — Small sheet metal, wedge-shaped tabs that are inserted into the tail pipe of some older turbine engines to reduce the nozzle opening and increase thrust. Adjustment of thrust is now done at the fuel control. Same as tai lpipe inserts. Used to “trim” a turbine engine.

micro- — One millionth (0.000001) of a unit.

microammeter — An electrical current measuring instrument capable of measuring current flow in millionths of an ampere.

microballoons — Microscopic-size phenolic or glass spheres used to add body with very little weight to a resin when used as a filler or potting compound.

microbes — Microscopic forms of animal life. They exist in water and feed on hydrocarbon aircraft fuel. Microbes form a water-entrapping scum on the bottom of jet aircraft fuel tanks.

microcircuit — An extremely small electronic component that has a large number of circuit elements made onto a single supporting material. *See also integrated circuit supporting material.*

microfarad — Abbrev.: mf. One millionth (0.000001) of a farad.

microfiche – A method of storing microfilmed information which stores from 24 to 288 individual frames, or pages of material on a single 4″ × 6″ sheet of photographic film. The microfiche film is read on a reader which enlarges the film and can make a full size printed copy of the desired page(s), if needed.

microfilm – Reproduction of printed material on 35mm, or 16mm photographic film; used to store vast quantities of written material in a small space. The microfilm is read on a reader which enlarges the film and can make a full size printed copy of the desired page(s), if needed.

microinch – One millionth (0.000001) of an inch.

micrometer – One millionth (0.000001) of a meter. A micrometer is also called a micron.

micrometer caliper – A precision measuring device having a single movable jaw, advanced by a screw. One revolution of the screw advances the jaw 0.025″.

micrometer setting torque wrench – A hand-operated torque wrench in which a preset torque is adjusted with a micrometer-type scale. When torque is reached, the handle of the wrench breaks over, indicating this torque to the operator.

micro-microfarad – A unit of capacitance equal to one millionth of a millionth of a farad. Micro-microfarad has been changed to picofarads.

micron –

[1]One millionth (0.000001) of a meter, or one thousandth (0.001) of a millimeter (0.000001 meter or 1×10^6 meter). A micron is also called a micrometer.

[2]Pressure measurement in a column of mercury: One micron of pressure is equal to 0.001 millimeter of mercury (1×10^6 meter of mercury) at 0°C.

[3]One micron is normally used to denote the effectiveness of a filter.

Micronic filter – A disposable element filter used in hydraulic or pneumatic systems that filters particles as small as one micron.

microorganism – An organism of microscopic size, normally bacteria or fungus.

microphone – A device for converting sound waves to electric impulses.

microprocessor – A small central processing unit (CPU) for a microcomputer.

microscope — An optical instrument used to magnify extremely small objects so that they can be seen by the human eye.

microsecond — One millionth (0.000001) of a second.

microshaver — An adjustable metal-cutting tool used for shaving the heads of a countersunk rivets so they can be absolutely flush with the surface of the skin.

microshaving — A process in sheet metal work in which the head of the countersunk rivet is shaved to absolute smoothness with the surface of the skin.

Microswitch — A type of electrical switch which is used to open or close a circuit with an extremely small movement of the actuator.

mid-flap — The middle flap on a triple-slotted segmented flap.

mid-span shrouds — The lugs on fan blades which contact each other to provide a circular support ring; giving strength and vibration reduction.

mid-span weight — A weight placed in the mid-span area of a helicopter rotor blade to add inertia to the blade.

mid-wing airplane — An airplane having one main supporting surface which is located in the center of the fuselage.

mil — Commonly used to represent one one-thousandth (0.001) of an inch.

MILSPEC — Term identifying Military Specifications.

mild steel — Steel that contains between 0.05 and 0.25% carbon.

mildew — A gray colored parasitic fungus growth that forms on organic matter.

mildewcide — An additive to dope or sealers used on organic materials to inhibit the growth of mildew.

mile — Abbrev.: mi. One statute mile is equal to 5,280 ft.

Military Standards — Abbrev.: MS. The standards used for aircraft hardware which was originated by the U.S. military services in order to maintain a high degree of quality standards in the manufacture, repair and maintenance of aircraft.

mill bit — A tool used with a router to remove metal and honeycomb core for repairs to bonded structure.

mill file – A single-cut file, tapered slightly in thickness and in width for about one-third of its length.

milli- – One one-thousandth (0.001) of a unit.

milliameter – An electrical current measuring device calibrated to read in milliamperes. 1000 milliamperes = 1 ampere.

milliampere – One-thousandth (0.001) of an ampere.

millibar – A unit of barometric pressure equal to approximately 0.75 millimeters of mercury. Standard sea level atmospheric pressure is equal to 1,013.2 millibars.

milling machine – A metal-working machine tool with a movable table that feeds the work into a rotating milling cutter.

millivolt – One-thousandth (0.001) of a volt.

mineral-based hydraulic fluid – A petroleum-based hydraulic fluid consisting essentially of kerosine and additives to inhibit corrosion and to minimize foaming. It is dyed red and is identified as MIL-H-5606.

miniature screw – A miniature screw is a screw less than 0.06″ in diameter, having a slotted head, and threaded for assembly with a preformed internal thread.

minimum fuel – The minimum fuel specified for weight and balance purposes when computing an adverse loaded center of gravity. It is no more than the quantity of fuel necessary for one half hour of operation at rated maximum continuous power.

minimum-flow stop – Refers to a fuel control design which prevents the power lever from shutting off fuel. A separate shutoff is provided in this case.

minor alteration – Any alteration that is not considered to be a major alteration.

minor axis of an ellipse – A straight line that passes through the center of the ellipse and is perpendicular to the major axis.

minor fastener diameter – The diameter of a threaded fastener measured at the thread root.

minor repair – Any repair not considered to be a major repair.

minuend – The number from which the subtrahend is subtracted.

minus – A term that means a negative value. Minus values are indicated by using a short dash in front of the value (–4). A minus sign is used in electricity to indicate a negative condition.

minute –
[1]Measurement: An angular measurement equal to 1⁄60 of a degree in a 360° circle or 21,600 minutes.
[2]Time: A unit of time that is equal to 1⁄60 of an hour or 60 seconds.

mirror image – An object that is an exact duplicate of the original but reversed as if the object were viewed through a mirror.

misalignment – A condition which exists when two mating surfaces do not meet or match as they should.

miscible – The ability of a material to combine or mix with another material.

misfire – A failure of an explosive charge. As in the misfire of a rocket or an engine.

misfiring – The interruption of even firing of the cylinders of a reciprocating engine.

mist – Tiny droplets of water suspended in the air.

mist coat – A very light spray coat of thinner or other volatile solvent with little or no color in it.

miter – A term used to describe a cut to the edges of a board or a surface in such a way that they will match or fit together.

miter box – An angle cutting holding device used for cutting wood or metal to form a mitered joint.

miter square – A small square used for marking the ends of wood or metal for other than right angle cuts.

mixed exhaust – On a turbofan, a design which allows the primary and secondary airstreams to mix prior to leaving the engine. A sound alteration feature of more modern engines. Same as forced exhaust mixer.

mixer –
[1]A system of bellcranks that prevents the cyclic inputs from changing the collective inputs on a helicopter control system.
[2]A circuit in which two frequencies are combined to produce sum and difference frequencies. *See also heterodyne and beat frequencies.*

mixture – A combination of matter composed of two or more components not bearing fixed proportions to one another.

mixture control — The primary carburetor control for adjusting the fuel-air mixture ratio. It may be either a manual control or it may be a combination of both. In the case of the combination, the pilot adjusts for a particular ratio, and the automatic control maintains that ratio by compensating for temperature and pressure variations of the atmosphere.

mobile charges — Electrons in a semiconductor material that drift within the material from one electrical charge to another.

mobile test stand — An engine run-in stand which is portable and may be used at other than a fixed location.

Möbius loop — A single-sided surface that is twisted in the form of a loop.

mockup — A full-size reproduction of a part or assembly used to determine whether or not all of the components will fit as they are designed. It is also used as an expedient for instruction when the real object is impractical to use.

mode — The manner of doing some operation.

modem — A modulator-demodulator device that is used to couple two computers together by the use of telephone lines.

modification — The change in the design or configuration of an original unit.

modify — To change something such as an alteration or redesign of an original unit, or the changing of a schedule from the original date or time.

modular maintenance — A maintenance procedure which allows replacement of major assemblies, called modules, in a minimum amount of time and expense. The removed module is returned to a repair facility and bench tested and repaired as needed.

modular structure — Standardized units built up as modules.

modulate — To change. This normally refers to a radio carrier being changed by the voice or by a tone.

modulated anti-skid system — A form of anti-skid brake system that senses the rate of deceleration of the wheels to maintain pressure in the brakes that will hold the tire in the slip area, yet not allow a skid to develop. It does this by modulating, or continually changing, the pressure in the brake.

modulated continuous wave — A radio code transmission that consists of sending modulated audio frequency tones in a series of short and long burst.

modulation — The changing of frequency or amplitude by superimposing an audio frequency on a carrier frequency.

modulator — That portion of a transmitter circuit which modulates the carrier wave.

moisture separator — A device used in a pneumatic system to separate moisture from the air.

moisture-proof — Property of an object which resists absorption of moisture.

mold line — In metal layout, a mold line is a line used in the development of pattern used for forming a piece of sheet metal. It is that part of formed part that remains flat, and is formed by the intersection of the flat surfaces of two sides of a sheet metal part.

mold line dimension — The distance from the edge of metal to a mold point, or between mold points.

molecule — The smallest particle of an element or compound that is capable of retaining the chemical identity of the substance in mass.

molybdenum — A metallic element similar to chromium which is used as an alloying agent for most aircraft.

moment — The product of the weight of an object in pounds and the distance from the center of gravity of the object to the datum or fulcrum (the point about which a lever rotates) in inches. Moment is used in weight and balance computations and is expressed in pound-inches. The formula used is: Moment = distance × force.

moment index — The moment divided by a constant such as 200, 1,000, or 10,000. Its use is to simplify weight and balance computations by eliminating large and unwieldy numbers.

momentum — The tendency of a body to continue in motion after being placed in motion.

moment-weight number — An identification number or letter indicating a measurement of both weight and center of gravity and used on rotating airfoils for balancing purposes.

monel — An alloy of iron with 67% nickel, 30% copper, and 3% aluminum. It is extremely resistant to corrosion.

monkey wrench – A name for a type of adjustable wrench that has one fixed jaw and one movable jaw.

monocoque – A stressed-skin type of construction in which the stiffness of the skin provides a large measure of the strength of the structure. No truss or substructure is required.

monolithic casting – A casting that is formed as a single piece.

monomer – A chemical compound that can be polymerized.

monoplane – An airplane with one main supporting surface wing sometimes divided into two parts by the fuselage.

monopropellant – A type of rocket engine propellant in which the fuel and the oxidizer are both part of a single property.

monorail – A single rail used to carry cars or objects.

monospar wing – A fundamental wing design which incorporates only one main longitudinal member in its construction.

monostable – The condition of a device which has one stable condition. When disturbed from this, it will return to its original state.

monostable multivibrator – An electronic circuit that tries to maintain a condition of on or off. When it is disturbed from this position, it will automatically return to its stable condition.

Morse code – A system of dots and dashes used in aviation to identify ground radio facilities.

mothball – Surplus airplanes, parts, or equipment that have been preserved in special packings and stored until they are needed.

mothballed – A term used for parts, machinery, or equipment that has been prepared and placed in storage.

motion – The act of changing place or position.

motor bypass – A device in a hydraulic system that prevents a hydraulic motor from receiving excessive fluid. The fluid bypasses the motor.

motor over – The process of rotating the engine with the starter for reasons other than for starting.

motoring – Rotating a turbine engine with the starter for reasons other than starting.

mounting lug – A lug used to secure an accessory, cylinder, etc.

mounting pad – Provision made on the accessory section of an aircraft engine for attaching such accessories as magnetos, generators or alternators, and fluid pumps.

muff – A shroud placed around a section of the exhaust pipe. The shroud is open at the ends to permit air to flow into the space between the exhaust pipe and the wall of the shroud. This heat can be used for carburetor heat or cockpit and cabin heat. An alternate air valve is operated by means of the carburetor heat control in the cockpit.

mule – A term used to describe an auxiliary hydraulic power supply that can supply fluid under pressure to the aircraft hydraulic system when the engines are not running. The mule is normally used to operate and test the landing gear and flight control systems.

multi-spar wing – A fundamental wing design which incorporates more than one longitudinal member for support.

multi-engine – An aircraft having more than one engine.

multimeter – A piece of electrical test equipment consisting of one meter movement and several shunts, multipliers, and other circuit elements to allow the meter to be used as a voltmeter, ohmmeter, milliammeter, and ammeter. Rectifiers make it usable for AC and well as DC.

multiple-disk brake – A form of aircraft brake in which a series of discs, keyed to the wheel, mesh and rotate between a series of stationary discs keyed to the axle. The brakes are applied by hydraulic clamping action squeezing the discs together.

multiple-spar (multi-spar) wing – A type of airplane wing structure that uses several spanwise structural members to give the wing its strength.

multiplex communications – A method of two-way communications in which two sites can transmit and receive on the same frequency and at the same time.

multiplicand – The number that is to be multiplied by another.

multiplier – A number by which another number is multiplied. To multiply or extend the range of something.

multiplier resistor – The resistor in series with a voltmeter movement used to multiply or extend the range of the meter.

multiplier tube – An electron tube designed to amplify or multiply very weak electron currents by means of secondary emission.

multivibrator — A form of oscillator which produces its output by having one or another transistor or vacuum tube conduct. When one conducts, the other is shut off. Conduction alternates between the two.

Mumetal — An alloy of 14% iron, 79.5% nickel, 1.5% chromium, and 5% copper.

muriatic acid — Commercial hydrochloric acid (HCl).

mushroom head — A flared head which forms on a pounding tool such as a punch or chisel when it is hammered. This is a dangerous condition, and it must be ground away.

mutual inductance — The inductance of a voltage in one coil due to the field produced by an adjacent coil. Inductive coupling is accomplished through the mutual inductance of two adjacent coils.

Mylar — A polyester film.

Mylar capacitor — A capacitor that uses mylar film as a dielectric.

N

nacelle – The streamlined enclosure on the wing or fuselage of an aircraft which houses the engine.

NAND gate – A not and logic device, which will not have a voltage at its output only until a voltage appears at all of the inputs.

nano – One billionth (0.000000001) of a unit.

nanovoltmeter – A sensitive voltmeter that measures voltages as low as one nanovolt.

nap – The short fiber ends which protrude from the surface of a fabric. When the fabric is doped, these fibers become stiff and must be sanded off. This is called "laying the nap".

naphtha – A volatile and flammable hydrocarbon liquid used chiefly as a solvent or cleaning agent.

narrowing grinding – Removing part of the top edge of the valve seat in the cylinder of a reciprocating engine.

NAS drawings and specifications – Dimensional and material standards for aircraft fasteners developed by the National Aircraft Standards Committee. All drawings and specifications are prefixed by NAS.

natural aging – Solution heat-treated aluminum alloy material that has been allowed to harden at room temperature following heat treatment.

naturally aspirated engine – A reciprocating aircraft engine that is not supercharged, but whose induction air is forced into the cylinders by atmospheric pressure only.

nautical mile – A measure of distance used primarily in navigation. It is equal to 6,076 ft. and is one minute of latitude at the equator.

navigation lights – Lights on an aircraft consisting of a red light on the left wing, a green light on the right wing, and a white light on the tail. FARs require that these lights be displayed in flight during the hours of darkness.

Navigational Aids – Abbrev.: NAVAID. Short for navigational aids. NAVAIDS are electronic systems installed around airports and along airways to help pilots navigate by IFR, NAVAIDS include VOR, ILS and DME.

neck —

[1]A specialized form of a portion of the body of fasteners near the head to perform a definite function, such as preventing rotation, etc.

[2]A reduced diameter of a portion of the shank of a fastener which is required for design or manufacturing reasons.

needle and ball indicator — *See turn and slip indicator.*

needle bearings — An anti-friction bearing made of hardened steel. The bearing consists of a series of small diameter rollers that ride between two hardened and polished steel races. One race is pressed into the housing, and the other race is pressed onto the rotating shaft.

needle valve — A tapered end fluid control needle valve that fits into a seat or recess to control or restrict a flow of fluid through an orifice.

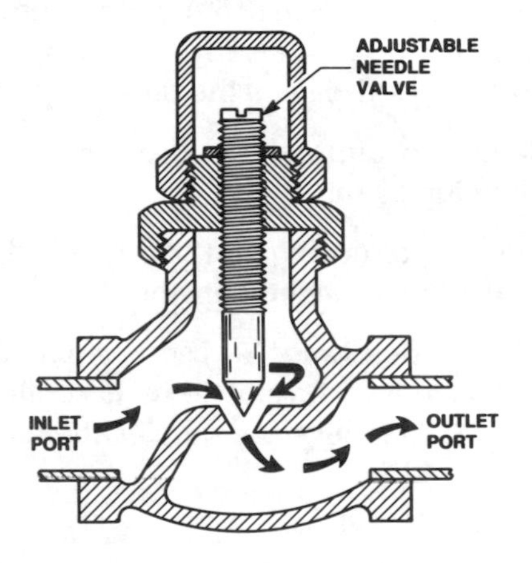

negative —

[1]A condition in which there is an excess of electrons.

[2]An accepted name for the terminal of a battery or power source from which the electrons flow.

negative acceleration — Deceleration; slowing down.

negative battery terminal — The terminal of a battery from which the electrons leave and enter the circuit.

negative condition — One in which there is an excess of electrons.

negative dihedral — A downward inclination of a wing or other surface. It is the downward angle that is formed between the wings and the lateral axis of the airplane.

negative electrical charge — An unbalanced electrical condition caused by an atom having more electrons than protons.

negative electrical condition – A condition in which there are more negative charges than there are positive charges.

negative electrical resistance – A decrease in current through a device when there is an increase in voltage.

negative feedback – Information or a signal that is fed back from a circuit or device that tends to decrease the output.

negative ion – An atom that has more electrons than protons spinning around the nucleus.

negative moment – A force that causes a rotational force on a body in a counterclockwise direction.

negative pressure – Pressure that is less than atmospheric pressure.

negative pressure relief valve – A valve in an aircraft pressurization system that prevents the outside air pressure becoming greater than the pressure inside the cabin.

negative stagger – The placement of the wings of a biplane in which the leading edge of the lower wing is ahead of the leading edge of the upper wing.

negative static stability – A condition in which an object disturbed from a condition of rest will tend to move further away from its condition of rest.

negative temperature coefficient – A conductor or device that decreases in resistance as the temperature increases.

negative torque system – A system in a turboprop engine which prevents the engine encountering excessive propeller drag by controlling the pitch of the propeller blades.

negative vacuum relief valve – A relief valve used on pressurized aircraft that opens when outside air pressure is greater than cabin pressure.

negative value – A value less than zero.

neoprene – An oil-resistant synthetic rubber made by polymerizing chloroprene. Used in seals and locknuts, etc.

net thrust – Symbol: F_n. The effective thrust developed by a jet engine during flight, taking into consideration the initial momentum of the air mass prior to entering the engine.

neutral –
[1]The condition in which a gear, lever, or other mechanism is not engaged.
[2]An electrical condition that is neither positive nor negative.

neutral axis – An imaginary line through the length of a loaded beam at which point the forces of compression and tension are neutral.

neutral flame – A flame used in oxyacetylene welding which is neither carburizing nor oxidizing and which uses the correct ration of acetylene gas and oxygen.

neutral line – In sheet metal bending, the space unaffected by either compression on the inside curve, or by stretching on the outside curve.

neutral plane – An imaginary line drawn perpendicular to the resultant flux in a generator. For arcless commutation, the neutral plane should lie directly over the plane of the brushes.

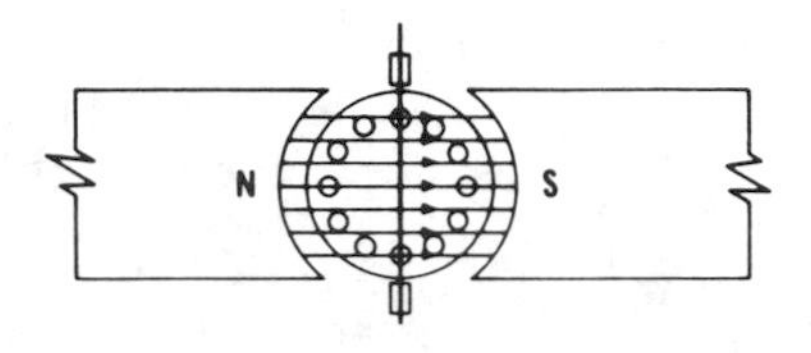

neutral position – The position of the rotating magnet of a magneto between the pole shoes. In the neutral position, no lines of flux flow in the magneto frame.

neutral static stability – The condition of an object in which, when once disturbed from a condition of rest, it has neither the tendency to return to a condition of rest, nor to depart further from it. It remains in equilibrium in the direction of disturbance.

neutralize – To make electrically balanced or inert by combining equal positive and negative quantities.

neutron – Uncharged particle in the nucleus of an atom. Its mass is essentially equal to that of a proton.

never exceed speed – The speed beyond which an aircraft should never be operated.

Newton's first law of motion – The law of physics which describes inertia. "A body at rest tends to remain at rest and a body in motion tends to continue to move at constant speed, along a straight line, unless it is acted upon by an external force."

Newton's metal – A low melting point metal that melts at approximately 95°C.

Newton's second law of motion — The greater the force acting on an object, the greater the acceleration. And the greater the mass, the less the object will accelerate.

Newton's third law of motion — The law of physics which describes action and reaction. For every action, there is an equal and opposite reaction.

nibble — To take small bites or quantity.

nibbler — A sheet metal cutting tool that cuts the metal by a series of small nibbles or bites.

Nichrome — An alloy of nickel and chromium. It is used for making precision wire-wound resistors.

nick — A sharp-sided gouge or depression with a V-shaped bottom which is generally the result of careless handling.

nickel — A silver-white, hard, malleable, metallic chemical element used for plating because of its high resistance to oxidation.

nickel silver — A metal alloy of copper, zinc, and nickel.

nickel-cadmium battery — A battery made up of alkaline secondary cells. The positive plates are nickel hydroxide, the negative plates are cadmium hydroxide, and potassium hydroxide is used as the electrolyte.

nimbostratus — A dark gray cloud layer that produces rain or snow.

nipple pipe fitting — A short piece of pipe fitting that is threaded on both ends.

nitrate dope — A finish for aircraft fabric, consisting of a film base of cellulose fibers dissolved in nitric acid, with the necessary plasticizers, solvents, and thinners.

nitric acid — A colorless, or yellowish, flowing, suffocating, caustic, corrosive, water-soluble liquid (HNO_3) with powerful oxidizing properties.

nitriding — A form of case hardening in which a steel part is heated in an atmosphere of ammonia (NH_3.) The ammonia breaks down, freeing the nitrogen to combine with aluminum in the steel to form an extremely hard abrasive-resistant aluminum nitride surface. Cylinder walls and crankshaft journals may be nitrided.

nitrogen — A colorless, tasteless, odorless, gaseous element forming nearly ⅘ of the earth's atmosphere.

noble — Chemically inert or inactive, especially toward oxygen.

noble gas — An inert gas such as neon, argon, krypton, and xenon.

nodal system — A vibration dampening system used by Bell Helicopter to reduce main rotor vibration.

noise — A general term for any loud or unusual sound that is annoying or excessive.

noise suppressor — A device installed in the tailpipe of a turbojet engine to slow the mixing of the exhaust gases with the surrounding air, thus decreasing the intensity of the sound.

no-load current — When a device, which is connected to an electrical circuit, is at idle.

nominal resistance of a thermistor — The true resistance of a thermistor at a particular reference temperature. Most manufacturers use 20°C as their reference temperature.

nominal size — The designation used for the purpose of general identification.

nominal value — A stated value that may not be a fact or an actual value.

nomograph-viscosity index — A chart as produced by ASTM which is used to plot the viscosity change of turbine oils with temperature change. *See also viscosity index.*

nonabrasive — Material which will not scratch or scar when rubbed on another surface.

nonabrasive scraper — A scraper which has no abrasive materials attached to it.

non-atomizing spray — The application of a material to a surface by a spray gun in which the material is fed in a solid stream rather than in tiny droplets.

non-atomizing spray gun — A spray gun that propels a solid stream from the spray nozzle.

nondestructive inspection — An inspection of aircraft parts, units, components, etc., without altering or destroying the physical or material properties and/or integrity. It is used to determine the continued serviceability for another inspection period. Also called nondestructive testing.

nondimensional number — A number that does not have a dimensional value such as Mach numbers.

nondirectional antenna — An antenna that has the ability to receive or transmit a signal with equal strength.

nonferrous metal — A metal which contains no iron.

nonflexible control cable — Control cable, called wire, made up of 7 or 19 strands of solid wire preformed into a helical or spiral shape.

nonlinear output — Output that is not directly proportional to the input.

nonlinear scale — The scale of an indicating instrument in which the numbers are spread out at one end and are bunched up at the other.

nonlinear system — Nonuniform in length, width, or output.

nonmagnetic — Metal that does not have the properties of a magnet; that cannot be magnetized or attracted by a magnet.

nonporous — Any material that does not allow a liquid to pass through it.

nonrepairable — Damage which cannot be repaired and which renders the part nonserviceable.

nonrepairable damage — Damage that requires the aircraft, or aircraft component, to be replaced.

nonrigid airship — An engine-driven, lighter-than-air aircraft, such as a blimp, that can be steered.

nonscheduled airline — An airline that does not operate according to a regular published schedule.

nonservo brakes — Brakes that do not use the momentum of the aircraft to assist in the application of the brakes. Nonservo brakes are applied by the pilot.

nonskid brakes — A feature found in high performance aircraft braking systems that provides wheel antiskid protection. The wheel consist of a skid control generator unit that measures the wheel rotational speed. As the wheel rotates, the generator develops a voltage and current signal. The signal strength indicates the wheel rotational speed. This signal is fed to the skid control box though the harness. The box interprets the signal and signals a solenoids in the skid control valves to release the brake pressure until the wheel begins to speed up allowing the wheel to continue to rotate without skidding.

nonstandard fastener — A fastener which differs in size, length, material, or finish from established and published standards.

nonstructural — That portion of an aircraft which does not carry any aerodynamic loads.

nontautening dope — A special formulation of aircraft dope to use on heat-shrunk polyester fabric. It provides the necessary fill for the fabric, but produces a minimum of shrinkage.

NOR gate — A not or logic device which will have a voltage at its output only when there is no voltage on any input.

normal category airplane — An airplane certificated for nonacrobatic operation.

normal heptane — *See heptane.*

normal operating speed — The operating speed is the velocity obtained in level flight at design altitude of the airplane at not more than 70% of normal rated engine power.

normal rated power — The highest power at which an engine can be operated continuously without damage to it; maximum continuous power at 100%.

normal shock wave — A shock wave formed ahead of an airfoil approaching the speed of sound. It is perpendicular to the path of the airfoil.

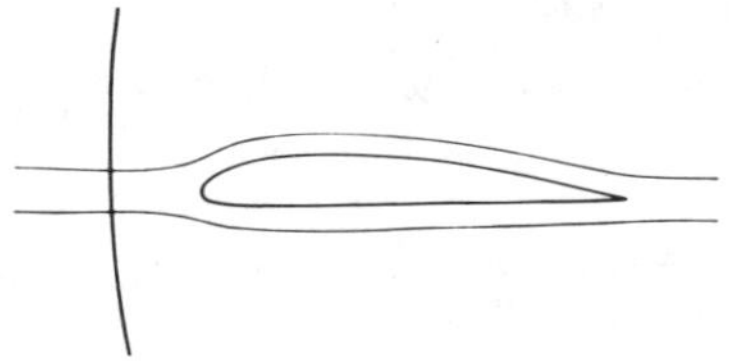

normalizing — A form of heat treatment in which the metal is heated to its critical temperature and allowed to cool slowly in still air. Normalizing relieves stresses in the metal.

normally closed relay — A relay switch consisting of a coil or solenoid, an iron core, and movable contacts controlled by a spring. Normally, closed relay contacts are held closed by the spring. When current flows through the solenoid, the contacts are opened by the magnetic pull of the electromagnetic coil.

normally open relay — A relay switch consisting of a coil or solenoid, an iron core, and movable contacts controlled by a spring. Normally, open relay contacts are held open by the spring. When current flows through the solenoid, the contacts are closed by the magnetic pull of the electromagnetic coil.

North Pole — The north-seeking pole of a magnet.

nose cone – A conical-shaped dome usually attached to the front portion of the fuselage to house the radar scanner or other electronic equipment. A nose cone can also be described as the front cover of propellers, intakes for jet engines, etc., for streamlining and directing the airflow.

nose gear – The forward auxiliary wheel in a tricycle landing gear.

nose heavy – A condition which exists on an aircraft in which the center of gravity is ahead of the forward limit.

nose rib – A false, or partial, wing rib extending back from the leading edge only to the main spar. Its purpose is to add smoothness to the leading edge of a wing.

nose section – The forward section of an aircraft.

NOT gate – A logic device having one input and one output. There will be no voltage on the output, when a voltage appears at the input.

notes – Specific instructions on an aircraft drawing.

nozzle – The tapered end of a duct. *See also nozzle, fuel.*

nozzle blades – Any of the blades of a nozzle diaphragm.

nozzle diaphragm – A ring of stationary blades in a turbine engine ahead of the turbine wheel. used to direct the flow of hot gases into the turbine in the direction of maximum efficiency.

nozzle, fuel – The pressure-atomizing unit that receives fuel under high pressure from the fuel manifold and delivers it to the combustor in a highly atomized, precisely patterned spray.

NPN transistor – A three-element semiconductor made up of a sandwich of P-type silicon or germanium between two pieces of N-type material.

N-strut – Struts of a biplane near the wing tips shaped in the form of the letter N.

N-type silicon – Silicon which has been doped with an impurity having five valence electrons.

nucleus – The center or core around which other parts are grouped. Center or core of an atom, consisting of positively charged protons and uncharged neutrons.

nucleus of an atom – In chemistry, it is the central portion of an atom.

null – An indicated low or zero point in a radio signal.

null balance – An electrical circuit in which two voltages have cancelled each other out.

numbers – A symbol used to assign a value that shows how many or the place in a sequence.

numerator – A term used to denote a part of a fraction that shows a portion of something that makes up a whole. In decimals, the numerator is the number to the right of a decimal point. In a fraction, it is the number above the line and signifies the number of parts of the denominator used.

nut – An internally threaded collar used to screw onto bolts or screws to form a complete fastening device.

nutation – The oscillation of a spinning gyroscope.

nutplate – A special form of nut which may be riveted to the inside of a structure. Bolts and screws can be screwed into the nutplate without the aid of a wrench.

nylon – A tough, lightweight, elastic polyamide material, used especially in fabrics and plastics.

O

oblique angle — An obtuse angle. Any angle having an angle other than a right angle.

oblique photography — Photographs taken with a camera pointed at any angle.

oblique shock wave — A shock wave attached to the bow and tail of an aircraft flying at a speed greater than the speed of sound. The sides of the oblique shock wave form the Mach cone.

oblique triangle — A three-sided, closed figure that does not contain a right angle.

oblong shape — An object that is longer than it is broad. An elongated circle or square.

observation aircraft — Military aircraft that fly behind enemy lines and observe the movement of troops or supplies.

obsolete — Something that is no longer in use or in practice.

obtuse angle — An angle greater than 90°. Also called an open angle.

obtuse triangle — A triangle that contains an angle greater than 90°.

occluded front — A weather front that overtakes another front — shuts it off.

O-condition — The soft, or annealed, tempered condition of an aluminum alloy.

octagon — An eight-cornered figure with each side having the same length.

octahedron — A solid design or figure with eight plane surfaces.

octal number system — The numbers used in digital electronics that uses a system base on eight units (0-7).

octane rating — The rating system of aviation gasoline with regard to its antidetonating qualities. Fuel with an octane rating of 87 is made up of a mixture of 87% isooctane and 13% heptane.

odd harmonics — The odd multiples of a frequency.

oersted — One oersted is equal to a magnetomotive force of 1 gilbert per square centimeter, and this is 79.577 ampere-turns per meter.

off-idle mixture — The fuel-air mixture ratio of an aircraft engine in the transition period between idle RPM and the main metering system.

offset rivet set — A rivet set used in a hand-held pneumatic riveting gun in which the head is offset from the center line of the shank. Offset rivet sets are used in locations where a straight set cannot be used.

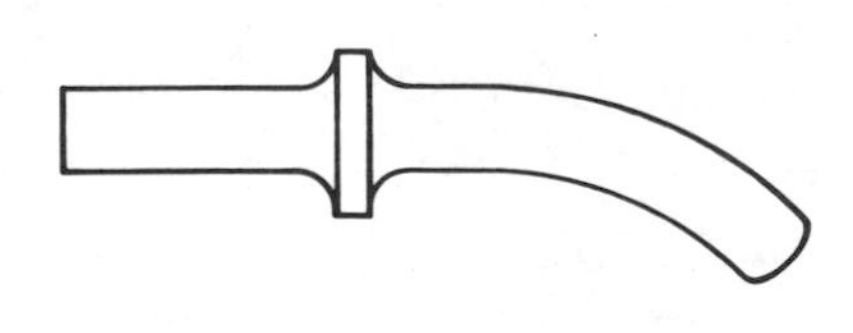

offset screwdriver — A screwdriver in which the blades are oriented at 90° to the shank. An offset screwdriver is used to turn screws where there is not enough clearance in line with the screw to allow a regular screwdriver to be used.

off-the-shelf item — Any item, part, or program that is standard that can be used rather than having a custom manufactured part.

ohm — Symbol: Ω. The unit of electrical opposition equal to the resistance of a circuit in which an electromotive force of one volt will maintain a current of one ampere.

ohmmeter — An electrical measuring instrument used to measure resistance in a circuit. An ohmmeter measures resistance by measuring the amount of current which flows when a known voltage is applied across the unknown resistance.

Ohm's law — The law which establishes the relationship between current, voltage, and resistance in an electrical circuit. The current in a circuit is directly proportional to the voltage causing it and inversely proportional to the resistance of the circuit.

ohms-per-volt — The measure of a voltmeters sensitivity. It is found by dividing the number one by the amount of current needed to deflect the meter pointer full scale.

oil canning — A condition of the sheet metal skin of an aircraft which is slightly bulged or stretched between rows of rivets. This bulge will pop back and forth in the manner of the bottom of an oil can.

oil control ring — The piston ring below the compression rings used to control the amount of oil between the piston and the cylinder wall of an aircraft reciprocating engine. It is usually a multi-piece ring and normally fits into a groove with holes to drain part of the oil back to the inside of the piston.

oil cooler –
[1]A heat exchanger used to cool the oil.
[2]A radiator used to maintain normal operating temperature of lubricating oil. Some coolers utilize air as a cooling agent, others use fuel.

oil dilution – The process of thinning engine oil by adding fuel to it at shutdown to make starting easier in cold weather.

oil film – A light coating of oil sufficient to prevent metal-to-metal contact or to protect metal parts from corrosion.

oil filter – A device for removing impurities and foreign matter from the lubricating oil used in an aircraft engine.

oil hardening – A process of hardening steel using oil as a quenching agent. When red hot steel is immersed in a bath of oil in quenching, the steel cools more slowly than by other methods, and it gives the steel a more uniform hardness.

oil inlet – The fitting on an aircraft engine through which the lubricating oil enters the oil system.

oil jet – A small nozzle opening which directs a stream of oil onto a point to be lubricated, such as bearings gears, etc.

oil outlet – The fitting on an aircraft engine from which the lubricating oil leaves the engine to return to the external reservoir.

oil pan – The removal part of the crankcase where engine lubricating oil is collected by gravity and stored. Oil in the pan is forced through passages in the engine by a pressure pump.

oil passages – Channels or holes in an engine through which lubricating oil flows to lubricate, seal, or cool the engine.

oil pressure indicator – An instrument which indicates the pressure of the oil in the engine lubricating system.

oil scraper ring – A piston ring located at the bottom, or skirt end, of a piston used to wipe the oil either toward or away from the oil control ring, depending on the design of the engine.

oil screen – A fine mesh screen in the engine lubrication system used to stop and hold impurities, preventing their passing through the engine and causing damage.

oil seal – A device used to prevent oil leaking from the engine past a moving shaft.

oil separator – A device used to separate oil from the discharge air of a wet-type air pump.

oil slinger — A rotating device used as a centrifugal impeller to direct oil flow, usually away from a bearing sump and towards a scavenge pump.

oil sump — A container built into the lower part of an aircraft engine which holds the supply of lubricating oil.

oil tank pressurizing valve — A check valve used to trap oil vapors in the expansion space above the oil surface to provide a pressurization effect of approximately 3-5 PSI within the oil tank.

oil temperature indicator — Indicates oil temperature of the inlet oil to the engine. System usually consists of the ratiometer type. Normal temperature is a good signal to the pilot that the engine is warmed up for takeoff.

oil temperature regulator — A control device which maintains the oil temperature within the desired operating range by either passing the oil through the core of the cooler or around the jacket of the cooler.

oil wiper ring — The bottom ring on a piston that is used to direct oil up between the piston and the cylinder wall for lubrication and sealing.

oilite bushing — A special type of friction bushing made of a bronze material impregnated with oil. Friction of the moving parts generate enough heat to bring the oil to the surface and provide the needed lubrication.

oleo strut — An aircraft landing gear shock strut which absorbs the initial landing impact by the transfer of oil from one chamber to another through a restricting orifice. Taxi shocks are absorbed with compressed air or by a spring.

OMNI — A phase comparison of two navigational signals transmitted from ground stations. The signals are in phase when they are received at a location that is directly magnetic north of the station. An instrument on the instrument panels tells the pilot when the needle gets off center from the pre-selected radial. If the aircraft gets away from the preselected radial, the needle moves out of center to show the pilot which direction they should turn in order to bring the aircraft back to the radial.

omni bearing selector — Abbrev.: OBS. The variable phase shifter in a VOR system with which the pilot selects the radial from the VOR station along on which he wishes to fly. The OBS shifts the phase of the course deviation indicator reference signal so the needle will center when the aircraft is on the selected radial.

omni station — The ground station of a VOR.

omnidirectional microphone — A nondirectional microphone that picks up sounds equally well from all bearings.

omnidirectional radio antenna — A type of nondirectional radio antenna that has consistent field power in all horizontal bearings.

omnirange navigation equipment — Abbrev.: VOR. A phase comparison type of electronic navigation equipment which provides a directional reference between the airplane and the ground station, which is measured from magnetic north.

on condition maintenance — A maintenance concept whereby some components of the engine remain in service as long as they appear airworthy at each inspection. The replace-on-condition concept is as opposed to replace after a "life-limited" time interval. In the case of engines themselves, this means no TBO is required.

one hundred and eighty degree ambiguity — An error inherent in radio direction finding systems, in which the system is unable to determine whether the station is ahead of the airplane or behind.

one-shot rivet gun — A special type of rivet gun that hits the rivet with a single hard blow for each pull of the trigger. It is used for rivets that are subject to becoming brittle if driven by ordinary rivet guns that deliver a continuous series of blows as long as the trigger is held.

on-speed — The condition in which the actual engine speed is equal to the desired engine speed as set on the propeller control by the pilot.

on-speed condition — A condition of a propeller governor system which maintains the required balance between all three control forces selected by the governor, by metering to, or draining from, the propeller piston the exact quantity of oil necessary to maintain the proper blade angle for constant-speed operation.

open angle — The angle through which metal has been bent that produces a "V" with less than 90° between the sides.

open circuit — An electrical circuit that is not complete or that provides a continuous path for electrons to flow.

open wiring — Any wire, wire group, or wire bundle not enclosed in conduit.

open-assembly time — The assigned time between the time when an adhesive is spread on to two surfaces to be joined, and the time the surfaces are clamped together.

open-center selector valve — A type of hydraulic selector valve used in open-center hydraulic systems that allows fluid to flow from the pump to the reservoir when the selector valve is place in neutral (when none of the actuating cylinders are receiving fluid under pressure).

open-circuit voltage — The measured voltage of a battery or generator when there is no load or flow of electrons in the circuit.

open-end wrench — A solid nonadjustable wrench with open parallel jaws on one or both ends.

open-tip turbine blades — A blade with no shroud attached at the tip. This blade can withstand higher speed-induced tip loading than the shrouded tip blade. Also called open perimeter tip.

operate — To use, cause to be used, or authorize the use of an aircraft for the purpose of air navigation.

operating center of gravity range — The distance between the forward and rearward center of gravity limits as specified in the Aircraft Specifications or Type Certificate Data Sheets.

operating pressures — The hydraulic or pneumatic pressures to which an object or system is subjected in normal operation.

operating relay time — The measured time from the time a relay control switch is closed until the relay contacts are completely closed.

operating time — The time that is measured from the time an engine, component, or unit begins to operate until the operating unit completes its operation or movement.

operating weight — Term used by aircraft operators to include the empty weight of the aircraft and items always carries in the aircraft, such as crew, water, food, etc.

operational checks — Checks made to a unit to determine that it is operating properly within the specifications outlined by the manufacturer.

Operations Limitations Manual — *See flight manual.*

opposed-type engine — *See horizontally-opposed engine.*

opposite polarity ignition system — An older type of turbine ignition system which used no storage capacitors.

optical micrometer – A precision measuring device used to measure the depth of scratches or fissures in the surface of a material by measuring the change in focus of a complex lens. The lens is focused on the surface of the material which is used as a reference, and then it is focused at the base of the damage. The amount of change in the lens focus is converted into a measure of the depth of the damage.

optical pyrometer – A temperature measuring instrument that is used to indicate the temperature of molten metal inside a furnace.

optional equipment – Aircraft equipment that is approved for installation in an aircraft, but is not required for airworthiness.

OR gate – A logic device that will have a voltage on its output any time a voltage appears at any one of its inputs.

orange peel – A defect in a painted surface that resembles the skin of an orange. It may be caused by incorrect paint viscosity, wrong air pressure, incorrect spray gun setting or an improper distance between the spray gun and the work.

orbital electron – An electron spinning around the nucleus of an atom as distinguished from free electrons that move from one atom to another.

ordinate – A line that represents the Y-axis of a graph used to fix a point.

organic brake linings – Organic material reinforced with brass wool which is attached to solid metal backings used for single-disk brakes.

organic fabric – *See organic fibers.*

organic fibers – Fibers of natural origin such as cotton or linen and used in the manufacture of aircraft fabric covering materials.

organic lining – The friction material used in spot-type, single disk brakes. It is a composition material in which brass- or copper-wool or particles of brass are embedded to control the coefficient of friction.

orifice – A small hole which is of a specific size to meter or control the flow of a fluid.

orifice check valve – A component in a hydraulic or pneumatic system that allows unrestricted flow in one direction, and restricted flow in the opposite direction.

original skin – The skin or metal covering which was originally used in the manufacture of an airplane.

O-ring — A type of sealing device used in a pneumatic or hydraulic system which has a circular cross section and is made in the form of a ring.

ornithopter — An aircraft designed to produce lift by the flapping of its wings.

oronasal oxygen mask — An oxygen mask covering only the mouth and the nose of the wearer.

orthographic projection — The projection of a single view of an object in which the view is projected along a line perpendicular to both the view and the drawing surface.

oscillate — To swing back and forth with a consistent force or rhythm.

oscillator —
[1]An electronic device which converts DC into AC.
[2]An electronic circuit which produces AC with frequencies determined by the inductance and capacitance in the circuit.

oscillograph — A device for producing a graphical representation of an electric signal mechanically or photographically.

oscilloscope — An electrical measuring instrument with which recurring voltage and current changes may be observed on a cathode ray tube similar to a small television tube.

Otto cycle — A constant-volume cycle of events used to explain the energy transformation that takes place in a reciprocating engine. In this type of engine, four strokes are required to complete the required series of events or operating cycle of each cylinder. Two complete revolutions of the crankshaft (720°) are required for the four strokes; thus, each cylinder in an engine of this type fires once in every two revolutions of the crankshaft.

out of phase — A condition of cyclic values in which two waves, such as voltage and current, do not pass through the same point at the same time.

outer flame — The enveloping, almost transparent flame that surrounds the bluish-white inner flame, or cone, in oxyacetylene welding.

outer liner — The annular and can-annular combustion liner outer shell as opposed to its inner shell. Formerly used to refer to the outer case of a can-type combustion chamber.

outflow valve — The valve in a pressurized aircraft cabin which maintains the desired pressure level inside of the cabin by controlling the amount of air allowed to flow out of the cabin.

out-of-rig — A condition of aircraft flight control rigging that prevents the aircraft from being flown without the aid of the pilot touching the controls, because the controls are not properly adjusted.

out-of-round — Eccentrically shaped because of wear.

out-of-track — A condition of a helicopter rotor or the propeller of an airplane in which the tips of the blades do not follow the same path in their rotation.

out-of-trim — A condition in an aircraft in which straight and level hands-off flying is not possible because of an aerodynamic load caused by an improperly adjusted trim device.

output — Power or energy produced by a device or a signal delivered from a device.

outside caliper — A measuring device having two movable legs, used to determine the distance across an object. Once the distance has been established, the actual dimension may be made, using a steel scale or a vernier caliper.

outside skin — The outer surface of an aircraft.

oval binding head screw — Obsolete term for a truss head screw.

oven head screw — Obsolete term for truss head screw.

overall efficiency — The product of multiplying propulsive efficiency and thermal efficiency.

overbalance — The displacement of the trailing edge of a control surface above the horizontal position when performing a static balance test.

overboost — A condition in which a reciprocating engine has exceeded the maximum manifold pressure allowed by the manufacturer.

overcontrol — Any movement of a control device in excess of that needed for a given condition.

overhang —

[1]One-half the difference in span of any two main supporting surfaces of an airplane. The overhang is positive when the upper of the two main supporting surfaces has the larger span.

[2]The distance from the outer strut attachment to the tip of a wing.

overhaul — To restore an aircraft, engine, or component to a condition of normal operation.

overhead cam — The cam of an aircraft reciprocating engine, located above the cylinder head, which operates the valves directly without the aid of pushrods.

overhead valve — A valve located in the upper part of an aircraft reciprocating engine cylinder head.

overheat warning system — A fire warning system that warns the pilot of an overheat condition that could lead to a fire.

overinflation valve — A relief valve that opens to relieve excessive air pressure used in some of the larger aircraft wheels that mount tubeless tires.

overlapping — To lap over or extend beyond.

overload — To apply a load in excess of that for which a device or structure is designed.

overrunning clutch — Generally, a pawl and ratchet arrangement used on various types of starters. It is designed such that it permits the starter to drive the engine but not the engine to drive the starter. The ratchet on the engine side will slip around within the pawls if normal disengagement does not occur.

oversized stud — A stud having a greater diameter than standard, but with the same number of threads, and same pitch as a standard stud.

overspeed —
[1]A condition in which an engine has produced more RPM than the manufacturer recommends.
[2]The condition in which the actual engine speed is higher than the desired engine speed as set on the propeller control by the pilot.

overspeed condition — A condition of the propeller operating system in which the propeller is operating above the RPM for which the governor control is set. This causes the propeller blades to be in a lower angle than that required for constant-speed operation.

overspeed governor — A fuel limiting device normally configured to the fuel control and usually flyweight operated.

overtemperature —
[1]Condition in which a device has reached a temperature above that approved by the manufacturer.
[2]Any exhaust temperature that exceeds the maximum allowable for a given operating condition or time limit.

overtemperature warning system — A warning system that warns the pilot of an overheat condition. If the temperature rises above a set value in any one section of the overheat sensing circuit, the sensing device turns on a cockpit light indicating the location of the overtemperature.

overvoltage protector — A type of electrical circuit protection device used to protect components from damage caused by high voltage surges. If the overvoltage is excessive, the overvoltage protector opens and protects the component.

oxidation — A chemical action in which a metallic element is united with oxygen. Electrons are removed from the metal in this process.

oxide — A chemical combination in which oxygen is combined with another element.

oxide film — A layer, coating, or metallic oxide on the surface of a material.

oxidized — Combined with oxygen.

oxidizer — Any substance that causes another to unite with oxygen.

oxidizing flame — An oxyacetylene welding flame in which there is an excess of oxygen passing through the torch. An oxidizing flame is noticed by a sharp-pointed inner cone and a hissing noise made by the torch.

oxyacetylene — Type of gas welding using oxygen and acetylene.

oxy-gas welding — *See oxyacetylene.*

oxygen — One of the basic elements. In the free gas state, it is always O_2 because two atoms of oxygen must combine to form one molecule of oxygen gas. Oxygen is a colorless, odorless, tasteless, gaseous chemical element forming about 21% of the earth's atmosphere. It will not burn but supports combustion and is essential to life processes.

oxygen bottle — Special, high-strength steel cylinder which is used to store gaseous oxygen under pressure.

oxygen cell corrosion — A type of corrosion that results from a deficiency of oxygen in the electrolyte.

oxygen concentration cell corrosion — A type of corrosion that forms between the lap joints of metal that trap and hold moisture.

oxygen manifold — A device for connecting several oxygen masks onto one oxygen supply.

oxygen mask – Small face mask with special attachments for breathing oxygen.

oxygen plumbing – Tubing and fittings used in the oxygen system to connect the various components.

oxyhydrogen – A type of gas welding using oxygen and hydrogen.

ozone – A variety of oxygen which contains three atoms of oxygen per molecule rather than the usual two. The major portion of ozone in the atmosphere is formed by the interaction of oxygen with the suns rays near the top of the ozone layer. It is also produced by electrical discharges (lightning storms). Ozone is important to living organisms because it filters out most of the sun's ultraviolet radiation.

ozonosphere – The stratum of the earth's atmosphere that has a high concentration of ozone that absorbs ultraviolet radiation from the sun. The ozonosphere is approximately 20 to 30 miles above the earth.

P

pack carburizing — A heat treatment method for case-hardening steel.

package — A complete assembly unit.

packing — A hydraulic seal that prevents leakage of fluid between two surfaces which move in relation to each other.

packing ring — An O-ring used to confine liquids or gases, preventing their passing between a fixed body and a movable shaft.

paint — A form of aircraft finish which is a mechanical suspension of pigments in a vehicle. The paint protects and improves the appearance of the aircraft.

paint drier — A substance added to paint to improve its drying properties.

paint stripper — A chemical material which penetrates the paint film and loosens its bond to the metal. Once the stripper penetrates the paint, the paint softens and wrinkles and can be easily wiped or washed away.

Pal nut — A thin, pressed-steel check nut used to screw down over an ordinary nut to prevent it from coming loose or from backing off.

pan — A hollow, depression or shallow metal object used to hold oil.

pancake landing — An aircraft landing procedure in which the aircraft is on an even plane with the runway. As the aircraft reduces speed and lift, it drops to the ground in a flat or prone attitude.

panel — Any separate or distinct portion of the surface of an aircraft structure.

pan-head screw — A machine screw or sheet metal screw having a large-area, slightly domed head.

pants — Streamlined covers for airplane wheels.

paper electrical capacitor — An electrical component that uses two strips of metal foil for its plates and strips of waxed paper as its dielectric.

parabola — A plane curve that remains equally distant from a fixed point.

parabolic light reflector — A light reflector, such as a light bulb, whose surface is in the form parabola or curve.

parabolic microphone — A sensitive microphone mounted in a way so that it is aimed at the source of the sound to be picked up.

parachute — A large cloth device shaped like an umbrella used to retard the fall of a body or object through the air.

par-al-ketone — A heavy, waxy grease used to protect control cables and hardware fittings from corrosion on seaplanes.

parallax — The apparent displacement of an object as viewed from two different locations not in line with the object.

parallel —

[1]Lines which run in the same direction and which will never meet or cross.

[2]Having more than one path for electron flow between the two sides of the electron source.

parallel access — A computer term used to describe a method of accessing data simultaneously, or all data at the same time.

parallel circuits — Two or more complete circuits connected to the same two power terminals.

parallel lines — Two or more lines extending in the same direction and at the same distance apart at every point.

parallel of latitude — Any of the imaginary lines on the surface of the earth that are parallel to the equator and representing degrees of latitude on the earth's surface.

parallel resonant electrical circuit — A circuit that is made up of an inductor and a capacitor connected in parallel. Also called a tank circuit.

paralleling — Controlling the output of more than one generator so that they will equally share the load.

paralleling generators — An operational procedure in which the output voltages of multi-engine aircraft electrical system generators are adjusted to share the electrical load equally.

paramagnetic material — A material that has a permeability that is less than that of a ferromagnetic material.

parameter — A quantity or constant whose value varies with the circumstances of its application.

parasite drag — Drag caused by the friction of the air flowing over a body. Parasite drag increases as airspeed increases.

parasol wing airplane — An airplane having one main supporting surface mounted above the fuselage on cabane struts.

Parkerizing — A method of treating metal parts by immersing them in a solution of phosphoric acid and manganese dioxide. Used to protect the surface from rusting.

Parker-Kalon screws: — Abbrev.: PK screws. Self-tapping sheet metal screws. PK screws are made of hardened steel, and has sharp, coarse threads.

parking brake — A mechanical or hydraulic brake system used to prevent the aircraft from moving from its parked position.

part number — An identification number assigned to a particular part or assembly by the manufacturer.

part power trim check — Trimming with the power lever against a trim stop or rig pin, then checking the EPR or N_1 speed against a trim curve for ambient conditions. If the correct values are not present, adjustment of the fuel control called trimming is required.

partial pressure — Pressure exerted by a single gas in a mixture of gases, such as oxygen in air.

partial-panel flight — Flight by reference to instruments using the turn-and-slip indicator, clock, and airspeed indicator, instead of the artificial horizon and directional gyro.

particle — A small piece of any substance or matter.

parting agent — A material or substance used to cover a mold and prevent resin adhering to it.

Pascal's law — A basic law of fluid power which states that pressure in an enclosed container is transmitted equally and undiminished to all points of the container, and acts at right angles to the enclosing walls.

passenger mile — The amount of air travel of one passenger, pilot or crew member.

passivating — A treatment of corrosion-resistant steels after welding. Much of their corrosion resistance is due to the buildup of an invisible oxide film on the surface. After welding, this buildup will normally take place in service, but can be hastened by immersion in cold 20% nitric acid for approximately 30 minutes. This also removes any surface foreign particles.

passive electrical circuit — An electrical circuit that does not contain any source of electrical energy such as a battery or generator.

passive electrical component — An electrical component, such as resistors, capacitors and inductors, that produces no gain in the circuit.

patch — A small piece of material used for strengthening, reinforcing, or to cover a hole or weak spot in a structure.

pattern —
[1]A model, guide or plan that is used to form or make things.
[2]The flight pattern an aircraft must follow when approaching the airport for landing and when leaving the airport after taking off. Aircraft operating from the airport must follows the same flight pattern in order to reduce the danger of an in-flight collision.

pawl — A pivoted stop in a mechanical device which allows motion one way but prevents it in the opposite direction. It is commonly used in a ratchet mechanism.

payload — That part of the useful load of an aircraft that is over and above the load that is necessary for the operation of the vehicle.

P-band radar — A radar frequency range between 225 and 390 MHz.

peak alternating current — The greatest amount of current that flows in one direction between the zero current level and the greatest point of deviation as measured on an oscilloscope.

peak inverse voltage — The maximum voltage which may be applied safely to an electron tube in the direction opposite to normal current flow.

peak value — The maximum value of AC or voltage measured from the zero reference line.

pedestal grinder — A type of grinder which is mounted on a pedestal. and stands on the floor of the shop.

peen — To round over or flatten the end of a shaft or rivet by light hammer blows.

peened surface — A surface marked as from an impact with a blunt instrument. Caused by careless handling or concentrated load sufficient to permanently deform the metal surface.

pencil compass — A drawing instrument used to draw circles or arcs in which the mark is made with a pencil.

pendulum — A body which is suspended from a fixed point but is free to swing back and forth or to oscillate.

pendulum valves — Gravity-operated air valves over the discharge ports of the rotor housing of a pneumatic gyro horizon. When the gyro tilts, the pendulum valves change the airflow from the housing and cause a precessive force which erects the gyro.

pentagrid converter — A five-grid electron tube which serves as a mixer, local oscillator, and first detector in a superheterodyne radio receiver

pentode —

[1]Five-element vacuum tube.

[2]An electron tube containing five electrodes: cathode, plate, control grid, suppressor grid, and screen grid.

perfect dielectric — A dielectric that has no conductivity. A perfect dielectric is an insulator that returns all of the energy that is used to establish an electric field in it when the electrical field is removed.

perforate — A series of holes in material such as paper.

performance chart — A chart detailing the aircraft performance that can be expected under specific conditions.

performance number — The anti-detonation rating of a fuel which has a higher critical pressure and temperature than isooctane. Isooctane treated with varying amounts of tetraethyl lead is used as the reference.

period — The time required for one cycle of AC.

period oscillation — The amount of time needed to complete one cycle of an oscillation.

periodic event — Any regularly repeated event that is with the same amount of time between the events is said to be a periodic event.

periodic inspection — Any inspection that is repeated regularly. An annual or 100-hour inspection of an aircraft is a periodic inspection.

periodic vibration — A vibration that has a regularly recurring resonance.

periphery — The outside of a circular or curved figure.

Permalloy — An alloy of iron and nickel used in the manufacture of permanent magnets.

permamold crankcase — An engine crankcase which has been pressure molded in a permanent mold. It is thinner and more dense than a sand-cast crankcase.

permanent ballast — A permanently installed weight in an aircraft to bring its center of gravity into its allowable limits. *See also ballast.*

permanent magnet — A ferrous metal or alloy of ferrous metals usually containing nickel and cobalt, in which the magnetic domains are aligned, and tend to remain aligned. Lines of magnetic flux join the poles of the permanent magnet so that an electrical current may be generated when these lines of flux are cut by a conductor.

permanent set — A mechanical deformity caused by excessive stress placed on a material.

permeability — The ability of a material to accept and concentrate lines of magnetic flux.

Perminvar — Special alloy used for permanent magnets.

perspective — Technique of representing on a plane surface the special relationship of objects as they appear to the eye.

petrolatum-zinc-dust compound — A material used inside an aluminum terminal lug when swaging the lug onto aluminum wire. The zinc dust abrades the oxides from the aluminum, and the petrolatum prevents its reformation.

petroleum — A substance containing chemical energy and used as a fuel for most of our aircraft engines. It is a natural hydrocarbon product which was at one time plant or animal life, but was buried under billions of tons of earth. It is obtained as a liquid from deep wells.

phantom line — Thin lines made up of a long dash and two short dashes, used to show an alternate position or a missing part in an aircraft drawing.

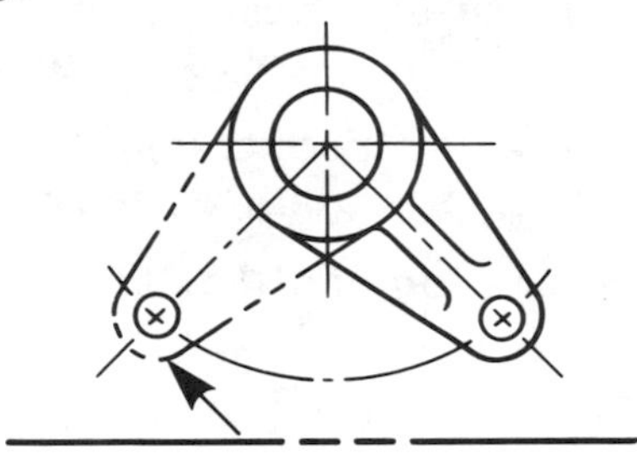

phase —

[1]The time difference between an event in the voltage waveform and the equivalent event in the current waveform in an AC circuit.

[2]A section or a distinguishable part of a maintenance program or inspection.

phase angle – The number of degrees of generator rotation between the time the voltage passes through zero and the time the current passes through zero in the same direction.

phase lock – A method of modulating the phase, condition, or state of one electronic oscillating device so it will exactly follow that of another oscillating device.

phase shift – The difference in time between similar points of an output and an input electrical wave form.

phenol-formeldehyde resin – A type of thermosetting plastic resin that is reinforced with cloth or paper to make molded plastic objects.

phenolic plastic – Plastic thermosetting phenolic-formaldehyde resin material, reinforced with cloth or paper.

Phillips-head screw – A form of recessed-head screw designed to be driven with a special cross-pointed screwdriver whose point has two distinct tapers.

phonetic alphabet – Standard words (lexicon) and combinations of words that are used for each of the letters in the alphabet during radio transmission. Examples of standard phonetic alphabet now in use are: ALFA, BRAVO, CHARLIE, DELTA, ECHO, FOXTROT, GOLF, HOTEL, INDIA, LIMA, NOVEMBER, OSCAR, PAPA, ROMEO, SIERRA, TANGO, UNIFORM, VICTOR, WHISKEY, X-RAY, YANKEE, ZULU.

phosgene – A colorless gas with an unpleasant odor which is produced when Refrigerant-12 is passed through an open flame. Causes severe respiratory irritation.

phosphate ester-base hydraulic fluid – A synthetic, fire-resistant hydraulic fluid used in high-pressure hydraulic systems of modern jet aircraft. It is identified by the specifications MIL-H-8446.

phosphate film – A dense, insoluble, inorganic film deposited on the surface of a metal which has been treated with a conversion coating.

phosphor coating – A coating for the numerals on instrument dials and for the pointers. It may be caused to glow by exciting it with ultraviolet light rays.

phosphorescent paint – A paint that absorbs energy from natural light or from ultraviolet light and which continues to glow after the natural light is removed.

phosphoric acid etchant – That constituent of a conversion coating which microscopically roughens the surface of the metal being treated and deposits a phosphate film.

photo cell — An electronic device which become conductive or produces a voltage when struck by light.

photocathodes — An electrode within an electron tube that releases electrons after it has been exposed to light.

photochemistry — That branch of chemistry having to do with the effects of light on chemical reactions.

photoconductive cell — A type of photoelectric cell that changes its resistance as the amount of light that strikes it changes. One of its uses is in photographic light meters.

photodiode — A semiconductor diode which can be caused to conduct in its reverse direction by the application of light to its junction. Photoconductive.

photoelectric characteristics — The changes (positive or negative) produced in any of the electrical characteristics of a material when exposed to light.

photoelectric material — Any element that emits electrons when exposed to light. Elements such as alkaline metals, cesium, lithium, and rubidium, are photoelectric.

photoelectricity — Electricity produced by the action of light on certain photoemissive materials.

photoemissive characteristic — Characteristics that cause a material to emit electrons when it is exposed to light.

photoemissivity — *See photomissive characteristic.*

photon — A particle of radiant energy.

photonegative characteristics — The characteristics of a material that has an increase in resistance when expose to light.

photosensitive — The property of emitting electrons when struck by light. Photoemissive.

phototransistor — A transistor which can be forward-biased into conduction by the application of light to its emitter-base junction.

photovoltaic cell — A solid-state electrical component that produces a voltage when exposed to light.

phugoid oscillations — Long-period oscillations of an aircraft along its longitudinal axis. It is often called hunting.

physical properties — Those properties of a body which may be determined by methods other than chemical, including weight, strength, and hardness.

physical tables — Tables listing the physical properties and characteristics of materials.

physics — A form of natural science which deals with matter and energy and their interaction in the various fields of mechanics.

pi filter — A network consisting of two capacitors and an inductor net arranged in the form of the Greek letter pi (π). It is essentially a capacitor-input filter followed by an L-filter.

piano hinge — A continuous metal hinge consisting of hinge bodies attached to both the fixed and movable surfaces. A hard steel wire connects the two bodies and serves as the hinge pin.

pickling —
[1]The treatment of a metal surface by an acid to remove surface contamination.
[2]Preparing an aircraft engine for longtime storage.

pick-off — That portion of a device or a system which removes a signal from a sensor.

pickup or scuffing — The buildup or rolling of metal from one area to another usually caused by insufficient lubrication or clearance.

pico- — One-millionth (0.000001) of a unit.

picofarad — Abbrev.: pf or µf. One-millionth (0.000001) of a microfarad.

pictorial diagram — The simplest kind of diagram for the maintenance technician. It may be either a line drawing, sometimes with shading to emphasize shapes, a parts blow-up, or even a photograph of a piece of equipment, which provides information concerning the overall appearance of a unit. The shapes, relative sizes and location of components, interconnecting wires, cables, etc.

pictorial drawing — A drawing which shows the object as it appears to the eye as in a photograph.

piezoelectric effect — The property of certain crystals enabling them to generate an electrostatic voltage between opposite faces when subjected to mechanical pressure. Conversely, the crystal will expand or contract if subjected to a strong electrical potential.

piezoelectric transducer — An electrical device which enables a mechanical movement to generate an electrical signal.

piezoelectricity — Electricity that is produced when certain crystalline materials such as quartz is subjected to mechanical pressure.

pig iron — Crude iron as it is reduced from the iron ore in a blast furnace.

pigment — A powder or paste mixed with a paint finish to give the desired color.

pigtail — A piece of wire that sticks out of a component which allows the component to be installed or tied.

pilot —
[1]A person licensed to operate an airplane, ship, or balloon in flight.
[2]A part that guides another part in its movement.

pilot chute — A small parachute attached to the canopy of the main parachute which pulls the main canopy out of the parachute pack so that it can open.

pilot hole — A small hole drilled or punched in sheet metal that is smaller than the bolt or rivet to be used. Used to temporarily fasten the sheet metal together while the final drilling and riveting is done.

pilot light — Electrical equipment light indicating power is on.

pin — A pin is a straight cylindrical or tapered fastener designed to perform an attaching or locating function.

pin contacts — Electrical connector contacts, called the male contacts, that are in the form of a set of metal pins in one-half of a connector. These pins fit into sockets, female contacts, in the other half of a connector.

pin jack — A single female receptacle that will accept and hold a small metal pin that is attached to the end of a wire or a test lead. Pin jacks are used on test equipment such as multimeters.

pin punch — A long punch with straight sides. Used to remove bolts and rivets from tight-fitting holes.

pin spanner — A form of semi-circular wrench having pins which fit into holes around the edge of a circular nut.

pinhole — A defect in a finish which appears as a tiny hole. It is caused by a bubble in the paint film.

pinion — A small cogwheel whose teeth fit into a larger gear.

pinion gear – A small gear on a shaft driven by either a sector gear or a toothed rack.

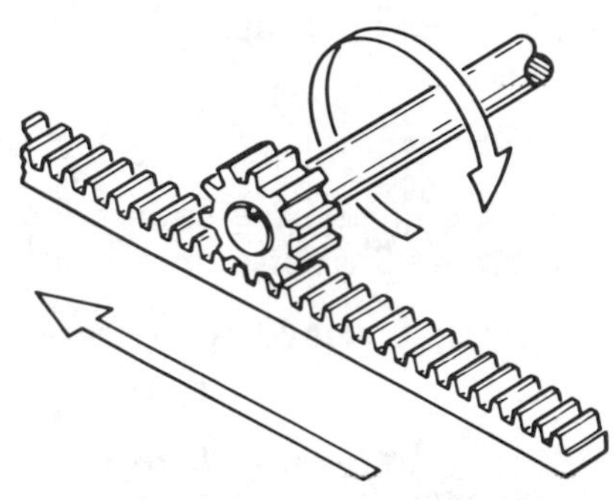

pinked edge – The edge of a fabric material which has been cut into a series of small Vs to prevent the material raveling.

pinked-edge fabric – *See pinked edge.*

pinked-edge tape – A surface tape whose edges have been cut into small Vs.

pinking shears – A special form of scissors which cuts the fabric in a series of small Vs.

pint – A measure unit of volume equal to ½ qt., ⅛ U.S. g.on, 29-7/8 cu. in., or approximately 4.73×10^{4} cu. m.

pipe threads – Tapered threads on a hollow pipe or a piece of round stock. The taper provides the seal.

piston – A cylindrical member which moves back and forth within a steel cylinder used to compress the fuel-air mixture and to transmit the force from the expanding gases in the cylinder to the crankshaft.

piston displacement – Total volume swept by the pistons of an engine in one revolution of the crankshaft.

piston engine – A term most commonly referring to a reciprocating engine.

piston fuel pump – Sometimes used as a main fuel pump in place of the more typical spur gear pump. This pump is capable of delivering fuel at higher pressures than other types and can also vary its output per revolution. A variable displacement-type pump.

piston insulator – Composition insulators between the actuating pistons and the pressure plate used to prevent heat transferring from the pressure plate into the piston where it would likely damage the seals and the fluid.

piston pin – *See wrist pin.*

piston pin boss – The enlarged area inside a piston to provide additional bearing area for the wrist pin.

piston pump – A pump in a fluid power system which uses pistons to move the fluid.

piston ring grooves – The grooves in the circumference of a piston into which the piston rings fit.

piston rings – Rings made of a special gray cast iron which fit into grooves in the periphery of a piston to form a seal between the piston and the cylinder wall.

piston skirt – The lower portion of a piston.

piston-type pump – A type of hydraulic fluid pressure pump in which fluid is moved by pistons that move up and down in the cylinders of the pump.

pitch –
[1]The rotation of an airplane about its lateral axis.
[2]Distance between the centers of adjacent rivets in the same row.
[3]The distance measured between corresponding points on two adjacent threads.

pitch axis – The lateral axis of an aircraft or the axis about which the aircraft pitches in a nose up or nose down attitude.

pitch, bolt threads – Distance from any point on the thread of a screw to corresponding point on an adjacent thread.

pitch of a propeller – Zero-thrust pitch. The distance a propeller would have to advance in one revolution to give no thrust. Also called "experimental mean pitch".

pitch of screw thread – The distance from the center of one thread to the center of the next thread.

pitch ratio of propeller – The ratio of the pitch to the diameter.

pitch rivet layout dimension – The distance between the centers of adjacent rivets that are installed in the same row.

pitot pressure – Ram or impact pressure used in the measurement of airspeed.

pitot static system – The pressure system for airspeed indicators, altimeters, and vertical speed indicators. It consists of the pitot tube and a static port, along with all of the necessary tubing and moisture traps.

pitot tube – An open-ended tube that faces directly into the relative airstream of an aircraft and picks up the ram, or pitot, pressure to be used in an airspeed indicator.

pitot-static tube – A combination tube which has the pick-up for the pitot pressure as well as openings that pick up undisturbed, or static, air pressure.

pitting – The formation of pockets of corrosion products on the surface of a metal.

pitting corrosion – A form of metal corrosion in which small, localized pits filled with the salts form on the surface of the metal.

pivot trunnion – A bearing surface on the top of the landing gear on which the gear rides when folding into the aircraft.

placard – A notice placed in or on the aircraft depicting pertinent information relating to the aircraft, its operation, particular component limitations, etc.

plain bearing – A simple form of bearing used to support an aircraft engine crankshaft or camshaft and which is designed to take only loads that are perpendicular to its face.

plain flap – A form of wing flap in which a portion of the trailing edge of the wing folds down to increase the camber of the wing without increasing the wing area.

plain nut – A simple hex nut which has no provisions for locking.

plain overlap seam – A seam used for machine sewing of aircraft fabric in which the edge of one piece of fabric laps over the edge of the other. and one or more rows of stitches are made to hold the pieces together.

plain rib – A former rib used to give an airfoil its shape.

plain washer – A simple flat washer used to provide a smooth bearing surface for a nut or to shim between a surface and a nut.

plan view – The view from the top of an object.

plane of rotation — The plane in which a propeller or a helicopter rotor rotates. It is perpendicular to the crankshaft or the rotor shaft.

plane of symmetry — A vertical plane which passes through the longitudinal axis of an aircraft and divides the aircraft into two symmetrical sides.

planetary gears — A reduction gearing arrangement in which the propeller shaft is attached to an adapter holding several small planetary gears. These gears run between a sun gear and a ring gear, either of which may be driven by the crankshaft, and the other is fixed into the nose section. Planetary gears are efficient and do not reverse the direction or rotation between the two shafts.

planform — The outline of a wing as viewed from aboave. The shape you would see when looking at a top view of a blueprint, or "plan"..

plan-position indicator — Abbrev.: PPI. A radar system component for presenting a maplike display of the search area on the screen of a CRT.

Plante cell — A secondary cell in which the pole pieces are formed of sheets of lead and lead dioxide. The electrolyte is a dilute solution of sulfuric acid.

plaque — The base for plates of nickel-cadmium batteries. It is powered by nickel, formed under heat and pressure onto a close mesh nickel screen.

plasma coating — Process of applying a thin coating of highly wear-resistant material on the surface of turbine engine parts. The process is accomplished by extremely high surface heat and melting together of the base and coating materials.

plastic — Commonly used term to denote any of the thermosplastic or thermosetting polymers used in modern aircraft construction.

plastic range — A stress range in a material in which, though the material does not fail when subjected to force, the material does not completely return to its original shape, but will be deformed.

plasticizer — A chemical used in a lacquer finish to give the film its flexibility and resilience.

plate —
[1]The electrode in a vacuum tube which serves as the anode receiving the electrons from the cathode.
[2]The active element in a storage battery.

plate resistance – In an electron tube, the ratio of a change in plate voltage to a change in plate current with grid voltage constant.

plate saturation – The condition in an electron tube when the plate will no longer attract electrons as fast as they are emitted by the cathode.

plating – A process in which one metal is used to cover another using a process of electrical deposition. Specifically, chromium and cadmium are useful metals for covering steel.

platinum – A hard, gray metallic element with an extremely high melting point. It is used for the electrodes of fine-wire spark plugs.

platinum spark plug – A type of fine-wire electrode spark plug that can operate at very high temperatures.

play – A commonly used term to mean relative movement between parts. As in the case of flight controls, play is the amount of movement of a flight control without causing it to move.

P-lead – A colloquial term for the primary lead of an aircraft magneto connected to ignition switch.

plenum – An enlargement of a duct or an enclosing space in an aircraft engine induction system or air conditioning system. Used to smooth out the pulsations in the flow of the air.

plenum chamber – An enclosed volume of air in which the air held at a slightly higher pressure than that of the surrounding air.

Plexiglas – A transparent acrylic plastic material used for aircraft windshields and side windows.

pliers – Small, pincher-like hand tools used for holding small objects or for bending or cutting wire.

plies – Sheets of material which are laminated together.

plumb – Anything that lines up with the plumb line is said to be plumb or in plumb.

plumb bob – A heavy, pointed weight which is used on the end of a string or line to establish a location directly below a point.

plumb line – The straight line of a string to which a plumb bob is attached and hung. Anything that lines up with the plumb line is said to be plumb or in plumb.

plumbing – Tubing and fittings or connections used for transmitting fluid within an aircraft.

plumbing connection — Threaded connections which join sections of tubing, or which are used to connect the tubing to a component.

plunger — A part of a machine that works with a plunging motion.

ply rating — A load rating for aircraft tires which relates to the strength of cotton plies. For example, a 20-ply rating nylon tire has the same load rating as a tire with 20 cotton plies.

plywood — Layers of wood glued together so that the grain in each layer is placed 45° or 90° to the others.

pneudraulic — A combination of air and hydraulic pressure.

pneumatic altimeter — A form of altimeter that measures height above a given pressure level. Its calibration is based on a specified lapse rate or change in pressure with height.

pneumatic drill motor — An air motor equipped with a chuck to hold twist drills.

pneumatic fire detection system — A newer type of system using a gas filled continuous tube. The gas expands and acts on a diaphragm to close an electrical circuit and show a warning light in the cockpit.

pneumatic starter — Starting motor operated by air pressure.

pneumatic system — The power system in an aircraft which is used for operating landing gear, brakes, wing flaps, etc. with compressed air as the operating fluid.

pneumatic-mechanical fuel control — A type of fuel control which utilizes pneumatic and mechanical forces to operate its fuel scheduling mechanisms.

pneumatics — That system of fluid power which transmits force by the use of a compressible fluid.

PNP transistor — Three-element semiconductor device made up of a sandwich of N-type silicon or germanium between two pieces of P-type material.

pod — An enclosure housing a complete engine assembly.

pointer — A thin strip of movable metal rod, sometimes called hands or needles, that is moved by an analog instrument mechanism over a calibrated scale.

point-to-point wiring – Previously, the universally used method of building electronic units. Electronic components were mounted directly on the chassis and interconnected by means of wires that were integral parts of the components leads or by means of insulated hook up wire.

polarity – The property of an electrical device of having two different types of electrical charges: positive (deficiency of electrons) or negative (excess of electrons).

polarization – A degradation in chemical cell performance, particularly in the case of Leclanché cells, caused by gas formation and the resulting insulation of portions of the pole area.

pole – The designation given to the ends of a magnet.

pole shoes – The field assembly part of an electric generator or motor.

poles of a magnet – Refers to the north and south poles of a magnet where magnetic lines of flux leaves the south pole and enters the north pole.

polishing – The process of producing a smooth surface by rubbing with fine abrasive wheels, belts or compounds.

polyester fiber – A synthetic fiber noted for its mechanical strength, chemical stability, and long life. It is used to make woven fabric for covering aircraft structures.

polyester resin – A synthetic resin, usually reinforced with fiberglass cloth or mat, and used to form complex shapes for aircraft structure.

Poly-Fiber – A fabric woven from polyester fibers.

polyethylene – A lightweight, thermoplastic resin material that has good chemical resistance. Polyethylene resins are used for making containers for liquids and sheets of protective covering material.

polyethylene plastic material – A lightweight, thermoplastic resin material that has very good chemical and moisture-resistant characteristics. It is used for plastic sheeting and containers.

polymer paint – A fast drying, water-based paint that contains vinyl or acrylic resins. When the water in the paint evaporates, it leaves a waterproof film of the plastic resin.

polymerization – The process of joining two or more chemicals with molecules of similar structure forming a more complex molecule with different physical properties. In this chemical reaction, the material essentially jells.

polymid – Translucent plastic material commonly called nylon.

polyphase alternating current – Three-phase AC electricity that is produced by more than one set of generator windings.

polyphase electric motor – An induction motor that operates on two-phase or three-phase AC.

polystyrene – A transparent plastic used to make cell cases for some nickel-cadmium batteries.

polyurethane enamel – A two-component, chemically cured enamel finishing system noted for its hard, flexible, high-gloss finish.

polyvinyl chloride – A thermoplastic resin used in the manufacture of transparent tubing for electrical insulation and fluid lines which are not subject to any pressure.

polyvinyl chloride – Abbrev.: PVC. A popular, low cost, wire insulating material.

pontoon – Floats attached to a land airplane landing gear to allow it to operate from water.

pop-open nozzle – Afterburner nozzle which pops full open at idle for the purpose of efficient engine operation at very low thrust.

poppet valve – A circular-headed, T-shaped valve used to seal the combustion chamber of a reciprocating engine, and at the proper time, either admit the fuel-air mixture into the cylinder or conduct the burned exhaust gases out of the cylinder.

pores – Small holes or openings on the surface of metals.

porosity – The condition of a material having small pores or small cavities throughout the material.

porous chrome – A plating of hard chromium on bearing surfaces. The surface of the plating consists of tiny cracks in which lubricant can adhere to reduce sliding friction.

porous chrome plating – An electrolytically deposited coating of chromium on the walls of aircraft engine cylinders. The surface contains thousands of tiny cracks which hold oil to provide for cylinder wall lubrication.

porous salt – The type of residue normally left on the surface of a metal which has been attacked by corrosion.

porpoising — Hunting, or oscillating, along the longitudinal axis of the aircraft normally caused by an incorrectly functioning automatic pilot.

port side — The left-hand side of an aircraft or ship as one faces the nose of the aircraft or bow.

position error — The error in an airspeed indicator caused by the static source not being exposed to absolutely still air.

position lights — *See navigation lights.*

positive — Symbol: +. A condition of electrical pressure caused by a deficiency of electrons.

positive acceleration — An increase in the rate of change of velocity.

positive electrical charge — An electrical condition that is caused by a deficiency of electrons.

positive ion — An atom that has fewer electrons than protons.

positive static stability — The condition of stability of an aircraft which causes it, when disturbed from a condition of straight and level flight, to tend to return to straight and level flight.

positive terminal — The terminal of a battery or power source where electrons enter the source after they have passed through the external source.

positive-displacement pump — A type of fluid pump which moves a specific amount of fluid each time it rotates. Examples of positive-displacement pumps are gear pumps, gerotor pumps, and vane pumps.

positron — The positive counterpart of an electron which has the same mass and spin characteristics as an electron, but it has a positive electrical charge.

post exit thrust reverser — *See mechanical blockage thrust reverser.*

pot life — The usable life of a resin. The time before it begins to thicken after the catalyst and accelerator have been added.

potential — A term for electrical pressure or voltage caused by dissimilar metals in an acid solution or an electrolyte.

potential difference — Abbrev.: pd. The difference in voltage that exist between two terminals or two points of differing potential.

potential drop — A drop in voltage in an electrical circuit caused by the resistance of current flow through a known resistor.

potential energy – That energy possessed by an object because of its position, configuration, or the chemical arrangement of its constituents.

potential hill – The apparent voltage produced in a semiconductor device by the depletion area. Before electrons can flow in the device, there must be enough voltage applied to overcome the potential hill.

potentiometer –
[1]A variation resistor having both ends and its wiper in the circuit and used as a voltage divider.
[2]An instrument used for measuring differences in electrical potential by balancing the unknown voltage against a known variable voltage.

potentiometer ohmmeter – An ohmmeter circuit in which resistance is measured by placing a known voltage across a standard resistor; then the circuit is opened and the unknown resistor is placed in series. The voltage drop across the standard resistor is read and displaced on the meter as ohms.

potentiometer variable resistor – *See potentiometer.*

potted circuit connector – An electrical circuit connector which is protected by encapsulating it with an insulating potting compound.

potting compound –
[1]A resin having filler capability, used to fill the cells when making minor repairs to damaged honeycomb panels.
[2]A non-hardening, rubber-like material used to moisture proof and protect the wires in certain electrical plugs.

pound – A measure of mass equal to approximately 0.454 kg.

pour point –
[1]The lowest temperature at which a fluid will pour without disturbance.
[2]The lowest temperature at which oil will gravity flow.

powder metallurgy – A development which makes use of powder rather than ingots. A process used to produce superalloys for high heat strength-type turbine components and for bearing material (sintered bearings).

powdered-iron core – A molded, magnetic powdered iron mixed with a binder. Used when high permeability and low eddy current losses are desired.

power — Symbol: P
[1]The time-rate of doing work. Force times distance, divided by time. Power can be expressed in terms of foot-pounds of work per minute, or in horsepower. One HP is 33,000 ft.-lbs. of work per minute.
[2]The basic unit of electrical power is the watt, and 746 watts of electrical power is equal to one mechanical horsepower. In electrical problems, power is the product of the voltage (E) times the current (I) ($P = E \times I$). Power in watts delivered to a circuit varies directly with the square of the applied voltage and inversely with the circuit resistance.

power brake control valve — A special form of pressure regulator between the aircraft hydraulic system and the brake cylinders. The amount of pressure applied to the brakes is directly proportional to the force the pilot puts on the brake pedals.

power control system — A form of control system in which the normal movement of the controls is assisted by the use of hydraulic or pneumatic actuators to reduce the amount of force the pilot must apply.

power enrichment system — A carburetor subsystem for a reciprocating engine which increases the fuel mixture when the engine is operating at full power.

power factor — The ratio of the resistance of an electrical circuit to the circuit impedance measured by a wattmeter.

power frequency — Frequency of AC electricity used for heat and light. Commercial power frequency in the U.S. is 60 Hz, and aircraft power frequency is 400 Hz.

power lever — The cockpit lever which connects to the fuel control unit for scheduling fuel flow to the combustion chambers of a turbine engine. Sometimes called the power control lever or throttle.

power lever angle — Abbrev.: PLA. A protractor on the fuel control showing movement of the power lever at full travel in degrees.

power loading — The ratio of the maximum gross weight of an aircraft to the brake horsepower produced by the engines.

power overlap — The time that two or more cylinders of an engine are simultaneously on the power stroke. The more cylinders an engine can have on the power stroke at one time, the greater the power overlap and the smoother the operation.

power pump —
[1]An engine-driven or electric motor-driven pump used to move fluid to produce hydraulic or pneumatic pressure.
[2]A hydraulic pump which is driven by the aircraft engine or by an electric motor.

power recovery turbine — Abbrev.: PRT. A form of power recovery device used on the Wright R-3350 engine. The exhaust gases spin a series of small turbines which are clutched to the crankshaft by fluid-coupling devices.

power section — That portion of a radial engine on which the cylinders are mounted.

power stroke — The movement of the piston of an aircraft reciprocating engine when the piston is forced down by the expanding gases. This is the only time work is accomplished by the engine.

power supply — The part of an electronic circuit which supplies the filament and plate voltages for the operation of the circuit.

power transformer — An electrical power supply transformer that changes voltages to that needed for the operating unit.

power turbine — Also referred to as free power turbine. A turbine rotor not connected to the compressor.

powerplant — The complete installation an aircraft of the engine, propeller, and all of the accessories and controls needed for its proper operation.

powerplant technician — A person who holds a certificate from the FAA authorizing him to perform maintenance or inspection on the powerplant, including the propeller, of certificated aircraft.

Pratt truss — A type of truss structure in which the vertical members carry only compressive loads, and the diagonal members carry only tensile loads. A Pratt truss is used for most fabric-covered wings.

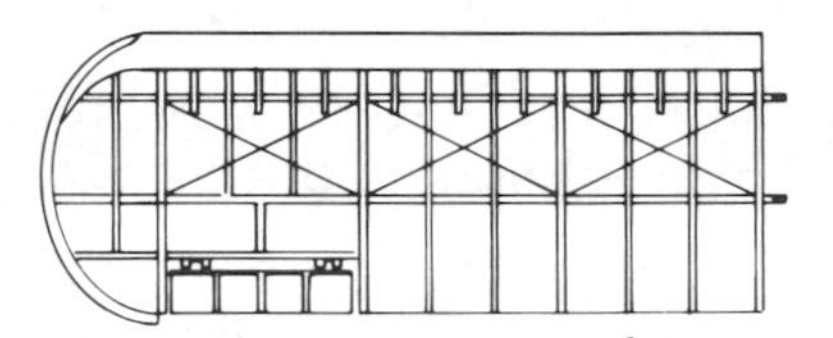

preamplifier — A type of electronic circuit component that amplifies an extremely weak input signal to a value that is strong enough that it can be used in other amplifiers.

precession – One of the characteristics of a gyroscope which causes an applied force to be felt, not at the point of application, but 90° from that point in the direction of rotation.

precious metal – Highly valued metals because they are scarce. Examples of precious metals are gold, silver, and platinum.

precipitate – To condense out of or to separate from a mixture.

precipitation – A general term that includes rain, hail, snow, and sleet.

precipitation hardening – Hardening caused by the precipitation of a constituent from a supersaturated solid solution.

precipitation heat treatment – A step in the heat-treating process of aluminum in which the metal, after having been heated to its critical temperature and quenched, is raised to an elevated temperature and held for a period of time. This artificially ages the metal and increases its strength.

precision measuring instruments – Instruments capable of making exact measurements.

predrilling – The drilling of a hole slightly undersize to enable reaming of the hole to the proper size later.

pre-exit thrust reverser – A thrust reverser system installed forward of the exhaust nozzle.

preflight inspection – Inspection of the aircraft before takeoff to determine that all systems are functioning properly for the intended flight.

preformed control cable – Steel aircraft control cable whose individual strands have been formed into a spiral before the cable was woven. This relieves the bending stresses within the cable and prevents the strands spreading out when the cable is cut.

preignition – Ignition occurring in the cylinder before the time of normal ignition. Often caused by a local hot spot in the combustion chamber igniting the fuel-air mixture.

pre-installation checks – Checks made on a unit before installation of the unit.

preoiling – Forcing lubricating oil through all of the oil passages in the engine before the engine is started.

press brake — A type of sheet metal bending tool in which the sheet is place on the bed with the sight line directly under the edge of the clamping bar with the correct bend radius die. The clamping bar is brought down to hold the sheet firmly in place. A bending leaf is raised until it bends the metal to the proper angle.

press fit — A tight interference fit between machine parts which requires one part to be pressed into the other.

press-to-test light — A type of light that is tested by pressing the light fixture to complete the circuit to ground. If the light illuminates, the bulb is good.

pressure — Force per unit area.

pressure altitude — That altitude in standard air that corresponds to the existing air pressure. It is the altitude above standard sea level pressure of 29.92 in. of mercury or 1013.2 millibars.

pressure capsule — The portion of the structure which is subjected to pressurization, comprising the cabin and cockpit.

pressure carburetor — A type of fuel metering system which senses the relationship between impact air pressure and venturi pressure to provide a metering force for the fuel.

pressure casting — A method of casting metal parts by forcing molten metal into permanent molds.

pressure controller — That portion of a turbocharger control system which maintains the desired manifold pressure.

pressure-demand oxygen regulator — A type of oxygen regulator that is capable of furnishing 100% oxygen under pressure to force the oxygen into the lungs of the wearer.

pressure demand oxygen system — A demand oxygen system which supplies 100% oxygen at sufficient pressure above the altitude where normal breathing is adequate. Also known as pressure breathing system.

pressure fed gun — A paint spray gun in which the material is fed to the gun by air pressure on the pot or cup holding the material.

pressure gage snubber — A unit installed in the pressure gage line which causes the needle of the pressure gage to give a steady reading.

pressure fed spray gun — A paint spray gun in which the material being sprayed is fed into the gun under pressure from a pressure device.

pressure jig – A means of putting pressure on a repaired bonded structure by air pressure inside a bladder between the surface and a jig.

pressure line – Tubing carrying hydraulic fluid under pressure from the pump to the selector valve or the control valve.

pressure plate –
[1]A heavy, strong plate in a multi-disc brake which receives the force from the brake cylinders and transmits it to the discs.
[2]A heavy, stationary disc in a multiple disc brake. It is provided with a wear surface on one side only, and the pistons press against its backside to force the disk stack over against the back plate.

pressure port – The opening in a device through which pressure is introduced.

pressure pot – A container holding the material to be sprayed. An agitator keeps the material in motion, and a regulator maintains the proper air pressure on the material to feed it to the gun.

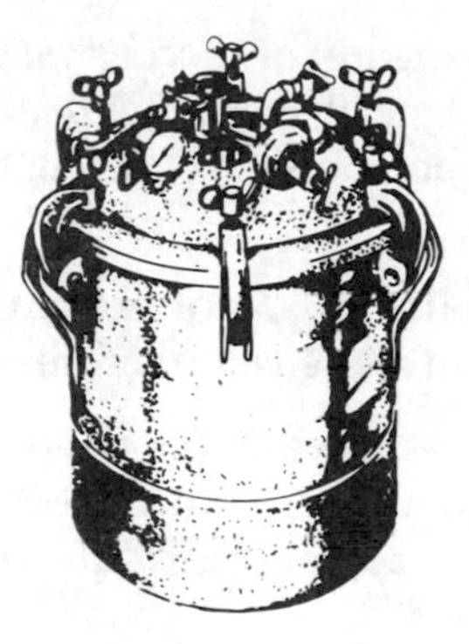

pressure ratio – One pressure divided by another, used to describe certain engine functions. *See also compressor pressure ratio; engine pressure ratio.*

pressure ratio controller – In turbocharged engines, controls the maximum turbocharger compressor discharge pressure (34 + or – .5 in. Hg to critical altitude (18,000 ft.).

pressure reducing valve – A device which reduces the pressure of a liquid or a gas from a high value to a fixed lower value.

pressure regulator – A device which maintains a constant output pressure from a constant displacement pump by bypassing a portion of the fluid back to the inlet side of the pump.

pressure relief valve – A pressure control valve that bypasses fluid back to the return manifold or reservoir in the event the pressure rises above a predetermined level.

pressure seal – A seal installed in a pressure bulkhead to permit a cable to pass through it.

pressure sensing switch – An electrical switch which will open or close when a predetermined pressure is reached in a system.

pressure tank – Another name for accumulator.

pressure transducer – *See pressure transmitter.*

pressure transmitter – A mechanical-to-electrical device in which an electrical signal is generated which is proportional to the pressure being sensed. The electrical signal is then transmitted to an instrument on the instrument panel.

pressure vessel – A pressurized aircraft that is sealed and pressurized in flight.

pressurization – A means of increasing the supply of oxygen in the cabin of an airplane flying at high altitude by increasing the air pressure in the cabin to that of an altitude which requires no supplemental oxygen.

pressurization controller – A controller which maintains the pressure in an aircraft cabin at the value required.

pressurized aircraft – Any aircraft in which the cabin area is sealed off and is pressurized with air from a cabin supercharger. The cabin can be pressurized to a pressure that compares with an altitude of approximately 8,000 ft.

pressurized ignition system – An ignition system that is pressurized with compressed air to keep the high voltage at the correct electrode.

pressurizing and dump valve – A valve used with a dual line duplex fuel manifold and duplex nozzle system. The pressurizing valve opens at higher fuel flows to deliver fuel to the secondary manifold. At engine shutdown, the dump portion opens to drain fuel overboard from the manifolds. Same as pressurizing and drain valve.

prestretching – A means of preventing an aircraft control cable stretching in operation by applying a load of 60% of its breaking strength for a specified period of time to the cable before it is installed in the airplane.

pretrack – A method used by Sikorsky to preset the track of a rotor blade prior to installation.

preventive maintenance – Simple or minor preservative operations and the replacement of small standard parts not involving complex assembly operation as listed in Appendix D of FAR Part 43.

prick punch – Tool used to place reference marks on metal.

primary air – That portion of the compressor output air used for actual combustion of fuel. Sometimes this term is used to refer to the amount of air flowing through the basic engine portion of a turbofan engine.

primary airstream – The air which passes through the core of the engine.

primary cell – An electrical device which generates electron flow by converting some of its substance into ions which free electrons. Some of the material is destroyed in the process. Primary cells are not rechargeable.

primary circuit – The main circuit in a magneto ignition system. It consists of turns of wire on which the primary current flows. *See also primary current.*

primary controls – Movable surfaces which cause an aircraft to rotate about its three primary axes. The primary controls of an airplane are the ailerons, elevators, and rudder.

primary current – The alternating or pulsating current that flows in the primary winding of a transformer and induces a current into the secondary winding. *See also primary winding; primary circuit.*

primary exhaust nozzle – On a turbofan, the hot exhaust nozzle. On an afterburner, the inner exhaust nozzle.

primary fuel –
[1]Fuel sprayed into the combustion chambers at low air flow.
[2]Refers to a duplex fuel nozzle and the fuel that initially flows on starting; usually from the center orifice. Also called pilot fuel.

primary structure – Those portions of the airplane, whose failure would seriously endanger the safety of the airplane. It includes the wing structure, controls, engine mounts, etc.

primary winding – The winding in a magneto coil through which the current induced by the rotating magnet flows. The breaker points are in series with the primary winding. *See also primary circuit; primary winding.*

prime coats – The first coats of an aircraft finish used to bond the topcoats to the base material.

primer –
[1]A material which provides a sandwich between the topcoats of a finishing system and the metal to provide a good bond.
[2]Small hand pump used to spray raw gasoline into an engine cylinder to provide fuel for starting.

primer fuel system – A low output system for engine starting. Used primarily where the main fuel system uses vaporizing tube nozzles. Also called starting fuel system.

primer surfacer – *See prime coats.*

priming a pump – The act of replacing air in a pump with hydraulic fluid.

principal view – The view in an orthographic drawing which shows the most detail of the object.

print – A copy of a formal engineering drawing.

print tolerance – A notation on an aircraft drawing of the tolerance allowed on the finished part described by the drawing.

printed circuit board – Abbrev.: PCB. Modern replacement for the electronic component chassis consisting of a plastic, fiberglass, or other insulating board with bonded copper strips for component interconnection. The components are generally mounted directly on the board by means of solder joining.

priority valve – A pressure-actuated hydraulic valve which allows certain actuations before others. This type of valve is used to assure that the wheelwell doors will be opened before the landing gear extends.

probe – A sensing device that extends into the airstream or gas stream for measuring pressure, velocity or temperature. In the case of pressure, it is used to measure total pressure.

procedures – List of instructions using step-by-step methods.

process annealing – Heating a ferrous alloy to a temperature close to, but below, the lower limit of the transformation range and then cooling in order to soften the alloy for further cold working.

Production Certificate – A certificate issued by the FAA to allow the production of a type-certificated aircraft, aircraft engine, or component.

profile drag – That portion of the drag of an aircraft caused by the air flowing over the surface of the craft.

program – A list of events or procedures. In computers, a program is a series of instructions that tell the computer exactly how it is to receive data, store it, process it, and deliver it to the user.

program flowchart – A pert or gnatt chart which show the steps that are taken in the execution of a program.

programmable calculator – An electronic calculator that can be programmed by the user and is stored in its memory.

programmable read-only memory – Abbrev.: PROM. An integrated circuit memory device for a digital computer. Read-only memory cannot be written to or changed.

prony brake – Device used to measure the usable power output of an engine on a test stand. It consists largely of a hinged collar, or brake. which can be clamped to a drum splined to the propeller shaft. The collar and drum form a friction brake which can be adjusted by a wheel. An arm of known length is attached to the collar and terminates at a point that bears on a scale. As the propeller shaft rotates, the force is measured on the scale. This force multiplied by the lever arm indicates the torque produced by the rotating shaft.

proof load – A load applied to a structure that does not cause permanent deformation. A testing measure to insure the structure will be airworthy.

proof pressure test – A series of tests to show that a pressure capsule will withstand the pressure exerted upon it in service.

prop blast – The colloquial term for the rush of air generated by a propeller.

propeller – A device for propelling an aircraft that, when rotated, produces by its action on the air, a thrust approximately perpendicular to its plane of rotation. It includes the control components normally supplied by its manufacturer.

propeller anti-icer – A system within an airplane which meters a flow of alcohol and glycerine along the leading edge of the propeller blades to prevent the formation of ice on the blades.

propeller blade – That part of a propeller which forms the airfoil and converts the torque of the engine into thrust.

propeller blade angle – The acute angle between the chord of a propeller blade and the plane of rotation.

propeller blade pitch – The distance a propeller will advance if there isn't any slip.

propeller blade tipping — *See propeller tipping.*

propeller boss — The thick, central portion of a fixed-pitch propeller hub.

propeller brake — A friction brake used on turbopropeller engines to prevent the propeller windmilling in flight after it has been feathered and to prevent its rotating when the engine is ground idling.

propeller butt — The blade shank or base of the propeller blade that fits into the propeller hub.

propeller critical range — An operational range where engine speed will set up harmonic vibration in the propeller. Engines are usually placarded against operation in this speed range.

propeller diameter — Twice the distance from the center of the propeller hub to the blade tip.

propeller efficiency — Ratio of thrust horsepower to brake horsepower. On the average, thrust horsepower constitutes approximately 80% of the brake horsepower. The other 20% is lost in friction and slippage.

propeller hub — The central portion of a propeller to which the blades are attached and by which the propeller is attached to the engine.

propeller pitch — The acute angle between the chord of a propeller and a plane perpendicular to the axis of its rotation.

propeller slip — A condition of propeller aerodynamics which is the difference between the geometric and the effective pitch.

propeller spider — The foundation unit of a controllable pitch propeller. It attaches to the propeller shaft, and the propeller blades ride on bearings on the spider. The spider is enclosed in the propeller hub.

propeller synchronization — A condition in which all of the propellers have their pitch automatically adjusted to maintain a constant RPM among all of the engines of a multi-engine aircraft.

propeller thrust — The component of the total air force on the propeller which is parallel to the direction of advance.

propeller tipping — Thin sheet brass or stainless steel covering along the leading edge and around the tip of a wooden propeller, to protect the blade from erosion. The metal tipping is installed over a fabric reinforcement.

prop-fan — An advanced technology propeller, designed to operate at supersonic tip speeds, Mach 0.8 airspeeds and 20,000 to 35,000 ft. altitude. Several 6- to- 8-bladed propeller designs with curved tips are presently under development for use in future passenger liners. A design stated to have excellent TSFC characteristics.

propjet — Same as a turboprop.

proprietary reducers — Thinners or solvents for paints which are formulated according to and distributed under a trade name of a chemical manufacturer.

propulsive efficiency — External efficiency of an engine or kinetic energy to work.

propwash — The force of air blown rearward by the propeller.

proton — Positively charged particles in the nucleus of an atom.

prototype — The first unit of a design to be built.

prototype device — A working model of a design used to test its concept.

protractor — A device for measuring angles in degrees.

protruding head rivet — An aircraft rivet in which the head protrudes above the surface of the metal; universal-head, round-head, and flat-head rivets.

Prussian blue — A compound used in checking for valve contact with a valve seat.

psychrometer — An instrument for determining the relative humidity of the air by measuring both the wet-and dry-bulb temperatures.

P-type silicon — Silicon doped by an impurity having three valence electrons.

public aircraft — An aircraft used only in the service of a government or a political subdivision. This does not include any government-owned aircraft engaged in carrying persons or property for commercial purposes.

pucks — Colloquial term for brake linings used on disc brakes.

pull test — A fabric-strength test in which a sample strip, one inch wide, is pulled until it breaks. The strength of the fabric is determined by the measured amount it took to break the strip.

pulley — A simple machine in the form of a wheel grooved to accommodate a cable. It is used to guide cables and change direction.

pull-through rivet — A form of blind, mechanically expanded rivet in which the hollow shank is upset by pulling a tapered mandrel through it.

pull-up resistor — A resistor used to limit the current through a two-state device when the device is in its low resistance state, and to develop a potential difference when a two-state device is in its high resistance state.

pulsate — To expand and contract rhythmically, yet not change direction.

pulsating direct current — DC which has been chopped by a vibrator or chopper that changes from zero to maximum and then back to zero. This produces the changing current required for use in a transformer.

pulsation — A beat, or rhythmic throb.

pulse — A rhythmic throb in the voltage of an electrical circuit.

pulse amplifier — A wide-band electrical amplifier that is used to increase the voltage of alternating current.

pulse counter — A device that measures pulses of electrical energy that it receives in a specific interval of time.

pulse generator — An electronic circuit designed to produce sharp pulses of voltage.

pulse-echo — An ultrasonic non-destructive inspection used to detect the presence of internal damage or faults in a piece of aircraft structure.

pulse-echo method of ultrasonic inspection — Method of detecting metal thickness or indication of internal damage by introducing a pulse of ultrasonic energy into a part, timing its travel through the material and back to the point of injection.

pumice — An extremely fine natural abrasive powder used for polishing metal surfaces.

pump — A mechanical device used to move a fluid. A pump is not a pressure producing machine as pressure can be produced only when a flow of fluid is restricted.

punch —
[1]A short, tapered steel rod used for driving pins, bolts, or rivets from holes.
[2]A device for cutting holes in paper, thin metal, or gasket material by shearing the material between close-fitting male and female dies.

punch test – A test of the strength of aircraft fabric while on the airplane. A pointed, spring-loaded plunger is pushed into the fabric, and the amount of force required to penetrate indicates the strength of the fabric.

puncture – A hole that is pierced in a material.

purge – To cleanse a system by flushing.

push fit – A form of interference fit in which the parts may be assembled by hand-pushing them together rather than having to drive or press them.

push rod – The component in a reciprocating engine which transmits the movement of the cam to the rocker arm to open the valves of a reciprocating engine.

push to-test light – A light fixture for an indicator light which may be pressed to complete a circuit which will determine that the bulb is in operating condition.

push-button electrical switch – An electrical switch that is actuated on or off by a push button. Each time the button is pressed, it opens or closes the circuit.

pusher propeller – A propeller that fits onto an engine whose shaft points to the rear of the aircraft. The thrust pushes the aircraft through the air rather than pulling it.

push-pull amplifier – A type of electronic amplifier that has two output circuits whose output voltages are equal but 180° out of phase with each other. A push-pull amplifier is also called a balanced amplifier.

push-pull rod – A rigid control rod used to move a component by either pushing or pulling it.

pylon – The structure which holds a turbine engine pod to the wing or fuselage of the aircraft.

pyrometer – Temperature measuring instrument used to indicate temperatures that are higher than can be measured with a mercury thermometer.

Q

Q factor – The "figure of merit" or "quality" of an inductance coil. It is the ratio of the inductive reactance to the resistance of a coil.

Q factor of a coil – *See Q factor.*

QEC unit – *See quick engine change kit.*

Q-springs – A system used for transmitting artificial feel of the control surface movement of large, high-speed aircraft to pilot.

quad-clamp – Quick-Attach-Detach clamp used to attach accessories to their gearbox mounting pads.

quadrant – A mounting for aircraft engine operating controls in which the control handles are pivoted levers which move in an arc rather than push-pull rods in the instrument panel.

quality control – Abbrev.: QC. The management and inspection function which controls the quality, standards, and performance in the manufacturing of aircraft, aircraft engines, or components.

quantity – An amount or portion. The exact amount of a particular thing.

quantum – A quantity or packet of energy.

quantum theory – The theory that energy is absorbed or radiated in a discontinuously in step-by-step units called quanta.

quart – A liquid measurement of volume equal to ¼ gal; 2 pt. or 57.75 cu. in.

quarter-sawed wood – Lumber that is cut at 45° across the annual rings.

quarter-turn cowl fastener – Any of a series of quick-release cowling fasteners which require only ¼ turn to either fasten or release them.

quarter-wave antenna – The length of a radio antenna that is one quarter of the wavelength of the frequency for which the antenna is used.

quench hardening – Hardening a ferrous alloy by austenitizing and then cooling rapidly enough so that some or all of the austenite transforms to martensite.

quenching – Rapid cooling of a metal as part of the heat-treating process. The metal is removed from the furnace and submerged in a liquid such as water, oil, or brine.

quick engine change assembly – Abbrev.: QECA. An assembly made up of an engine, propeller, all accessories, along with all of the necessary cowling and the engine mount. Used to minimize the downtime when an engine change is required.

quick engine change kit – Abbrev.: QEC kit. Used to configure a basic engine for a particular airframe installation. A QEC kit consists of the engine with all of accessories and propeller attached for minimizing the time it takes to replace an engine.

quick-disconnect couplings – A type of fluid coupling designed for easy and quick connecting and disconnecting. A check valve is incorporated in both halves of the coupling to prevent the loss of fluid when they are disconnected.

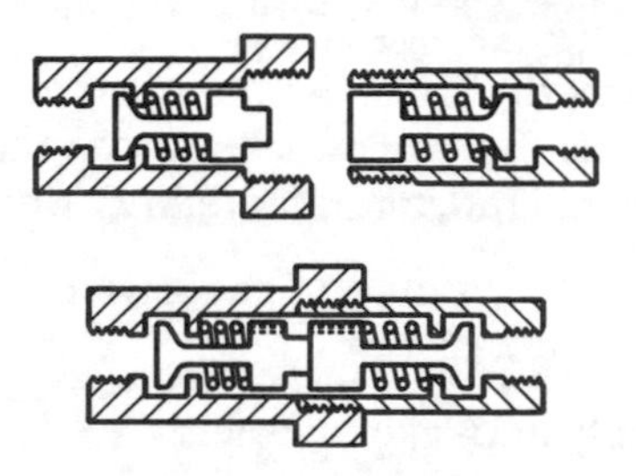

quick-turnaround installation – A liquid oxygen installation in which the converter is installed with quick disconnect fittings and mounts so it may be removed from the airplane for rapid servicing or replacement.

quill shaft – A hardened steel shaft with round cross section and splines on each end. Torsional flexing of the shaft is used to absorb torsional vibrations.

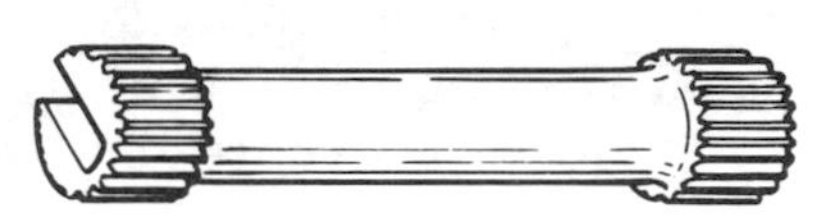

quotient – The number resulting from the division of one number by another.

R

rabbet – A groove that is cut into the edge of a piece of wood in such a way so as to fit another board to it to form a joint.

rabbet plane – A type of wood-working tool with a blade that extends to the outside edge of the body of the plane.

race – A hardened and polished steel surface on which a bearing is supported.

rack – A straight piece of metal that has teeth cut into one side.

rack and pinion – A set of gears arranged in such a way that a rotary motion of the pinion gear is changed into linear motion of the rack.

rack and pinion actuator – A form of rotary actuator in which the fluid acts on a piston on which a rack of gear teeth is cut. As the piston moves, it rotates a mating pinion gear.

radar – A term meaning radio detection and ranging, it is a form of pulsed electronic system in which a pulse of electromagnetic energy is transmitted from the aircraft. It travels to the target, bounces off, and returns to the aircraft. The display, representing the direction and distance of the target, is made on a cathode ray oscilloscope mounted in the aircraft instrument panel.

radar altimeter – A form of absolute altimeter that determines the distance of the airplane from the terrain below it by measuring the time required for a pulse of electrical energy to travel from the airplane to the ground and back.

radar beacon transponder – A piece of electronic equipment which receives an interrogation from ground radar and responds with a coded transmission that appears on the radar scope of the traffic controller.

radar mile – The time required for a radar pulse to travel a distance to any reflecting object within its range.

radial – Branching out in all directions from a common center. A line of radio bearing radiating out from a VOR station. Each VOR station has 360 radials. A radial is associated with the VOR. All directions associated with a VOR station are related to magnetic north. The radials are assigned numbers which pertain to their situation around the magnetic compass card. They are considered as being drawn away from the VOR station to a particular magnetic direction.

radial engine — A reciprocating aircraft engine in which all of the cylinders are arranged radially, or spokelike, around a small crankcase.

radial inflow turbine — A turbine wheel which receives its gases at the blade tips and guides the air inward and outward to the exhaust duct. It has a look similar to a radial outflow compressor. It is used extensively in APUs.

radial lead — An electrical component lead that protrudes from the outside of the component.

radial outflow compressor — *See centrifugal flow compressor.*

radial ply tire — A tire in which layers of metal or synthetic casing plies run straight across the tire. Completely encompassing the tire body, the plies are folded around the wire beads and back against the tire sidewalls.

radian — A unit of angular measurement equal to the angle between two radii, separated by an arc equal to the length of the radius. Radians are used in the measurement of angular velocity.

radiant energy — Any form of energy that is radiating from electromagnetic waves.

radiant heat — Any form of heating an area by placing the heating element installed inside a floor or wall.

radiation sensing detector — A type of fire detection system which utilizes heat-sensitive units to complete an electrical circuit when the temperature rises to a preset value.

radiation shield — A layer of aluminum foil used in an insulation blanket around turbine engines to prevent heat radiating from the engine into the structure.

radical sign — The mathematical sign "$\sqrt{\ }$" placed before a quantity to show its root is to be extracted.

radio altimeter — A radio altimeter is called a "radio" rather than "radar" because its transmissions are not pulsed as radar transmissions are. The transmission is a continuous wave, constant amplitude, frequency modulated pulse that measure the height of and aircraft above the terrain. The difference in the frequency of the reflected signals at any time is read on the RA indicator in feet above the ground.

radio control — A means of operating a unit from a point away from the unit by radio transmissions.

radio direction finding — A method of determining the direction from which signals are received. When an aircraft is within reception range of a radio station the radio receiver provides a means of fixing the position and the bearings are plotted on a chart to determine the position of the airplane with reasonable accuracy.

radio frequency — Frequencies of AC which produce electromagnetic waves which radiate from the conductor. It is above the audible frequency range and below the frequency range of heat and light.

radio-frequency cable — *See coaxial cable.*

radiography — The system of nondestructive inspection using X-rays or gamma rays to determine the condition of a part not visible without disassembly.

radium poisoning — Poisoning associated with handling the luminous materials used for marking the dials on some of the older aircraft instruments.

radius — One-half the diameter of a circle.

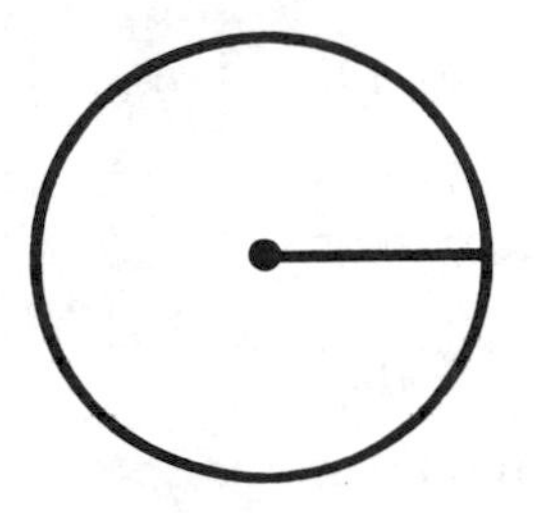

radius bar — The part of the cornice brake top leaf that has an accurately ground radius at its edge for bending sheet metal to obtain a specific bend radius.

radius block — A metal block around which sheet metal is bent to obtain a specific bend radius.

radius dimpling — A form of cold dimpling of thin sheet metal in which the cone-shaped male die is forced into the recess of the female die with either a hammer blow or a pneumatic rivet gun. After the dimpling process has been completed, it can be riveted with flush rivets.

radius gage — A precision gage having accurately cut inside and outside radii, used to measure the radius of a bend.

radome — The protective covering of a radar antenna mounted on an aircraft. A radome is constructed of lightweight honeycomb core material that is electrically transparent, so it will not interfere with the transmitted and received pulses of electrical energy.

rag wing — A common slang term used to refer to a fabric covered airplane.

rain — Water droplets formed by the condensation of moisture in the atmosphere.

rain gage — An instrument that is used to for measuring the amount of rain that has fallen in a given locale.

rake — A pressure sensor, usually small holes in a stationary engine component which act as a total pressure probe.

ram air pressure — Pressure slightly above ambient. Caused by the forward motion of the aircraft.

ram air temperature rise — The increase in temperature caused by the ram compression of the air as an aircraft passes through the air at a high rate of speed. The rate of temperature increase is proportional to the square of the speed of the aircraft.

ram drag — Described as gross thrust minus net thrust. Ram drag occurs by a decrease in momentum change rate of mass airflow through the engine as the aircraft speeds up. That is, V_1 increases rapidly and V_2 stays relatively constant.

ram pressure rise — Pressure rise in the inlet due to the forward speed of the aircraft.

ram ratio — The ratio of ram pressure to ambient pressure in a jet engine.

ram recovery — The increase in thrust as a result of ram air pressures and density on the front of the engine caused by air velocity.

ram recovery point — The point at which the suction condition in the inlet returns to the value of ambient pressure. This occurs from the effect of air ramming in the inlet. This point is said to occur near 160 MPH on most aircraft.

ram temperature rise — Inlet temperature rise due to inlet ram pressure rise.

ramp — The apron or paved surface around a hangar used for parking aircraft.

ramp weight – The total weight of the aircraft while on the ramp. It differs from takeoff weight by the weight of the fuel that will be consumed in taxiing to the point of takeoff.

Rankine Temperature – Absolute temperature scale using degrees Fahrenheit with minus 460°F as absolute zero. Used in many engine performance calculations.

rapid decompression – Almost instantaneous loss of cabin pressure in aircraft with pressurized cockpit or cabin.

rarefied air – "Thin air" such as that found at higher altitudes.

rasp – A type of rough file with raised points used for scraping soft materials such as wood or plastic.

ratchet – A mechanism which consists of a toothed wheel and a bar, or pawl, which allows the wheel to rotate in one direction but prevents its backward motion.

ratchet coupling – A type of toothed wheel with a pawl that keeps the wheel from ratcheting backwards.

ratchet handle – A special handle for turning socket-type wrenches. A ratcheting mechanism in the handle allows the socket to be turned in one direction while the handle is moved with a back-and-forth movement. Reversing the catch in the ratchet body, the same back-and-forth movement of the handle will rotate the socket in the opposite direction.

ratchet ring – A ring whose teeth slope in one direction, allowing movement in one direction only.

rate gyro – An instrument that is not affected by roll or pitch of the aircraft but aligns itself with the fore and aft line of the airplane. A rate gyro measures the unambiguous rate of rotation of an airplane about its vertical axis through 360°.

rate of burning – Time required for a fuel-air mixture to burn or to release its heat energy.

rate of climb – Rate at which vertical motion occurs.

rate of yaw – The rate in degrees per second at which the airplane rotates about its vertical axis.

rate pressure – The pressure in the outflow valve control which is established by reference pressure being restricted by the outflow valve control rate control valve.

rate signal – Any signal proportional to a rate of change.

rated horsepower — The maximum horsepower an engine is approved to produce under a given set of circumstances.

rated maximum continuous power — *See maximum except takeoff power.*

rated thrust — The manufacturer's guaranteed thrust as specified on the Type Certificate.

rate-of-climb indicator — A rate of pressure change indicator which indicates the rate at which the airplane is climbing or descending in feet per minute. It is also called a vertical speed indicator.

rate-of-temperature rise indicator — A thermocouple type of fire detection system whose operation is dependent on a rapid rate of temperature rise. It will not indicate an overheat condition that has developed slowly.

ratio — The relationship between one number and another expressed as a fraction. A proportion.

ratiometer indicator — A form of DC remote-indicating system whose pointer movement is determined by the ratio of current-flow between two resistors, or portions of a special variable resistor.

rattail file — A type of round, long tapered file used for filing holes.

rawhide mallet — A mallet made of rawhide, wound into a tight cylinder Used to form sheet metal without scratching it.

RC circuit — A circuit containing both resistance and capacitance.

RC time constant — The time required to charge a capacitor to 63.2% of its full-charge state through a given resistance.

reach — The length of the shell thread of a spark plug. For 18 mm spark plugs, long-reach plugs are threaded for 13⁄16" and short-reach plugs for ½".

reactance — The opposition to the flow of AC made by an induction coil or a capacitor.

reaction — A response to a stimulus or force that is caused by another action.

reaction engine — Jet engines which receive thrust only from reaction to expelling of hot gases.

reaction turbine – A stator vane and rotor blade arrangement, whereby the vanes form straight ducts and the blades form convergent ducts. The rotor is then turned by reaction to the squirting action of gases leaving the trailing edge of the blades. Design common to turbine type accessories such as air-starters.

reactive current – Current in an AC circuit which is not in phase with the voltage.

reactive metal – A metal, such as aluminum or magnesium, which reacts with oxygen to form corrosion.

reactive power – The product in an AC circuit of the total voltage times the total current.

reactor – A device in an AC circuit used to add reactance to it.

reader – A microfiche film or microfilm optical reader that magnifies the information on the film so it can be read, and if needed, to make a hard copy of the page(s).

real power – The power in an AC circuit which is the product of the voltage and the current in phase with the voltage. It is the voltage times the current times the power factor.

ream – Enlarging and smoothing a drilled hole with a precision cutting tool called a reamer. A hole is reamed in preparation for close-tolerance parts.

reamer – A specially designed sharp edged cutting tool used for enlarging or tapering drilled holes in preparation for close-tolerance parts.

rebreather bag – A bag connected to an oxygen mask so that oxygen delivered at fixed rate becomes mixed with a portion of the expired air, allowing rebreathing of a portion of each expired breath.

rebreather oxygen mask – *See rebreather bag.*

reciprocating engine – An engine which converts the heat energy from burning fuel into the reciprocating movement of the pistons. This movement is converted into rotary motion by the connecting rods and crankshaft.

reciprocating motion – A back and forth motion.

reciprocating saw – A power-driven, metal cutting saw with a blade which is driven back and forth across the material being cut.

recirculating fan – A fan in an aircraft cabin comfort system which circulates the air in the cabin without taking in any outside air.

reclaimed oil — Used lubricating oil that is restored to a useful state by a process to remove the impurities, and it is sold for reuse.

rectangle — A closed plane figure with four sides and four right angles. The opposite sides are parallel.

rectifier — A device which changes AC into DC. It is an electron check valve.

rectifier bridge — A form of rectifier using four diodes arranged in a bridge circuit.

rectify — To change. Usually this refers to changing AC into DC.

recurring — Happening again and again at regular or frequent intervals.

recurring Airworthiness Directive — An Airworthiness Directive that requires compliance at regular hourly or calendar time periods.

red line — A mark on an aircraft instrument which indicates the maximum allowable operating condition.

red rust — A form of nonmagnetic iron oxide.

red-line condition — The maximum condition of an operating unit that is allowed without exceeding its limit. Most aircraft instruments are marked with a red line on the instrument glass to let the operator know about the limit.

reduction factor — A constant by which a moment index is divided to simplify weight and balance computations.

reduction gear train — The gear arrangement of an operating unit in which the output shaft turns slower than the input shaft. An example of this is a gear arrangement in an aircraft engine which allows the engine to turn at a faster speed than the propeller

reduction gears — The gear arrangement in an aircraft engine which allows the engine to turn at a faster speed than the propeller.

Reed and Prince screw — A form of recessed-head, cross-point screw which is driven by a special cross-point screwdriver whose tip has a single taper.

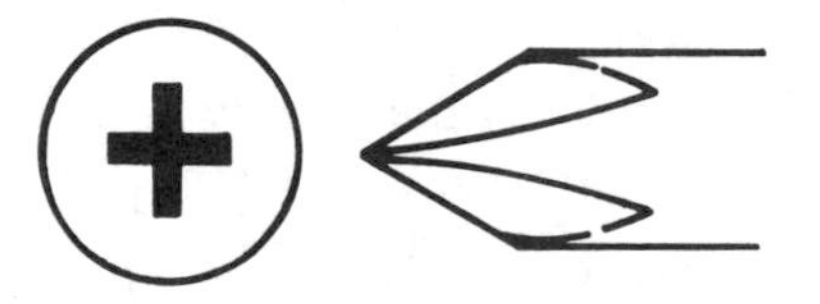

reed valve – Thin, leaf-type valves located in the valve plate of a reciprocating-type air conditioning compressor to control the inlet and outlet of the refrigerant.

reface – To resurface an object to remove the pits or wear marks.

reference datum – An imaginary vertical plane at or near the nose of the aircraft from which all horizontal distances are measured.

reference designator – Generally consisting of a combination of letters and numbers, the reference designator is used to uniquely identify a component or assembly. Reference designators are stackable, that is, a full reference designator identifies the major assembly, subassemblies, and finally, the component.

reference dimensions – Dimensions on an installation drawing necessary to show the relationship between two parts.

reference junction – One of the two junctions in a thermocouple system; the one which is held at a constant or stable temperature to serve as a reference for the measuring junction.

reference pressure – The pressure in the outflow valve control which is established by cabin air pressure flowing the cabin air filter and orifice and is metered by the reference pressure metering valve.

refining – A crude petroleum oil process in which crude oil is broken down into all of its different parts.

reflector – A light reflected from a polished shiny surface to a shaded surface.

refrigerant – A fluid which is used in an air conditioning system to absorb heat from the cabin and carry it outside the airplane where it can be transferred to the outside air.

refrigerant 12 – Commonly referred to as R-12. Dichlorodifluoromethane, a chemical compound used in most aircraft air conditioning systems.

regeneration – A process in an electrical circuit in which part of the output is fed back to the input with such a phase that it causes amplification.

registration certificate – The document in the airplane which contains the name and address of the person to whom the airplane is registered.

Reid vapor pressure – A measure of the pressure required above a liquid to hold the vapors in the liquid at a given temperature.

reinforce — To strengthen by the use of an additional material.

reinforced shell — A type of aircraft construction in which the outer skin is reinforced by a complete framework of structural members.

reinforcing tape — A narrow woven cotton or polyester tape used over the fabric reinforcing it at stitching attachments

rejuvenation — The restoring of resilience to a dope film by opening up the film with potent solvents and allowing the plasticizers in the rejuvenator to replace those which have migrated from the dope.

rejuvenator — A finishing material consisting of potent solvents and plasticizers used to restore resilience to weathered and cracked dope film.

relative bearing — Between two objects is the direction measured in degrees clockwise from a reference point on the first object to the second object.

relative humidity — The ratio of the amount of water vapor in the air to the amount of water vapor required to saturate the air at the existing temperature.

relative motion — *See relative movement.*

relative movement — The movement of one object with relation to another.

relative wind — The velocity and direction of the airflow with reference to the body which it is flowing.

relaxation oscillator — A form of electronic oscillator which produces a sawtooth waveform by charging a capacitor then rapidly discharging it.

relay — An electrically operated remote switch whose contacts are closed or opened by an electromagnetic field.

relay rack (open or closed) — A standard support system for ground equipment permitting the mounting of units with standard 19" wide panels. The open rack consists of two upright supports. The closed rack is actually a freestanding cabinet.

relief — The amount one plane surface of a piece is set below or above another plane, usually for clearance or for economy in machining.

relief hole — A hole drilled in a flat sheet metal part to allow intersecting bends to be made. The relief hole prevents the metal buckling.

relief map — A map that shows the mountains and valleys, shading, or color of land forms the way the land actually looks.

relief tube — An installed urinal that drains overboard. The discharge area around these tubes is an area highly susceptible to corrosion.

relief valve — A valve which limits the pressure in a system by releasing it at a preset value.

reluctance —
[1]Relative difficulty with which magnetic domains may be aligned.
[2]Opposition to magnetic flux. The opposite of permeability.

remanufactured engine — An engine assembled by the engine manufacturer or his authorized agent, built up of used parts which are held to the new parts' dimensional limits. The engine is given zero time records and usually the same warranty and guarantee as a new engine.

remote control — The control of aircraft and missiles from a remote location.

Rene metal — A nickel chromium superalloy produced for General Electric Co. and used in the manufacture of gas turbine engines.

repair — To restore to a condition of practical operation or to the original condition.

repairman license — A licensed issued by the FAA to persons employed in a specific job requiring special qualifications of a repair station, or by a certified commercial operator or a certified air carrier that is required to provide a continuous airworthiness maintenance program. The repairman license is valid only as long as the person holding the license is employed by the company for which the certificate was issued.

repeater indicator — An instrument that repeats the information produced by a master indicator.

repeating decimal — A fraction that cannot be expressed as a definite number is called a repeating decimal. The fraction ⅓ is a repeating decimal (0.333333333).

repulsion — The force that tends to cause objects to move away from each other.

required equipment — Items of equipment on an airplane that are determined by the FAA to be necessary for the aircraft to be considered to be airworthy.

reserve — To keep back, or set apart, for use in an emergency or for some special purpose.

reservoir – A tank in which fluid is stored for a system.

residual – The remainder, or anything that is left over.

residual charge – The remaining electrical charge that is left on capacitor plates after it has been discharged.

residual fuel or oil – The fuel or oil that is trapped in the lines and is not usable. In weight and balance computations, residual fuel and oil are considered to be part of the airplane's empty weight.

residual magnetic flux – *See residual magnetism.*

residual magnetism – Magnetism that remains in the core of an electromagnet after the magnetizing current no longer flows in the coil.

resin – A family of natural or synthetic fluids or semi-solid materials which may, by the addition of appropriate catalyzers, be changed into a solid.

resistance – Symbol: R. The opposition to the flow of electrons offered by a device or material. Opposition by resistance causes a loss of power.

resistance welding – *See electrical resistance welding.*

resistive current – Current in a circuit which is in phase with voltage.

resistivity – The ability of a material to resist the flow of electrons. It is the opposite of conductivity.

resistor – An electrical circuit element used to provide a voltage drop by dissipating some of the electrical energy in the form of heat.

resistor color code – A color code marking system used to identify the resistance value of carbon resistors. The resistor is marked by either three or four color bands. The first color band (nearest the end of the resistor) indicates the first digit. The second color band indicates the second digit of ohmic value. The third color band indicates the number of zeros to be added to the two digits derived from the first and second bands. If there is a fourth color band, it is used as a multiplier for percentage of tolerance.

resistor power dissipation rating – The amount of power that a resistor may safely dissipate in the form of heat under controlled conditions.

resistor spark plug – A composition resistor installed in the barrel of most shielded spark plugs. The resistor limits the current which is stored in the capacitive effect of the shielding and minimizes electrode erosion.

resolution – The ability to set and then reset a variable resistor to a specific resistance.

resonance – A frequency in an AC circuit in which the capacitive reactance is equal to the inductive reactance.

resonance method of ultrasonic inspection – A method of detecting material thickness or indications of internal damage by injecting variable frequency ultrasonic energy into a material. A specific frequency of energy will produce the greatest return in a given thickness of material. When the equipment is calibrated for a specific thickness and this thickness changes, an aural or visual indication is given.

resonant frequency – The frequency of a source of vibration that is exactly the same as the natural vibration frequency of the structure.

resonate – To vibrate at a certain frequency. A mechanical system is said to resonate when its natural vibration frequency of the force applied. The amplitude of an object in its vibration will increase immensely as that frequency is reached and will be less on either side of that frequency.

respirator – A device worn over the mouth and nose, to prevent the harmful inhaling of dangerous substances.

restart – The act of starting an engine after it has once been operating and then shut down.

restrictor – An orifice for reducing or restricting the flow of a fluid.

resultant flux – The flux in a magnetic circuit of an aircraft magneto that is the resultant of the flux of the rotating permanent magnet and the flux which surrounds the primary windings when primary current is flowing.

resultant lift – The vector sum of the magnitude and direction of all of the lift forces produced by an airfoil.

retard – To slow or delay the progress of something.

retard breaker points – *See retard points.*

retard points – An auxiliary set of breaker points in a magneto equipped with the Shower of Sparks starting system. These points are operative only during the starting cycle and open later than the run, or normal, points. This provides a late, or retarded, spark.

retarder – A slow-drying solvent used to prevent blushing or to provide a more glossy finish by allowing the finish a longer flow-out time.

retentivity – The ability of a material to retain its magnetic properties.

retirement schedule — A list of parts and times of a limited life. This list will contain the part, serial number, the time installed, and the removal time.

retort — A heated container with a long tube used to distill substances.

retract — To pull into the structure of an airplane any device that causes parasite drag.

retractable landing gear — A landing gear that folds into the aircraft structure to reduce parasite drag.

retraction test — That portion of an aircraft inspection in which that airplane is put on jacks and the landing gear cycled through its retraction and extension sequence.

retread — Tire recapping. Tires which meet injury limitations can be recapped. Retreading or recapping means reconditioning of a tire by renewing the tread, or renewing the tread plus one or both sidewalls.

retreating blade — Any blade that is located in a semicircular part of the rotor disc where the blade direction is opposite to the direction of flight.

retrofit — Term used to indicate a major modification, or, in other words, to go back and modify existing equipment.

return to service — The completion of all applicable maintenance records and forms after maintenance has been performed on an aircraft that will allow the aircraft to be flown legally.

revalidate — To reconfirm something.

reverse bias — The polarity relationship between a power supply and a semiconductor that does not allow conduction.

reverse idle — A power lever position where the thrust reversers are deployed, but engine power is at idle.

reverse pitch — An angle to which the propeller blade may be turned to provide reverse thrust from the propeller.

reverse riveting — A process of driving aircraft rivets in which the manufactured head is bucked by holding it in a rivet set supported in a bucking bar and upsetting the shank with a flush rivet set.

reverse-current relay — A relay incorporated into a generator circuit to disconnect the generator from the battery when battery voltage is greater than generator voltage.

reverse-flow annular combustor – A combustor design which forms an S-shaped path in which the gases flow from the diffuser to exhaust. This design shortens the entire engine length because the liner is coaxial to the turbines rather than in front of them as in a conventional annular combustor.

reversible-pitch propeller – A propeller system that has a pitch change mechanism that includes full reversing capability. When the pilot moves the throttle controls to reverse, the blade angle changes to a pitch angle that produces a reverse thrust which slows the airplane down during a landing.

reversing mechanism – A linkage that reverses the direction of movement between two parts.

revision – A change in dimensions, design, or materials in drawings.

revision block – That portion of an aircraft drawing which contains a record of all of the revisions made, and keys these changes with symbols to the location of the changes on the drawing.

revolutions per minute – Abbrev.: RPM. The number of complete revolutions of a body in one minute.

revolved section – A detail on an aircraft drawing in which the external view of a part shows the shape of the cross section of the part as though it were cut out and revolved.

Reynold's Number – A measure of the dynamic scale of a flow. Its usual form is the fraction $\frac{\rho Vl}{\mu}$ in which ρ is the density of the fluid, V is the velocity of the fluid, 1 is a linear dimension of the body of the fluid, and μ is the coefficient of viscosity of the fluid.

rheostat – A variable resistor having only two terminals. It is normally used in a circuit to drop voltage by dissipating some of the energy as heat.

rhomboid – A parallelogram whose angles are oblique and only the opposite sides are equal in length.

rib – The structural member of an airfoil which gives it the desired aerodynamic shape.

rib lacing – *See rib stitching.*

rib stitching – Attachment of fabric to an aircraft structure with rib-stitching cord. A series of loops around the structure and through the fabric are each secured with seine knots.

rib stitching cord – A strong cotton, linen, or polyester fiber cord used to stitch, or lace, the fabric, to the aircraft structure.

rib cap — *See cap strip.*

rich flameout — A condition of turbine engine operation in which the fire goes out in the engine because the fuel-air mixture is too rich to support combustion.

rich solvent — Slow-drying solvent.

rigging — The final adjustment and alignment of an aircraft, and aircraft flight control systems to provide the proper aerodynamic reaction.

rigging fixture — Special manufactured template designed to measure control surface travel.

rigging pins — Pins which may be inserted into control system components to hold the controls in their neutral position for rigging the control cables and rods.

right angle — An angle of 90° whose sides are perpendicular to each other.

right-hand thread — A thread is a right-hand thread if, when viewed axially, it winds in a clockwise and receding direction. All threads are right-hand threads unless otherwise designated.

rigid airship — A type of dirigible that has a rigid framework covered with cloth fabric, and with a cabin suspended underneath the frame that houses the crew and engine pylons for driving it through the air

rigid rotor — A helicopter rotor capable of only changing pitch.

rigid tubing — Fluid lines made of thin wall aluminum alloy, copper, or stainless steel which are used in an aircraft when there is no relative movement between the ends of the tube.

rigidity — The opposition of a structure to bending.

rime ice — A milky-appearing ice having a rough surface. It is formed by the instant freezing of moisture on the surface as an airplane flies through certain cloud formations.

ring and tube assembly — The ring of outer combustion chambers on a multiple can-type combustor engine.

ring cowl — A type of streamlined covering over the cylinders of a radial engine.

ring gear – One of the gears in a turboprop negative torque signal and prop reduction gear system. When a predetermined negative torque is applied to the reduction gearbox, the stationary ring gear moves forward against the spring force due to a torque reaction generated by helical splines. In moving forward, the ring gear pushes two operating rods through the reduction gear nose.

ring grooves – *See piston grooves.*

ringworms – A circular pattern of cracks in a brittle dope finish that results when a blunt object presses against the fabric.

rip panel – The air dump panel of a free floating balloon that can be opened by pulling a rip-cord to dump the air in the from the balloon in the event of an emergency and for landings.

ripple – A small periodic variation in the voltage level of a DC power supply.

ripsaw – A saw with course teeth used for cutting wood along the direction of its grain.

rivet – A small metal pin with a specially formed head on one end. It is used to fasten sheet metal parts together by upsetting the shank to form a clamping head.

rivet cutter – A tool used to cut rivets to the required length.

rivet gage – The transverse pitch or distance between rows of rivets.

rivet gun – A hand-held pneumatic riveting hammer used to vibrate aircraft rivets against a heavy bucking bar to form the upset head.

rivet pitch – The distance between the center of the rivet holes in adjacent rows.

rivet set – The tool which fits into a rivet gun used to hammer against the manufactured head of a rivet so the bucking bar may form the upset head on the opposite side of the skin.

rivet snap – Another name for rivet set.

rivet spacing – *See rivet pitch.*

rivet squeezer – A heavy, tong-like clamping machine used to squeeze the ends of a rivet to form an upset head.

riveting burr – A riveting burr is a small plain washer which is assembled with a small rivet before installation to provide a large area of contact on the part.

Rivnut — A patented, hollow, blind rivet manufactured by the B.F. Goodrich Company, in which the inside of the shank is threaded. The upset rivet may be used as a blind nut.

robot — A automated android mechanism built to do routine manual labor.

robotics — The research and development of the technology that deals with the design, application, construction, and maintenance of androids.

rocker arm — A pivoted arm on the cylinder head of an aircraft reciprocating engine. It is actuated by the push rod to push open the intake or exhaust valve.

rocker arm boss — That portion of the cylinder head of an aircraft engine which provides support for the rocker arm shaft.

rocket — A flight vehicle propelled by ejected expanding gases generated in the engine from self-contained propellants and not dependent on the intake of outside substances.

rocket assisted takeoff — Abbrev.: RATO. An auxiliary means of assisting an aircraft when taking off. The system consists of small rockets fastened to the aft structure. When a heavily loaded aircraft rotates for takeoff the jets are fired. The added boost provides extra thrust needed for takeoff. Once airborne the rockets are jettisoned.

rocket fuel — Any fuel specifically developed for rocket engines.

rocket ship — Any space craft that uses rocket engines for propulsion.

rocketry — The study, theory, and development of rockets.

rocking shaft — A shaft or rod in an instrument that changes the direction of a movement, usually by ninety degrees.

Rockwell hardness tester — A machine used to determine the hardness of a material by using a calibrated weight to press either a diamond pyramid or a hardened steel ball into the material. A dial indicator measures the depth of penetration and indicates it by a Rockwell number.

roll — The motion of the aircraft about the longitudinal axis. It is controlled by the ailerons.

roll pin — A pressed-fit pin made of a roll of spring steel. The spring force tending to unroll the pin holds it tight in the hole.

roll threading – Applying a thread to a bolt or screw by rolling the piece between two grooved die plates, one of which is in motion, or between rotating grooved circular rolls.

roller bearings – A form of antifriction bearings using hardened steel rollers between two hardened steel races.

root – Supporting base or structure as in the wing root that is connected to the fuselage.

root mean square – Abbrev.: rms. The value of a sine wave AC which is 0.707 times the peak value of one alternation.

rosette weld – A weld made through a small hole in a piece of steel tubing to weld an inner tube to the outer tube to prevent relative movement.

rosin – A light yellow-colored resin. Resin is the remaining material after oil of turpentine has been distilled from crude turpentine.

rosin core solder – A soft solder made chiefly of tin and lead alloys. used primarily for soldering copper, brass, and coated iron in combination with mechanical seams; that is, seams that are riveted, bolted, or folded. Rosin core solder is a hollow wire filled with rosin. During the soldering process, the solder is melted along with the rosin.

rosin joint – The name used to describe a soldered electrical connection in which the rosin, not the solder, is holding the connection. Rosin joints are unacceptable connections.

rotary breather – A rotating set of vanes or a centrifuge device through which oil laden air from the vent subsystem passes. Deaeration takes place and air only exits back to the atmosphere.

rotary pick-off – A device that is rotated by some object whose movement is to be measured. The pick-off generates a signal proportional to the amount of movement.

rotary radial engine – A form of aircraft engine popular in World War I, in which the propeller was attached to the crankcase and the pistons were attached to an offset cam mounted on the airframe. When the engine ran, the cylinders, crankcase, and propeller all spun around.

rotary solenoid – An electromagnet whose movable core is rotated by current through the coil.

rotating wing – A term often used to describe the rotors of a helicopter.

rotation – The act of a body turning about an axis.

rotor —
[1]The rotating element in an alternator. It is excited by DC, and the interlacing fingers on the two faces of the rotor form the alternating north and south poles.
[2]The rotating blades of a helicopter.
[3]The portion of a turbine compressor that spins.
[4]Either compressor or turbine. A rotating disk or drum to which a series of blades are attached.

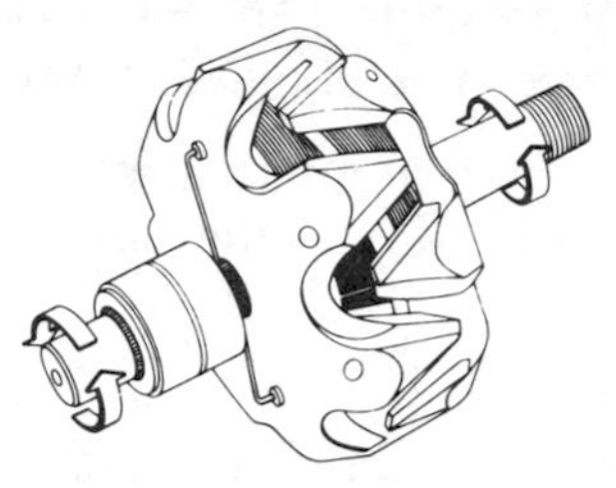

rotor brake — A device used to stop the rotor blade on shutdown. This may be either a hydraulic or mechanical mechanism.

rotor disc — That area within the tip path plane of the rotor of a helicopter. It is often called the disc area.

rotorcraft — A heavier-than-air aircraft that depends principally for its support in flight on the lift generated by one or more rotors.

round files — Hand files that are circular in cross section and tapered in length. Often called rat-tail files.

round head — A fastener with a semi-elliptical top surface and flat bearing surface.

round off —
[1]To make something become round.
[2]In mathematics, to change a fraction to the closest whole number.

round-nose pliers — Pliers used to crimp sheet metal to shrink it.

router — A metal-working machine using a high-speed, rotary-edged cutting tool. Often used to remove damaged honeycomb core from a bonded aircraft structure.

routine service items — Inspection items listed in a progressive inspection that require only a visual inspection to determine their condition.

roving — A slightly twisted roll of glass fibers used to reinforce resins in making molded parts.

rubber cement — An adhesive made from rubber that has not gone through the vulcanization process but has been dissolved in solvent.

rudder — The movable vertical control surface used to rotate the airplane about its vertical axis. The pilot operates the rudder by the movement of the foot pedals in the cockpit.

rudder pedals — Foot-operated controls in an airplane that move the rudder.

ruddervator — A pair of control surfaces on the tail of an aircraft arranged in the form of a "V". These surfaces, when moved together by the control wheel, serve as elevators, and when moved differentially by the rudder pedals, they serve as a rudder.

rumble — A combustor noise caused by choking and unchoking of the turbine nozzle from improper fuel scheduling.

run —

[1]Defect in a paint surface caused by too much finish being applied. The material gathers and attempts to flow off.

[2]To set something in motion, such as operating an engine. To run the engine.

run-in — The period of time an aircraft engine is operated to seat the moving parts after an overhaul.

running fit — A loose fit between moving parts which allows them to move freely.

runout — A term frequently used interchangeable with eccentricity but which normally refers to the amount which the outside surface of one component runs out with respect to the outside surface of another component. As such, it includes eccentricity, angularity and bow. The amount of runout is usually expressed in terms of Total Indicator Reading (TIR).

runout check — A dial indicator check for measuring the plane of rotation or a rotor shaft or disc.

runway — A strip of level ground, usually paved, used by airplanes for takeoffs and landings.

runway temperature — The air temperature immediately above a runway at approximately wing level.

runway visibility — The observable visual range as seen by a pilot along a runway.

rust — Oxidation of iron. A red, crusty product which forms on iron or steel when it unites with oxygen.

S

saber saw – A hand-held, electrically operated jigsaw that uses a short, stiff blade. It operates by a reciprocating motion.

sacrificial corrosion – A method of corrosion protection in which a surface is plated with a metal less noble than itself. Any corrosion will attack the plating rather than the base metal. *See also cathodic protection.*

saddled-mount oil tank – Sometimes called quarter saddle tank. A sheet metal externally mounted tank with a contour which allows the tank to mount around the curvature of a main case, e.g. compressor case.

safety belt – A belt designed specifically to fasten across the lap of an occupant of an aircraft to hold the occupants in their seat.

safety factor – A structural design feature. Safety factor is the ratio of the maximum load that a structural member is designed to support to the maximum probable load to which it will be subjected.

safety gap – A specifically designed space in a high-tension magneto that allows a spark to jump without damaging the magneto internal parts in the event that the spark plug lead is disconnected from the spark plug.

safety glass – A type of glass that is tempered in such a way that it breaks into small beads when it is shattered. Safety glass does not leave potentially dangerous jagged fragments when it is broken.

safety valve – A form of pressure relief valve that will open to relieve pressure in container under pressure in the event it rises above a predetermined value.

safety wire – Soft wire made of galvanized low-carbon steel, annealed stainless steel, or brass, that is used to prevent nuts and bolts from vibrating loose.

safety wiring – A method of fastening bolt or screw-heads together with soft wire to prevent their loosening.

safetying – Installation of any type of device that prevents the loosening of an attachment.

sailplane – A high performance glider.

sal ammoniac – Ammonium chloride.

salt – The result of the combination of an alkali with an acid. Salts are generally porous and powdery in appearance and are the visible evidence of corrosion in a metal.

sand casting – A form of casting in which the mold is made by shaping a special casting sand around a wood pattern. The pattern is removed, and molten metal is poured unto the mold. When the metal is hard, the mold is broken away.

sandbag – A bag made of a heavy canvas, or leather, filled with coarse sand and used as a die to form sheet metal by the bumping process.

sanding coat – A coat of surfacer or heavy bodied material which is applied and sanded off to fill small surface imperfections and thus provide a smooth surface for subsequent coats.

sandpaper – Abrasive paper made by bonding grains of sand to the surface.

sandwich construction – A form of bonded structure in which a core of material such as metallic or plastic honeycomb or end grain balsa wood is bonded between two face sheets of metal or fiberglass cloth. Sandwich constructed materials are used where high strength and light weight are required.

satellite – A man-made object that rotates around the earth. A small object that rotates around a larger object.

saturate – To cause something to be thoroughly soaked, or penetrated, to the extent that it cannot take any more.

saturated vapor – The condition of the vapor above a liquid in which no further vaporization can take place without an increase in its temperature.

sawtooth wave – The waveform produced by the relaxation oscillator in which the voltage rises slowly and drops off rapidly.

Saybolt Seconds Universal – Abbrev.: SSU. A device which measures the viscosity of lubricating oils by giving the seconds it takes for 60 cu. cm of oil to flow through its calibrating orifice. Aviation 80 engine oil has an SSU viscosity of 79.2, and Aviation 100 oil has an SSU viscosity of 103.0.

S-band radar – Radar frequency range between 1,550 to 5,200 MHz.

scale –

[1]A graduated measure.

[2]An oxide of iron sometimes formed on the surfaces of hot headed or forged fasteners.

scale effect – The change in any force coefficient, such as the drag coefficient, due to a change in the value of Reynold's number.

scale model – A smaller version of the original model that is made in the same proportions.

scalene triangle – A three-sided form in which all sides and angles are unequal.

scarf joint – A type of joint used for the construction or repair of a wooden aircraft or component. The two parts to be joined are cut with a taper of 1 inch in 10.

scarf patch – A flush repair to plywood skins where the slope must be shallower than 1 in 12, or about 12 times the thickness of the plywood.

scavenge – To remove an undesirable material from an area.

scavenger pump – A constant displacement pump in an engine that picks up oil after it has passed through the engine and returns it to the oil reservoir.

scheduled maintenance – Inspection and replacement of components that are planned in advance on a day, month, or operating hour basis.

schematic diagram – *See schematic drawing.*

schematic drawing – A form of graphical presentation or diagram used to explain the operation of a system without showing its mechanical details or physical layout.

Scintilla magneto – A Swiss-designed and built magneto. It is the forerunner of the current Bendix magneto.

scoop – Air inlet which projects beyond the immediate surface of an aircraft structure.

scope – A contraction of oscilloscope.

score – A deep scratch mark or line that is made across a piece of material that makes it possible to break the material along the line.

scoring – Deep scratches on the surface of a material which are caused by foreign particles between moving parts.

scraper ring – The bottom ring on a piston whose function is to scrape the lubricating oil away from the cylinder wall thereby preventing oil from getting into the combustion chamber of the cylinder.

scratching surface — Narrow, sharp, shallow markings or lines resulting from the movement of a particle or sharp pointed object across a surface. The most common cause is carelessness in handling and any part may be so affected. Note: scratching is considered not to be caused by engine operation to distinguish from scoring.

screech liner — A perforated liner within an afterburner, designed to combat destructive vibrations which cause metal fatigue and noise emissions.

screeching — A shrill, high-pitch noise that comes from a gas turbine engine caused by the instability of combustion in the engine.

screeding tool — A tool used to smooth out or level plastic resins used in bonded structure manufacturing or repair.

screen — A frame covered with a netting or mesh material used to cover an opening to prevent the entry of foreign objects.

screen grid — The electrode in a tetrode vacuum tube used to minimize the interelectrode capacitance between the plate and the control grid.

screen-type filter — A fluid filter whose element consists of a screen wire.

screw dowel — A screw dowel is a dowel pin provided with a straight or tapered thread.

screw pitch gage — A gage with a series of V-notches cut along one edge. Used to check the number of threads, or pitch, of a screw or bolt.

screwdriver — A hand tool used for turning screws.

scriber — A hardened steel or carbide tipped sharp-pointed tool, used to scratch lines on metal for cutting.

scribing — The process of marking a line on metal with the use of a scriber.

scroll combustor — A type used widely on auxiliary power units in conjunction with a radial inflow turbine. This combustor fits around the turbine nozzle which has its vane openings at its inner perimeter. The vanes direct the gases inward onto the turbine.

scroll shear — A floor- or bench-mounted sheet metal cutting tool which is used to cut irregular lines in a piece of sheet metal without having to cut to the edge of the sheet.

scrubbing — Excessive abrasive wear to a piston within a cylinder caused by detonation.

scud – Small patches of low clouds that usually form below heavier overcast clouds.

scuffing surface – A dulling or moderate wear of a surface resulting from a slight amount of rubbing. Usual causes are improper clearance and insufficient lubrication. Parts affected are rollers, rings, and steel parts bolted together.

scupper – A recess around the filler neck of a fuel tank. It collects any fuel spilled during the fueling operation and drains it overboard rather than allowing it to enter the aircraft structure.

sea level – A reference height used to determine a standard condition for the atmosphere and for altitude measurements.

seal – A component or material used to prevent fluid leakage between two surfaces.

sealant – A material used to form a seal between two imperfectly fitting surfaces. Sealants differ from gaskets in that they are usually liquid or semi-solids.

sealed compartments – The compartments in an aircraft structure which are sealed off and used as fuel tanks.

seam welding – A method of electrical resistance welding which forms a continuous line of weld instead of individual spots. *See also spot welding.*

seaplane – An airplane which is designed to land and take off from the surface of water.

seasoned lumber – Lumber that has been dried or had its moisture content reduced to a specified amount.

seat – A place or space on which something sits.

seat belt – *See safety belt.*

seated – A condition in which moving parts have worn together until there is a minimum of leakage past them.

secondary – Something at a second level of importance.

secondary air – That portion of compressor output air used for cooling engine parts and combustion gases.

secondary airstream – The air which passes through the fan portion of a turbofan engine.

secondary cell – A storage cell or an electrical device in which electrical energy is converted into chemical energy and stored until needed, then converted back into electrical energy. Secondary cells do not produce electricity; they merely store it.

secondary coil – Secondary winding of a transformer.

secondary control surfaces – Control surfaces, such as trim tabs, servo tabs, and spring tabs, which reduce the force required to actuate the primary controls.

secondary current – The current that flows in the secondary winding of a transformer. Secondary current is induced in the secondary winding of a coil by the collapse of the primary coil circuit current.

secondary emission – The emission of electrons from a surface when struck by high-velocity electrons from the cathode.

secondary exhaust nozzle – On a turbofan, the cold exhaust fan nozzle. On an afterburner, the aft or outer exhaust nozzle. In this instance, it is made up of moveable flaps which change the geometry of the nozzle in different modes of engine operation.

secondary fuel – Refers to the duplex fuel nozzle and the fuel that flows at higher power settings from the secondary orifice. Also called the main fuel.

secondary voltage – The voltage in a circuit that is produced across the secondary winding of a transformer.

secondary winding – Output winding of a transformer.

second-class lever – A lever commonly used to help in overcoming big resistances with relatively small effort. The second class lever has the fulcrum at one end; the effort is applied at the other end. The resistance is somewhere between these points. A wheelbarrow is an example of the use of second class levers.

section line – The crosshatching used in a cutaway section of an aircraft drawing to identify the material.

sectional view – Details in an aircraft drawing which are obtained by cutting away part of an object to show the shape and construction at the cutting line.

sectioning – Marking a drawing by suitable crosshatching or other symbols to indicate the material in the cutaway view of a part.

sector – A part of a circle that is bounded by any two radii and the arc included between the two radii.

sector gear – A portion of a gear that appears to have been purposely cut from a whole gear wheel. It consists of the hub as a pivot point and a portion of the rim with the teeth. Sector gears generally drive a smaller pinion gears for small angular movements.

securing strap – A strap used to secure an oil or fuel tank to the airframe or engine.

sediment – Any matter that settles to the bottom of a liquid or container.

seesaw rotor – A term used for a semirigid rotor.

seesaw rotor system – *See semirigid rotor.*

segment – A division or a section. A part of a figure.

segmented rotor brake – A heavy-duty multiple disc brake used on large, high speed aircraft. Stators are keyed to the axle and contain high-temperature lining material. The rotors, keyed into the wheel, are made in segments to allow for cooling and for the large amounts of expansion encountered.

seize – Equipment failure in which moving parts fuse or bind because of friction, pressure, or excessive temperature.

SELCAL – A contraction of selective calling referring to an automatic signaling system used in aircraft to notify the pilot that his aircraft is receiving a call.

selective plating – A process of electroplating only a section of a metal part.

selectivity – The ability to select or choose from among several choices.

selector switch – A multi-pole switch that takes the place of several switches. It is used to connect a single conductor to one or more conductors.

selector valve – A hydraulic flow control valve which directs hydraulic pressure to one side of an actuator and connects the other side to the system return line.

selenium – A chemical element in the sulfur family that is used in photoelectric devices because it changes its conductivity with a change in temperature.

selenium rectifier – A rectifier using a thin coating of selenium on an iron disk to develop a unidirectional current-carrying characteristic. Electrons flow easily from the iron to the selenium but encounter high resistance in the opposite direction. A metal alloy is used to form the electrical connection with the selenium.

self-accelerating — The ability of a turbine engine to produce enough power to accelerate.

self-aligning bearing — A form of rod end bearing consisting of a ball fitted into a socket which maintains alignment between the operating control and the unit being controlled.

self-centering chuck — A drill motor chuck with jaws that connect together to move at the same time when loosing or tightening to hold the drill.

self-demagnetization — The process in which a magnet loses its magnetism if allowed to sit for long periods without a keeper bar.

self-excited generator — A generator whose field excitation is taken directly from the armature thereby increasing the generator output voltage.

self-extinguishing — The ability of a material to automatically stop burning as soon as the source of the flame is removed.

self-induction — The generation of a back voltage in a conductor by the action of expanding and contracting lines of flux cutting across the conductor in which the alternating current is flowing.

self-locking nut — A nut that is designed with a built-in locking device that grips the threads of a bolt when the nut is tightened in order to prevent the nut from loosening caused by vibration. Locking devices may be fiber or non-fiber depending on its intended purpose.

self-tapping screw — A form of wood or metal screw which cuts its own threads as it is screwed into sheet metal or wood.

Selsyn system — A DC-type of synchro remote indicator system.

selvage edge — The edge of a piece of fabric that is woven to prevent its raveling.

semi- — A prefix that means half of something, or a part of something

semiautomatic operation — A device that is partially automatic and partially manual in its operation.

semicircular — In the form of a half circle.

semiconductor — An insulating material treated with certain impurities which add free electrons to act as current carriers — carrying the flow of current in one direction and blocking it in the other.

semiconductor diodes — *See diodes.*

semimonocoque – A stressed-skin structure in which the skin is supported by a lightweight framework to provide extra rigidity. Most of the larger modern aircraft are of semi-monocoque construction.

semirigid rotor – A popular helicopter rotor system that has a rotor which makes use of a feathering axis for pitch change. In addition, the rotor is allowed to flap as a unit.

semispherical – In the shape of half a sphere. Dome-shaped.

sender – A measuring device located in a reservoir or tank in which a float mounted on an arm rides on the top of the liquid. The arm is free to float and is connected to a variable resistor. Any change in the fluid level sends a signal to an indicator on the pilot instrument panel showing the amount of fluid in the tank.

sense antenna – A type of non-directional radio antenna that picks up a signal with equal strength from all directions.

sensible heat – Heat added to a substance which causes a change in the temperature without changing its physical condition.

sensitive altimeter – A form of pneumatic altimeter in which a pointer makes a complete revolution for each thousand feet, and which has an adjustable barometric scale by which the instrument may be adjusted to the existing barometric pressure.

sensitive relays – A relay that operates electromagnetically with very small current. When the relay closes, it controls a larger current to operate other devices.

sensitivity – A measure of the ability of something to be sensitive or responsive to very small changes to external conditions.

sensitized paper – Chemically treated paper used for making prints of drawings by a photographic process.

sensor – A sensing unit used to actuate signal-producing devices in response to changes in physical conditions.

separation – Phenomenon in which the flow past a body placed in a moving stream of fluid separates from the surface of the body.

sequence valve – A mechanically actuated hydraulic valve that causes a sequential action of certain actuators. Wheelwell doors, for example, must open and contact a sequence valve before the landing gear can extend.

series circuit – A circuit in which there is only one path for electron flow from the source through the load back to the source.

series ohmmeter — An ohmmeter circuit in which the resistance is measured by determining the amount of current flow through the unknown resistor when a known voltage is placed in series with the meter and the unknown resistor.

series resonant circuit — A type of AC circuit that has a capacitor and an inductor connected in series.

series RLC circuit — An AC circuit in which a resistance, capacitance, and inductance are arranged so that all current must flow through all three elements.

series wound generator — A generator in which the field and armature are connected in series.

series-parallel circuit — A type of circuit that consists of groups of parallel components connected in series with other components.

series-wound motor — An electric motor that has the electromagnetic field coils connected in series with the armature. Series-wound motors are used as starter motors because of their high starting torque.

serrations — A formation resembling the toothed edge of a saw.

service bulletin — Information issued by the manufacturer of an aircraft. aircraft engine, or component that details maintenance procedures that will enhance the safety or improve the performance of the product.

service capacity — A measure of the amount of electrical energy that may be obtained from a chemical cell. Measured under controlled conditions and given in ampere-hours.

service ceiling — The height above standard sea level beyond which an airplane can no longer climb more than 100 ft./min.

service life — The expected length of time a unit, part, component, or piece of equipment is expected to operate satisfactorily.

service manual — A manual issued by the manufacturer of an aircraft, aircraft engine, or component and approved by the FAA. It describes the approved methods of servicing and repairing the component.

serviceable — Equipment or parts that are in a condition which allows them to be returned to operational status on an aircraft.

servicing diagram — Information furnished by the manufacturer of an aircraft showing the proper access to all of the items or components requiring servicing.

servo — A motor or other form of actuator which receives a small signal from the control device and exerts a large force to accomplish the desired work.

servo altimeter — An altimeter in which the aneroid mechanism moves a rotary pick-off whose signal is amplified to drive a servomotor which moves the drums and pointers.

servo fuel — An intermediate metered fuel in the Bendix RS fuel injection system used to control the opening of the flow control valve as a function of the airflow into the engine.

servo loop — An automatic control system that sends a signal to a servomoter to move a control device. The loop signal to a servomotor stops the servomotor when the control is moved the appropriate amount.

servo system — An automatic control system that senses changes in movement, such as lowering the flaps, and sends a feedback signal to the control motor to stop moving the flaps when the correct position is obtained.

servo tab — An adjustable tab attached to the trailing edge of a control surface. The tab moves opposite the direction of the control. Used to aid the pilot in moving the control. *See also balance tab.*

servomechanism — Automatic device controlling large amounts of power by the use of a small input. A feedback system allows it to produce only the required amount of control.

servomotor — A type of motor that receives a signal due to the action of the control system which causes a mechanical movement of a primary control. Servomotors have the ability to move in either direction when the current of the correct polarity is sent to turn the servomotor in the proper direction.

servo-type carburetor — A type of carburetor that uses the pressure drop across a servo a metering jet to control the amount of metered fuel, proportional to the amount of air, that is allowed to flow to the cylinders.

sesquiplane — A special form of biplane in which the area of one of the wings is less than one half of the area of the other.

set screw — Small headless screw used to secure a wheel, pulley, or knob onto a shaft.

setback — The distance between the mold line and the bend tangent line on a sheet metal layout. For 90° bends, setback is equal to the inside radius of the bend plus the thickness of the metal being bent.

settling with power — A helicopter operation whereby the main rotor blades are operating in its own downwash or vortex.

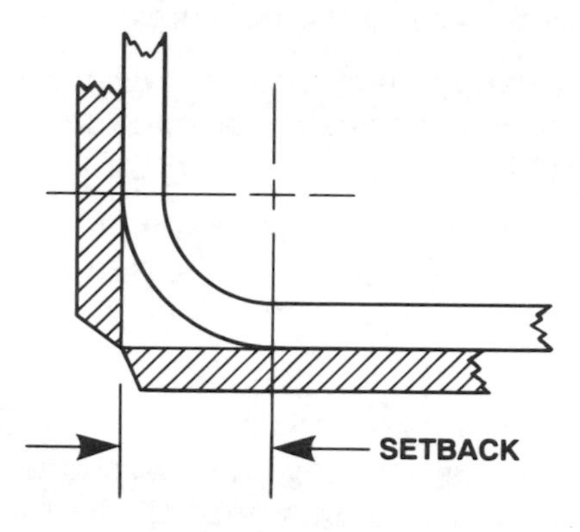

sewed seam — A seam in aircraft fabric made with a series of stitches joining two or more pieces of material.

sewed-in panel repair — A repair to fabric aircraft covering in which a panel extending from the leading edge to the trailing edge and from rib to rib is sewed in place. All of the seams are suitably reinforced with surface tape.

sewed-patch repair — A type of repair to aircraft fabric covering in which a patch is sewed into place and the seams are covered with surface tape.

sextant — An instrument used to measure the angular distance between the horizon and a navigational star to determine a position or location.

Seyboth fabric tester — A patented, hand-operated precision device for testing the relative strength of installed aircraft fabric by measuring the amount of force required to punch a hole in the fabric with a specially shaped punch.

shaded-pole motor — A low-torque AC induction motor whose rotating field is provided by the inductive action of shading poles on diametrically opposed pole pieces.

shaft horsepower — Abbrev.: SHP. Turboshaft engines are rated in shaft horsepower and calculated by use of a dynamometer device. Shaft horsepower is the exhaust thrust converted to a rotating shaft.

shaft runout – An inspection performed on a reciprocating engine crankshaft to determine the straightness of the crankshaft.

shank – A straight, narrow, essential part of a body such as the portion of a twist drill beyond the flutes, which holds the drill in the chuck.

shank of a drill – The part of the body of the twist drill that is round and smooth not including the tip and flutes.

shaving – A cutting operation in which thin layers of material are removed from the outer surfaces of the product.

shear – A stress exerted on a material which tends to slide it apart.

shear failure – The failure of a riveted or bolted joint caused by the rivets shearing rather than the sheet tearing.

shear nut – A thin nut used on clevis bolts to prevent the bolt from falling out. Shear nuts are only suitable for shear loads and must never be used for tensile loads.

shear pin – A specially designed pin used in the drive shaft of engine driven pumps to protect the accessory drive train if the pump should seize.

shear point – Refers to an intentionally weakened point on a shaft such as in dual element fuel pump. The shear point is designed to break away if one element becomes jammed, leaving the other element still functioning.

shear section – A narrow-necked down portion of a drive coupling designed to shear in case of pump seizure, preventing damage to either the pump or to the engine.

shear strength – The amount of force required to shear a pin, bolt, or rivet.

shear stress – *See shear.*

shears – A type of cutting tool, similar to scissors, used to cut sheet metal.

sheave – A wheel with a grooved center that is used as a pulley.

sheet metal – Metal of any thickness up to 1⁄8″. Metal of a greater thickness is called plate.

sheet metal drawing – A forming process in which sheet metal is pressed between dies to form the desired compound-curved shape.

sheet metal layout — The pattern of the sheet metal part before forming it or drilling holes in order to get a greater degree of accuracy in the finished part. Sheet metal layout consists of flat layout, duplication of patterns, or projection through a set of points.

shelf life — Period of time a material may be stored and remain suitable for use.

shell — The outer structure of an atom that is formed by the rotating electrons around the nucleus.

shell-type transformer — A transformer encased in steel to contain the magnetic lines of flux.

shielded cable — An electrical conductor encased inside a braided metal shielding. The shielding intercepts radiated electrical energy and conducts it to the ground rather than allowing it to cause radio interference.

shielded ignition cable — A type of electrical cable enclosed in a metal braid that is used to carry the high voltage from the distributor of the magneto to the spark plug. Its purpose is to prevent radio interference caused by electromagnetic radiation from the high voltage.

shielded spark plug — A spark plug completely encased in a steel shell. The radiated energy from the spark is conducted to the ground through the shielding, preventing radio interference.

shielded-arc welding — A type of gas welding in which a gas is used as a covering shield around the arc to prevent the atmosphere from contaminating the weld. The resulting weld is stronger, more ductile and more corrosion resistant.

shielding — Metal covers placed around electric and electronic devices to prevent the intrusion of external electrostatic and electromagnetic fields.

shim — A thin piece of metal placed between mating surfaces to control the clearance.

shim — Thin pieces of metal used to fill in a space between two objects in order to adjust a preload or the clearance between bearing parts.

shimmy — A rapid and violent oscillation of a nose wheel or tail wheel of an airplane caused by excessive wear in the support bearings.

shimmy damper — A hydraulic snubbing cylinder installed between the nose wheel fork and the landing gear structure. It is used to prevent or minimize shimmying of the nose wheel during takeoffs and landings.

shock absorber – A device built into the landing gear of an aircraft to absorb the energy of the landing impact.

shock loading – A form of stress loading of extremely short duration.

shock mounted – Any device that is attached to the airframe with shock mounts to prevent the transmission of vibration of one unit into another.

shock mounts – A shock absorbing attachment device used to mount an engine or instrument panel in an airframe in such a way that there will be a minimum of vibration transmitted.

shock stall – Turbulent airflow on an airfoil which occurs when the speed of sound is reached. The shock wave distorts the aerodynamic airflow causing a stall and loss of lift.

shock strut – *See oleo strut.*

shock wave – A compression wave formed when a body moves through the air at a speed greater than the speed of sound.

shop head – A term used for the upset head of an aircraft rivet.

shop head rivet – The head that is formed on the rivet when it is driven is called the shop head.

Shore scleroscope – A type of hardness tester used for testing the hardness of metal, plastic, and rubber.

short circuit – A path for electrons to flow from one level of potential to another, without completing a useful circuit.

short stack – A type of exhaust system for aircraft reciprocating engines in which a short exhaust pipe is attached to the exhaust port of the cylinder. Short stacks use no collector system.

short takeoff and landing – Aircraft that can takeoff and land in a distance of 1,500 ft. or less and clear a 50 ft. obstacle in the process.

short wave – Radio waves that are shorter than 60 meters.

shorting switch – A multi-pole switch in which one circuit is completed before another circuit is opened.

shot peening – A process used to strengthen metal parts by blasting its surface with steel shot.

shoulder bolts – A special form of bolt in which the threaded portion is smaller than the shank. It is used for the installation of plastic materials to prevent overtightening.

shoulder-wing airplane – An airplane having one main supporting wing surface mounted near the top, but not on top of the fuselage. Also called mid wing.

Shower of Sparks – The induction-vibrator-type starting system used in some Bendix magnetos. In this system, a vibrator directs pulsating direct current into the primary circuit of one of the magnetos when the points are closed. Two sets of points are used in the magneto with this system and when the retard points open, pulsating current flows to ground through the primary coil and induces a high voltage into secondary winding. As long as both sets of points remain open, a shower of sparks occurs at the spark plug.

show-type finish – A finish for aircraft having a glass-like appearance. It is achieved by many coats of dope, much sanding and rubbing.

shrink fit – An interference fit between parts in which the female part is heated and the male part is chilled, and they are assembled. When they reach the same temperature, they are essentially locked together.

shrinking – The act of compressing a material into a smaller volume or area.

shrinking block – A sheet metal forming tool which clamps metal to prevent its buckling as it is hammered on its edge to shrink it.

shrink-wrap – A method used to protect products from dirt and dust while they are held in storage. The part to be shrink-wrapped is covered with a film of transparent thermoplastic material. When heat is applied to the film, the film shrinks to encase the part.

shroud – Sheet metal cover or housing used to control the air or gas flow to a desired path.

shroud ring – A stationary removable air sealing ring positioned just outside the tip plane of rotating airfoils. Sometimes, it is an outer casing which acts as a shroud ring.

shrouded-tip turbine blade – A blade with tip platforms attached which fit one to another to form a circular support ring. Often, the shrouds have thin abradable rims attached at their outer edge to act as air seals.

shunt – An accurately calibrated resistor placed in parallel with a meter movement for measuring current. Current flows through the shunt and produces a voltage drop proportional to the amount of current. The ammeter movement measures this voltage drop and displays it in amperes.

shunt circuit – A circuit that has several paths for electrons to flow.

shunt ohmmeter – A type of ohmmeter circuit used for measuring low resistances. The unknown resistor is placed in parallel (shunt) with the meter and the resistance is measured by the amount of current the unknown resistor takes from the meter.

shunt-wound generator – A generator in which the field and armature are connected in parallel.

shunt-wound motor – A type of motor in which the electromagnetic field windings are connected in parallel with the armature.

shut-off valve – A flow-control valve which may be used to shut off or stop a flow of fluid.

shuttle valve – A valve mounted on critical components which directs system pressure into the actuator for normal operation but emergency fluid when the emergency system is actuated.

side bands – The bands of frequencies on each side of carrier frequency produced by modulation.

side slip – An uncoordinated flight condition in which the aircraft moves downward and toward the inside of the turn.

sight gage – A glass tube or window attached to a reservoir or tank to show the height or quantity of fluid in the container.

sight glass – A liquid level indicator located on the outside of a reservoir which provides a visual indication of the level of the liquid in the reservoir.

sight line – A mark on a flat sheet of metal set even with the nose of the radius bar of a cornice or leaf brake. This placement puts the bend tangent line at the beginning of the bend.

signal – The intelligence or directive portion of a radio wave.

signal generator — A test unit designed to produce reference electric signals which may be applied to electronic circuits for testing purposes.

silencer — A device used in cabin air distribution systems to minimize the noise caused by pulsations in the air delivered from the cabin supercharger.

silica gel — A desiccant used as a drying or moisture absorbing agent used in packaging.

silicon — A natural element having four electrons in its valence orbit. Silicon is used to produce semiconductor devices having excellent thermal characteristics.
[1]N-Type: Silicon which has been doped with an impurity having five valence electrons.
[2]P-Type: Silicon doped by an impurity having three valence electrons.

silicon carbide — An abrasive used in the manufacture of grinding stones and abrasive papers.

silicon controlled rectifier — A form of gated rectifier that allows current to flow only during that portion of the cycle after which the gate has been triggered by a positive pulse.

silicon glaze — A shiny, brown, glass-like deposit on the nose insulator of a spark plug that has been operated in sandy or dusty conditions. This glaze is an insulator at low temperatures, but at high temperatures it becomes conductive.

silicon steel — A steel alloy that contains silicon.

silicone rubber — An elastic material made from silicone elastomers. Used with fluids which attack other natural or synthetic rubbers.

silver — A white, precious, metallic chemical element that is very malleable and a good conductor of electricity.

silver solder — An alloy of silver, copper, and nickel used for hard soldering. It produces a joint that is stronger than soft solder but not as strong as some forms of brazing.

simple flaps — Wing flaps which are lowered by pivoting them about a point near their leading edge. They change the airfoil section of the wing but do not affect the wing area.

simple machine — Any device with which work may be accomplished. simple machines are used to transform energy, or can be used to change the direction of a force. The six simple machines include the lever, the pulley, the wheel and axle, the inclined plane, the screw, and the gear.

simple motion – Newton's law of motion. Objects in motion tend to stay in motion.

simplex communications – A method of communication in which only one transmitter location can transmit at a time while the other receives.

simplex fuel nozzle – A nozzle with one spray orifice and one spray pattern.

simplex nozzle – A form of fuel discharge nozzle for turbine engines which is fed from a single fuel manifold.

simulate – To have the characteristics or appearance of something that is real.

simulator – An enclosed housing that duplicates all of the controls, instruments, furnishings, and feel of an actual airplane cockpit. The simulated environment gives the pilot the same feel and indications that would be found in actual conditions.

sine – A trigonometric function which is the ratio of the length of the side opposite an angle, in a 90° triangle, to the hypotenuse.

sine curve –
[1]A graphic representation of the relationship between the angle and its sine.
[2]Curve showing the relationship between the voltage or current, and the angle through which a rotary generator producing the voltage, or current has turned.

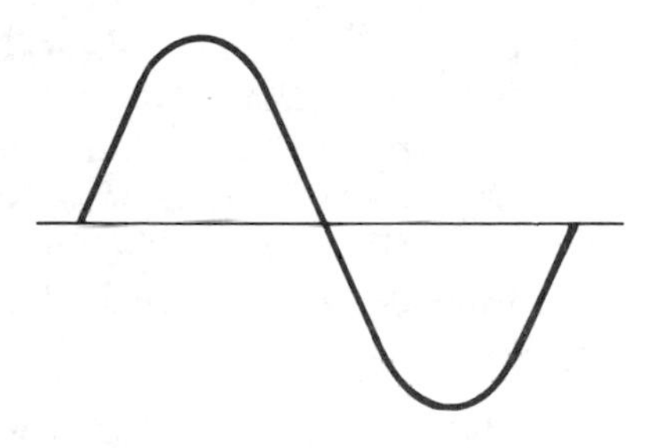

sine wave – The wave form of alternating current produced by a rotary generator. Its amplitude at any time is proportional to the sine of the angle through which the generator has turned.

single flare – A type of flare used for aircraft rigid tubing in which the end of the tube is flanged, but is not doubled back over itself as it is in a double flare.

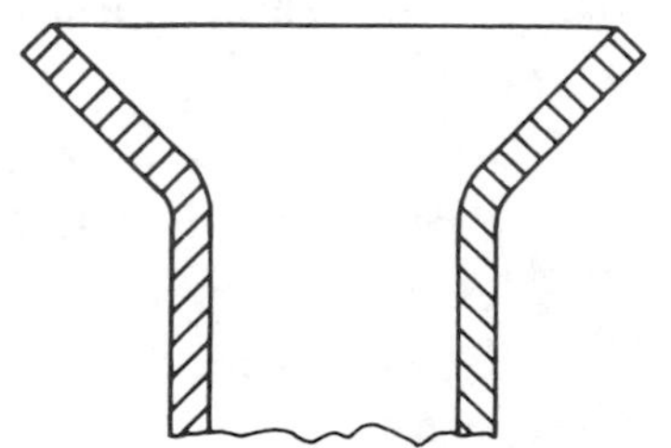

single spread – A method of applying adhesive to a bonded joint in which the adhesive is applied to one surface only.

single-acting actuator – A linear hydraulic or pneumatic actuator which uses fluid power for movement in one direction, and a spring force for its return.

single-axis autopilot – An automatic flight control device which controls the airplane only around the roll axis.

single-crystal turbine blade – A high temperature strength blade with no grain boundaries. It is manufactured by an advanced casting process which produces the blade from a single crystal of metal.

single-cut file – A type of hand file which has a single row of teeth extending across the piece at an angle 65-85°.

single-face repair – A repair to a bonded structure in which the damage extends through only one face sheet and into the core material.

single-loop rib-stitching – A method of attaching fabric covering to the ribs of an aircraft using only one loop of ribstitch cord for each stitch.

single-point fueling – Also called pressure fueling. A method of fueling the aircraft from a single point. It consists of a pressure fueling hose and a panel of controls and gages that permit one man to fuel or defuel any or all fuel tanks of an aircraft. The panel has valves connecting the various tanks to the main fueling manifold Fuel flows to each tank until the tank has reached the desired level or until it is full.

single-point grounding – A method of electrically grounding circuit by connecting all of the ground wires to a single point.

single-servo brakes – Brakes that use the momentum of the aircraft to wedge the lining against the drum and assist in braking when the aircraft is rolling forward.

single-sideband – A radio transmission in which only one of the sidebands in a signal is used.

single-spool compressor – A single compressor rotor design of the axial flow type.

sintered material – Form of powdered metal bonded to the rotating disk by heat and pressure. This material has become a coherent mass without melting.

sintered metal – A porous material made up by fusing powdered metal under heat and pressure.

siphon — A device that moves liquid from one container to a lower point. It consists of a flex tube in the liquid and a suction device that is used to start the flow of liquid. Atmospheric pressure on the surface of the liquid and gravity carries the liquid from the container to a point below the container.

siphon tube — A tube installed in a CO_2 fire extinguisher cylinder that assures that the CO_2 directed to the discharge nozzle will be in its liquid state.

Sitka spruce — A tall spruce tree that has needle-shaped leaves, drooping cones, and berry-like fruit. In the selection of wood for aircraft wood repairs, spruce is considered as the standard by which all other woods are compared.

sizing — Material used in the manufacture of some fabrics to stiffen the yarn for ease of weaving. Aircraft fabric must have no sizing.

sizing up a situation — A colloquial term used to denote the process of studying a situation to determine what must be done.

sketch — A simple drawing made without the use of drafting instruments and which shows a minimum of detail.

ski plane — An airplane whose wheels have been replaced with skis so it can be operated from snow or ice.

skid — A condition of uncoordinated flight in which the airplane moves toward the outside of the turn.

skid fin — A longitudinal vertical surface usually placed above the upper wing to increase the lateral stability.

skid shoes — Plates attached to the bottom of a helicopter skid-type landing gear protecting the skid.

skin — The smooth outer cover of the aircraft.

skin effect — The tendency of h-f alternating currents to flow in the outer portion of a conductor.

skin friction — Aerodynamic drag caused by the air flowing over the surface of the aircraft.

skin antenna — An antenna that is flush with the aircraft skin.

skip distance — The distance from a transmitter to the point where the reflected sky wave first reaches the earth.

skip welding — A welding technique used to prevent warpage of the material. Skip welding is a series of short welded beads with spaces between beads approximately as long as one of the beads. The remaining gaps are welded after the skip welds are completed.

skunk works — A top secret research and development project.

sky wave — That portion of a radio wave which is reflected from the ionosphere.

Skydrol hydraulic fluid — A synthetic, nonflammable, ester-base hydraulic fluid used in modern high-temperature hydraulic systems.

slab — An unfinished bar of metal which has a width greater than twice the thickness.

slag — A completely fused and vitrified matter caused by the flux used in welding. Slag separates from the metal in the process of its reduction in the melting process of welding.

slam acceleration — An improper operation of a turbine engine in which the power control lever is moved forward too rapidly. There is danger in this condition of a rich flameout because the fuel is metered before the airflow has picked up sufficiently.

slat — A movable auxiliary airfoil on the leading edge of a wing. It is closed in normal flight but extends at high angles of attack to duct air over the tip of the wing to delay the separation or stall.

slaved gyro — Directional gyro which is slaved to the output of a flux valve. In this way, a directional gyro can be given direction-seeking qualities.

sleeve — A tube or tube-like part fitting over or around another part.

slide rule — An ruler instrument with a central sliding piece both being marked with logarithmic scales. Used in making fast mathematical calculations.

sliding support — A type of duct support attached to a flexible bellows. The sliding support allows movement of the bellows while the duct is under pressure.

sling — A lifting attachment used to support and engine while it is being installed in, or removed from, an airplane or helicopter.

slinger ring — A tubular ring that is mounted around the hub of a propeller into which a deicing fluid is directed and is slung out onto the blades to prevent ice from forming.

slip —
[1]Propeller: The difference between the geometrical pitch and the effective pitch of a propeller. Slip may be expressed as a percentage of the mean geometrical pitch or as a linear dimension.
[2]Aircraft flight maneuver: A sideways aircraft maneuver used to descend at a steep angle or to compensate for excessive wind.

slip clutch mechanism — A typical installation on electric starter drives, designed to prevent sudden high torque to the engine. The clutch plates slip until torque on the engine side drops to the torque on the clutch side.

slip joint — A type of connection or joint in an induction system which remains airtight as the cylinders expand and contract with the temperature changes.

slip ring — A smooth circular ring used to put field current into a DC alternator.

slip stick — A slang term to describe a slide rule.

slippage mark — A painted mark between a tire and a wheel that will indicate any slippage between the two. If the mark of the tire and wheel are not lined up, the tire has slipped on the wheel and the wheel assembly must be removed.

slipping torque — The torque that must be produced by an aircraft engine starter before the clutch will slip.

slip-roll former — A metal working machine used to shape sheet metal into cylindrical and curved shapes.

slipstream — A stream of air pushed back by a revolving aircraft propeller.

sloshing sealing compound — A synthetic rubber sealant that is poured inside a metal fuel tank to seal the tank and prevent any fuel leakage.

slot — A fixed, nozzle-shaped opening near the leading edge of a wing, in front of the aileron, to duct the air down on the top of the wing, maintaining effective aileron control to high angles of attack.

slot-headed screw — A screw that has a single, straight groove cut across its head to fit the blade of a slot screwdriver.

slotted flap — A type of trailing edge wing flap in which a duct is formed when the flap is lowered. Air forced through the duct is held down on the surface of the flap, allowing greater extension before the air over the flap separates.

slotted nut — Similar to a castle nut. A hexagon nut that has grooves cut across its top to fit a cotter pin or safety wire which passes through a hole in the shank of a bolt.

slow-blow fuse — A special type of fuse which will take momentary overloads but will open the circuit when sustained excess current flows. Often used in motor circuits where the current is high until the motor begins to turn and generate back voltage.

sludge — A heavy, slimy deposit in aircraft lubricating oil which results from oxidation of the oil and contamination by water.

sludge chamber — Tubes or thin sheet metal chambers installed in the throws of an aircraft engine crankshaft. Sludge is slung into these chambers by centrifugal force and is held there until the engine is overhauled.

slug — A gravitational unit of mass to which a force of one pound would impart an acceleration of one foot per second per second.

slugging — A malfunction in a vapor cycle air-conditioning system in which liquid refrigerant is allowed to get into the compressor.

small aircraft — Aircraft having less than 12,500 lbs. certificated gross weight.

smaze — A mixture of fog vapor, smoke, dust (haze), and industrial smoke, with a lower moisture content than smog.

smile — The curved dimple around a rivet head caused by a rivet gun and rivet set not being held in a straight line with the head during the riveting process.

smog — A mixture of fog and industrial smoke that contaminates the air.

smoke detector — A system whereby the flight engineer of an aircraft can detect the presence of smoke in a baggage compartment before there is sufficient temperature change to actuate the fire warning system.

smolder — To burn without a flame.

snake drill — A long, flexible driving mechanism, one end of which is put in a drill chuck and the other designed to hold a twist drill.

snap ring — Small, spring-loaded, ring-type fastening device which fits into a groove either on the outside of a shaft or the inside of a hole. Spring tension holds the ring in place.

snap roll — An abrupt airplane maneuver in which the airplane does a single spin along its longitudinal axis while maintaining level flight.

snap-action electrical switch — A type of electrical switch that uses spring snaps to open the contacts when the switch is closed.

snips — Hand shears used for cutting sheet metal.

snow — Water vapor that changes directly into crystals of ice flakes when frozen in the upper air.

snubber — That portion of a hydraulic actuator which arrests the motion of the piston at the end of its stroke to cushion the stopping action.

soaking — Holding a metal at a specified temperature and time for the purpose of heat treating or annealing.

soap — A material mixed with water used for cleaning. Soap is produced by mixing an alkali and potash with a fat or oil.

soap bubble test — A method of testing for leaks in gas systems under pressure. A special non-flammable soap solution is brushed over the suspected fittings. If there is a leak, the escaping gas will cause the soap to bubble.

soapstone — A soft stone having a soapy feeling and composed essentially of talc. Used to mark steel parts prior to welding.

soaring — A type of glider flying without the use of an engine. The pilot locates and follows rising air to remain aloft, thereby slowing the aircraft descent.

socket wrench — A small cylindrical shaped wrench internally broached to fit the nut. It is equipped with a square hole in its top which fits a square drive on the wrench handle.

socket-head screw — A screw style that has a hex-shaped head.

sodium — A silver-white, metallic, alkaline chemical element with a waxlike firmness or thickness.

sodium bicarbonate — Also called baking soda. A white powdery crystalline compound. When mixed with water, sodium bicarbonate neutralizes spilled battery acid, and the acid on battery terminals.

soft magnetic material — A metal, such as iron, that readily accepts lines of magnetic flux and is also easily demagnetized.

soft solder — A physical alloy of lead and tin used to join non-structural metal parts or to increase the electrical conductivity of a twisted wire joint.

soft-faced hammer — A type of hammer which has a wood, plastic or rubber face on the head.

software — Programmed instructions that tell a computer what operations to complete.

solar cell — A silicon semiconductor device that converts solar energy into electricity.

soldered splice — A splice in electrical wiring made by twisting the wires together and then flowing soft solder over the joint. This splice is not recommended for aircraft use because of its tendency to be brittle.

soldering iron — An electrically heated, hand-held tool used to melt solder.

solderless connection — *See solderless splice.*

solderless splice — Terminal attached to an electrical conductor by crimping it onto the wire.

solenoid — A coil or wire with a movable core. In magnetic particle inspection, it is the coil used to longitudinally magnetize a part and to demagnetize parts using AC.

solid —
[1]A geometric figure which has three dimensions: length, width, and height.
[2]One of the three states of matter having a definite volume and shape and which is relatively firm or hard.

solid conductor — A wire made up of a single strand of metal covered with an insulating material.

solid fuel — Fuel such as wood or coal or one of the molded solid propellant materials used in rocket engines.

solid solution — A mixture where two or more elements or compounds may dissolve in each other at an elevated temperature but still remain in the solid state.

solidity — A helicopter rotor system in which the arc or circle of the sweep of the rotor blades is compared to a solid disk. Solidity is the ratio of the total blade are to the circle that is swept by the rotating rotor as it rotates.

solo flight — Refers to a single person piloting an aircraft in flight.

soluble — Any substance that can be combined with another substance. An example of this is mixing water and crystalline sugar.

solution – A state in which a base metal and alloying agents are united to form a single, solid metal.

solution heat treatment – A form of heat treatment of aluminum alloy in which the metal is raised to its heat treating temperature, held until it is uniform throughout, and then quenched. This process covers the alloying agents to be held in a solid solution, which increases the strength of the metal.

solvent – A liquid capable of dissolving another material.

sonic boom – A vibrational disturbance, heard as a loud noise, caused by an airplane moving faster than the speed of sound.

sonic cleaning – A method of cleaning parts by using high intensity sound waves in a cleaning fluid.

sonic frequencies – High frequency vibrational disturbances that the human ear can detect, normally considered to be the frequencies between 20 and 20,000 Hz.

sonic soldering – A method of soldering certain metals, such as aluminum, that are difficult to solder because of their quickness in building up surface oxides. To prevent the oxides from building up, the metal surface is coated with a flux to prevent any oxygen reaching it, and the tip of a sonic soldering iron heats the surface by vibrating at a sonic rate.

sonic speed – The speed traveled by sound though a medium.

sonic vibration – A high-frequency vibration caused by sound energy.

soot – A black residue that is created during oxyacetylene welding when the acetylene gas does not have enough oxygen in the mixture to completely burn the gas.

sound – A form of mechanical radiant energy transmitted by longitudinal pressure waves in the material medium. It is of such a frequency that it may be perceived by the human ear.

sound suppressor – Same as noise suppressor in reference to engines, or a device on a test cell used for sound attenuation of exhaust stack noises.

sound waves – A wave of motion in matter in the form of longitudinal wave motions. These waves are called longitudinal waves because the particles of the medium vibrate back and forth longitudinally in the direction of propagation.

source – The electrode of a field-effect transistor which compares to the emitter of an ordinary transistor.

south geographic pole – The south geographic pole is located at the 90th degree of south latitude. The south geographic pole of the earth is located at the southern end of the earth's axis.

space charge – The electric charge carried by the cloud of electrons in the space between electrodes of an electron tube.

space shuttle – An aerospace vehicle designed for carrying passengers and cargo to and from an orbiting satellite about the earth.

spacers – Devices or components used to take up space between two objects.

spaghetti – An insulating tubing slipped over wires.

spalling – A bearing defect in which chips of the hardened bearing surface are broken out.

span – Length dimension of a beam.

span loading – The ratio between the weight of an airplane and the span of its wing.

span of an airfoil – The length of an airfoil measured from tip to tip.

spanner – A special type of wrench that has a hook-shaped arm with a pin in its hooked end used for turning a ring-shaped nut.

spanner nut – Typical shaft retaining nut which can be tightened from notches in its face rather than from its outer surfaces as with a conventional hex-nut.

spanwise – From wing tip to wing tip.

spar – Main, or principle, spanwise structural member of a wing or other airfoil.

spar varnish – A phenolic modified oil which cures by oxidation rather than evaporation of its solvents, producing a tough, highly water-resistant film.

spark – Discharge of electrical energy of very brief duration between two conductors separated by air or other gas.

spark coil – A type of step-up transformer that produces a high voltage to the spark plug.

spark plug — A component in an aircraft engine which conducts high voltage electricity from the magneto into the combustion chamber of the cylinder. An accurately measured gap is provided for a spark to jump to ground to cause ignition for the fuel.

spark plug bushing — A bronze or steel insert in the cast-aluminum cylinder head of a reciprocating engine into which the spark plug is screwed.

spark plug resistor — A composition resistor installed in the barrel of most shielded spark plugs. The resistor limits the current which is stored by the capacitive effect of the shielding, minimizing the erosion of the spark plug electrodes.

spark suppressor — A device inside a magneto, such as a capacitor, that is placed across a set of contacts to keep the spark that forms from jumping across the contacts points as they open.

spark test — A common means of identifying various ferrous metals. In this test, the piece of iron or steel is held against a revolving grinding stone and the metal is identified by the sparks thrown off. Each ferrous metal has its own peculiar spark characteristics. The spark streams vary from a few tiny shafts to a shower of sparks several feet in length.

spark-ignition — A method of providing ignition of the fuel-air mixture inside the cylinder of a reciprocating engine by an electric spark.

spatula — A broad flat instrument used for spreading soft materials.

speaker voice coil — A small electromagnetic coil inside a telephone or radio speaker that moves and vibrates with a speakers voice and transmits this as sound waves.

special fastener — A special fastener is a fastener which differs in any respect from recognized standards.

specific fuel consumption — Number of pounds of fuel consumed in 1 hour to produce 1 HP.

specific gravity — The ratio of the weight of a given volume of a material to the same volume of pure water.

specific gravity adjustment — A fuel control adjustment which changes the fuel scheduling for use of different viscosity fuels.

specific heat — The ratio of the amount of heat required to raise the temperature of a body 1°, compared with the amount of heat required to raise the temperature of an equal mass of water 1°.

specific thrust – A ratio of mass airflow and net thrust. One means of comparison between engines.

specific weight – Density expressed in lb./cu. ft.

specifications – Data concerning dimensions, weights, performance, locations, etc.

spectrometric oil analysis – A system of oil analysis in which a sample of oil is burned in an arc, and the resulting light is examined for its wavelengths. This test can determine the amount of metals in the oil and can give warning of an impending engine failure.

spectro-photometer – A special device used to determine the way a surface reflects light waves of all frequencies. It is used to analyze paint pigments.

speed – The act of moving swiftly, or the rate of movement.

speed brakes – A type of control system that extends from the airplane structure into the slipstream to produce drag to slow the airplane down during descent.

speed handle – A crank-shaped handle used to turn socket wrenches.

speed of light – The speed at which light travels in a vacuum at a speed of 299,792.5 km/sec or 186,282 mil/sec.

speed of sound – The speed at which sound waves travel. At sea level, under standard atmospheric conditions, sound travels 760 MPH, 340 m/sec, or 1,116 ft./sec.

speed sensitive switch – An automatic, flyweight-operated sequencing switch driven by the engine gearbox. Used for completing electrical circuits to starting, ignition, fuel, etc.

speeder spring – The control spring used in a centrifugal governor to establish a reference force which is opposed by the centrifugal force of the spinning flyweights.

speed-rated engine – A gas turbine whose rated thrust is guaranteed to occur at a certain speed.

sphere – A geometric shape enclosed by a surface on which all points are an equal distance from an enclosed point called the center.

spherical – Having the form of a sphere.

spider – That portion of a propeller assembly used to support propeller blades.

spike – A transient condition in an electrical circuit when the circuit is first closed.

spillage – The movement of air from the bottom of the wing to the top; outward, and upward over the wing tip. It is the cause of wing tip vortices.

spin – A rotating stall in which the wing on the outside of the turn is flying while the wing on the inside is stalled.

spindle – Threaded part of a micrometer which is turned by the thimble and moves in and out of the frame.

spinner – Streamlined, bullet-shaped fairing that encloses the propeller hub assembly. The spinner streamlines the propeller installation and contributes to engine cooling.

spiral – A maneuver in which an airplane descends in a helix of small pitch and large radius, the angle of attack being within the normal range of flight angles. *See also spin.*

spiral flutes – Twisted grooves which run from one end of an object to the other. The grooves on a twist drill are spiral flutes.

spirit level – A measuring tool used to determine the relationship between a body and the horizon. The measuring element is a curved glass tube filled with a liquid, but having a single bubble. The position of the bubble in the tube is used to indicate the relationship.

spirit varnish – A wood finishing material made of resin dissolved in solvent. The varnish forms a hard, resin film on the wood when it dries.

splayed patch – A flush repair to a wood surface where the edges of the patch are tapered, but the slope is steeper than allowed in scarfing operations.

splice – A process whereby two ends of a material are joined together.

splice connectors – Devices, such as insulated solderless connectors, used for permanently connecting two ends of electrical wiring.

splice knot – A special form of knot used for joining two pieces of waxed rib-stitch cord. It will not slip as a square knot will.

spline – Any of a series of uniformly spaced ridges on a shaft, parallel to its axis and fitting inside corresponding grooves in the hub of a part.

split flaps — A form of wing flaps in which a portion of either the underside or the trailing edge of the wing splits and folds downward to increase the drag.

split lock washer — A heavy spring, steel lockwasher split at an angle across its face and twisted. Used with machine screws or bolts where the self-locking or castellated-type nut is not appropriate. The spring action of the washer provides enough friction to prevent loosening of the nut from vibrating.

split needled — A term used to describe the position of the two hands on the engine-rotor tachometer of a helicopter, meaning that the two hands are not superimposed.

split needles — A helicopter tachometer that has two needles: one shows engine speed, and the other shows the rotor system speed. When the clutch is fully engaged and the rotor is coupled to the engine, the two needles show as one. When the needles are split, it indicates that the rotor clutch is not fully engaged.

split steel lock washer — A heavy steel washer that is split and twisted to provide enough tension between the nut and the surface of the material to prevent the nut loosening.

split-lock keys — A form of split, tapered, cylindrical wedges used to lock the valve spring retainers to the stem of a poppet valve in an aircraft reciprocating engine.

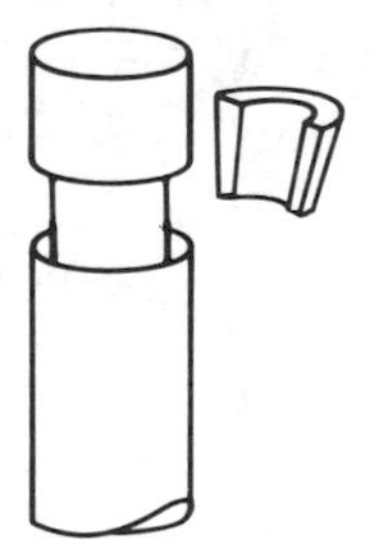

split-phase motor — An AC motor which utilizes an inductor or capacitor to shift the phase of the current in one of two field windings. This causes the resultant field to have a rotational effect.

spoiler — Any device used to spoil lift by disrupting the airflow over an aerodynamic surface.

spongy brakes — A brake malfunction caused by air in the hydraulic fluid. Since air is compressible, the braking action will not have a positive feel. It will feel as though there were a sponge or spring between the brake and the brake pedal.

sponson – A flange, or stub, projecting from the side of a flying boat hull to increase the beam of the hull and improve the lateral stability of the aircraft on the water.

spontaneous combustion – Self-ignition from the heat generated by exposure to oxygen. *See also spontaneous ignition.*

spontaneous ignition – A condition which exists when the temperature of a substance reaches its kindling temperature and self-ignites. No spark is required to start the fire.

spool – In an axial flow compressor, the spool shaped drum on which several stages of compressor blades are mounted.

spot check – The random selection and inspection of manufactured products. The parts that are checked represent the quality of all the parts that are manufactured.

spot face – Rotary tool used to remove a small amount of surface material around a hole.

spot facing – *See spot face.*

spot welding – A form of electrical resistance welding in which current is passed through sheets of metal stacked together. When the metal between the electrodes melts, it forms a button of metal, joining the sheets.

spotlight – A strong beam of brilliant light used to illuminate a particular area.

spot-type fire detection system – Bimetallic thermoswitches which will close to initiate a fire warning signal any time the temperature of the area in which they are located reaches a predetermined value.

spout – A projection on a pitcher by which a liquid is poured.

sprag clutch – A clutch which joins two rotating shafts. The clutch will ratchet and disengage when the driven shaft turns faster than the driving shaft.

sprag mount – An adjustable bracing system used on the Bell 47 series helicopters.

spray bar – An afterburner fuel nozzle which protrudes into the exhaust stream.

spray paint gun – An atomizing gun and reservoir device that sprays liquid paint or finishing material onto the surface that is being painted.

spray painting — A method of applying a finish to a surface by the use of an atomizing gun and a reservoir for the properly thinned paint.

spray strip — Metal strips that are mounted on the side of a flying boat hull used to divert water away from the aircraft.

spreader bar — The horizontal bar separating the floats of a twin-float seaplane.

spring coupling — A form of spring loaded device in a gear drive train which protects a system from excessive shock loads.

spring steel — Steel containing carbon in percentages ranging from 0.50 to 1.05%. In the fully heat treated condition, it is very hard, will withstand high shear and wear, and will have little deformation. Used for making flat springs and wire for making coil springs.

spring tab — An auxiliary airfoil set into a control surface which, under conditions of high control forces, will act as a servo-tab providing an aerodynamic assist for the pilot.

springback — The amount that metal springs back after it has been bent through a specific angle. This must be allowed for when making bends in the sheet metal.

spring-loaded — A condition in which one part is held in a particular relationship with another by the means of a spring. This usually allows some movement, but returns the parts to their original position.

spur and pinion reduction gear system — A type of gear system that is used for reduction gearing. In a prop reduction gearing system, the spur and pinion gearing consist of a large driving gear, or sun gear, splined to the shaft, a large stationary gear, called a bell gear, and a set of small spur planetary pinion gears mounted on a carrier ring. When the engine is operating, the sun gear rotates. Because the planetary gears are meshed with this ring, they also must rotate, and will walk, or roll, around it as they rotate, and the ring in which they are mounted will rotate the prop shaft in the same direction as the crankshaft but at a reduced speed.

spur gear — An external toothed gear.

square —
[1]A plane geometric shape having four equal sides with all four angles being right angles.
[2]The mathematical process of multiplying a number by itself.

square engine – An engine with cylinders whose bore and stroke dimensions are the same. The bore of the cylinder is its inside diameter. The stroke is the distance the piston moves from one end of the cylinder to the other, specifically, from TDC to BDC.

square file – A form of double cut file with a square cross section; tapered lengthwise. Used for filing slots in keyways.

square knot – A knot made up of opposite loops, each one enclosing the parallel sides of the other.

square mil – Abbrev.: sq. mil. An area equivalent to a square having sides 1 mil (0.001 in.) in length.

square root – The factor of a number which, when squared, will give the number.

square wave – The waveform of a multivibrator oscillator in which the leading edge and the trailing edge of the wave are both vertical.

squaring shears – A large, floor-mounted sheet metal tool used to make square cuts across sheets of metal.

squat switch – An electrical switch mounted on one of the landing gear struts. It is used to sense when the weight of the aircraft is on the wheels.

squealing brakes – A noise made by glazed brakes which are chattering at such a high frequency that the sound resembles a squeal rather than a hammering.

squeegee – A long handled rubber scraper blade used to remove liquid from a surface.

squeeler tip – A reduced thickness at the tip of rotor blades. This section is designed to wear away rather than to damage the shroud ring if tip loading forces cause contact.

squeeze bottle – A soft plastic bottle that can be squeezed to force the bottle contents out through its top.

squeeze riveter – Pneumatic or hydraulic riveting gun in which a set for the manufactured head and a smooth surface to form the upset head are mounted in the jaw of the large clamp. When the squeeze-gun is actuated, the jaws come together just enough to form the proper size upset head.

squelch – A circuit in a communications receiver which holds the output volume down until a signal is received.

squelch circuit – *See squelch.*

squib — A small explosive discharge plug sealed with a breakable disk combined with and explosive charge which is electrically detonated to discharge the contents of the fire extinguishing agent from the bottle when the pilot presses the push button discharge switch.

stabilator — A single-piece, horizontal tail plane which combines the functions of a stabilizer and an elevator.

stability — That property of a body which causes it, when its equilibrium is disturbed, to develop forces or moments tending to restore the original condition.

stabilizer — The fixed horizontal tail surface that has the elevators and rudder hinged to its trailing edge.

stabilizer bar — A dynamic component used on some Bell helicopters to insure rotor stability.

stable operation — An operating condition in which there is no appreciable fluctuation in any of the operational variables.

stable oscillation — An oscillation whose amplitude does not increase.

stage — In turbine engine construction, a single turbine wheel having a number of turbine blades.

stagger — The longitudinal relationship of the wings of a biplane. If the upper wing is forward of the lower wing, the airplane is said to have positive stagger.

stagger angle — Refers to blade twist design in an impulse-reaction turbine blade.

stagger wires — The wire between the cabane struts of a biplane. These wires are used to adjust the stagger.

staggered ignition — Dual ignition timed so that the two firing impulses do not occur at the same time.

staggered timing — A reciprocating engine ignition timing method that uses a dual ignition system.

stagnation point — The point on the leading edge of an airfoil where the airflow separates, some going over the surface and some below.

stagnation temperature — Temperature that results from friction of the airflow on a surface.

stainless steel — Steel, containing appreciable quantities of chromium and nickel and used for applications where its resistance to corrosion is important.

stake – Small, bench-mounted, anvil-type sheet metal tool.

staking – A term used to denote the swaging of terminals onto an electrical conductor.

stall – An aerodynamic condition in which the smooth flow of air has broken away from the upper surface of an airfoil, and the flow is turbulent, decreasing the amount of lift produced.

stall strip – A small triangular spoiler attached to the inboard leading edge of some wings to cause the center section of the wing to stall before the tips. This assures lateral control throughout the stall.

stall warning transmitter – A device which produces a signal to warn the pilot of an impending stall.

stalling angle – The angle of attack at which the flow of air over the wing ceases.

stall-warning system – A stall warning system warns the operator when the aircraft is approaching the critical stall angle.

standard – The degree of excellence required for a particular purpose.

standard barometric pressure – The weight of gases in the atmosphere sufficient to hold up a column of mercury 760 mm high (approximately 30″) at sea level (14.7 PSI). This pressure decreases with altitude.

standard cell – A cadmium-mercury cell that is made in a specially shaped glass container in which the two electrodes are covered with an electrolyte of cadmium sulfate. The voltage produced by a standard cell is 1.018636 volts at 20°C. A standard cell is also called a Weston standard or a Weston normal cell.

standard fastener – A standard fastener is a fastener which conforms in all respects to recognized standards.

standard sea level pressure – A standard value of pressure used as a reference for making aerodynamic computations. It is 14.7 lb./sq. in., 29.92 inches of mercury, or 1013.2 millibars.

standard temperature – 59°F.

standard-rate turn – A rate of change in aircraft direction of 3° per second, completing a 360° turn in two minutes. For higher speed jet aircraft, the standard-rate turn is 1½° per second, completing a 360° turn in four minutes.

standing waves — Stationary waves occurring on an antenna or transmission line as a result of two waves, identical in amplitude and frequency, traveling in opposite directions along the conductor.

standpipe — A vertical standing pipe in a tank or reservoir. It allows a space for a reserve of fluid between the top of the standpipe and the bottom of the tank. The reserve fluid is drawn from the bottom of the tank.

staple —
[1]Wire: A U-shaped piece of fine wire with sharp-pointed ends, driven into a surface to hold or fasten material to it.
[2]Textile material: The average length of textile material. Cotton fibers.
[3]A chief item: Any chief item, or part or element of a raw material.

stapler — A machine or tool used for driving staples through a piece of wood or a stack of paper for the purpose of binding them together.

starboard side — The right-hand side of an aircraft or ship facing the nose of the airplane.

stardust — Air impingement haze in an acrylic finish.

starter — A unit which uses electrical, pneumatic, or hydraulic energy to rotate an engine for starting.

starter engine component — Any device that uses electrical, pneumatic, or hydraulic energy to rotate an engine for it to start and run normally.

starter solenoid — An electrically operated switch which uses a small current controlled from the cockpit to close the high current-carrying contacts in the starter circuit.

starter-generator — A combined unit used on turbine engines in which the device acts as a starter for rotating the engine, and after it is running, internal circuits are shifted to convert the device into a generator.

statcoulomb — The amount of charge on each of two bodies 1 cm apart which causes them to exert a force of 1 dyne on each other. The statcoulomb is the charge resulting from the addition of approximately 2 × 109 electrons to a body.

state of charge — The condition or amount of charge of a battery.

static — Still; not moving. A condition of rest.

static balance –
[1]A condition of balance which does not involve any dynamic forces.
[2]When a body will stand in any position as the result of counter-balancing and/or reducing the heavy portions, it is said to be standing or static balance.

static charge – A charge of static electricity picked up by an airplane as it moves through the air.

static discharger – A device used to dissipate static electricity from a control surface before it can build up to a serious degree.

static electricity – Electrical charge which may be built up on a non-conductive surface by friction.

static flux – The concentration of lines of flux in the frame of a magneto due to the rotation of the magnet. At full register, the lines of flux are maximum, and at the neutral positions, the lines are minimum.

static friction – The friction on an object when an attempt is made to slide the object along a surface. The object must first be broken loose or started. Once in motion, it slides more easily.

static instability – The characteristic of an aircraft that, when it is disturbed from a condition of rest, will tend to move further from its original condition.

static interference – The noise in a radio caused by static electricity moving between two structures which have no common ground.

static port – A small hole, flush with the side of the aircraft, through which the static pressure is taken to operate the airspeed indicator, altimeter, and vertical speed indicator.

static pressure – Atmospheric pressure measured at a point where there is no external disturbance, and the flow of air over the surface is perfectly smooth.

static pressure pickup – A part of the static instrument system. Static pressure pickup is the location on the surface of an aircraft where the static air pressure is picked up from port holes or flush static ports on the outside surface of the airplane, and used to provide readings in the altimeter, airspeed indicator, and the vertical speed indicator. *See also static system.*

static radial engine — An engine with all of the cylinders radiating out from a small central crankcase. A single-throw crankshaft is used for each row of cylinders. All single-row radial engines have an odd number of cylinders, but two or four rows may be used if more power is required.

static RPM — The maximum RPM a reciprocating engine can produce when the aircraft is not moving through the air. The static RPM is lower than the RPM the engine will develop through the air because of the increased manifold pressure and power the engine can produce when forward movement rams air into the carburetor inlet.

static stability — The characteristic of an aircraft that, when disturbed from a condition of rest, will tend to return to the condition of rest.

static system — Plumbing that connects the altimeter, airspeed indicator, and vertical speed indicator to the outside static air source of the airplane. An alternate source is usually included in this system.

static temperature — A temperature measurement of air not in motion.

static test — A method of testing the structural integrity of an airplane structure to determine its ability to withstand loads that could possibly be encountered in flight.

static thrust — The thrust produced by a turbine engine that is not moving through the air.

static tube — A cylindrical tube with a closed end and a number of small openings normal to the axis, pointed upstream, used to measure static pressure.

static wick — A small device made of metal braid or graphite-impregnated cotton, attached to the trailing edge of a control surface to dissipate accumulated electrical charges into the air.

station web — A built-up section located at some point of applied force, such as attachments for wings, stabilizer, etc.

station — The location of a point within an aircraft identified by inches from the datum.

stator —

[1]The stationary part of an electrical machine such as a motor or alternator.

[2]The stationary portion of an axial flow turbojet compressor.

[3]The discs in a multiple-disc brake that are keyed to the axle and do not rotate.

stator case – Outer engine casing which houses either compressor or turbine stator vanes.

stator vane – Stationary vane, either compressor or turbine.

statute mile – A measure of land distance equal to 5280 ft. or 1.609 km.

stay – A cable or wire used as a structural member in a truss that is loaded in tension only.

steady-state condition – The condition of any system that exists when all of the measured values are stable.

steam – Vaporized water that has been changed by heating it.

steatite – The mass form of talc. Used for making ceramic insulating material for high-voltage systems.

steel – An alloy of iron that is hard and tough with small amounts of carbon and other alloying elements.

steel wool – An abrasive material made of long, fine, steel shavings and used for scouring steel parts.

steering damper – A hydraulically actuated device which is used to steer the nose wheel of an aircraft and to absorb shimmy vibrations.

stellite – An extremely hard, wear resistant metal used for valve faces and stem tips containing cobalt, tungsten, chromium, and molybdenum.

step – A break in the form of the bottom of a float or hull, designed to diminish resistance, to lessen the suction effects, and to improve control over longitudinal attitude.

step up coil – A form of transformer in which the secondary winding has more turns than the primary. The voltage in the secondary winding will therefore be stepped up.

step-down transformer – A device that steps down voltages and is made up of an iron core, a primary winding, and a secondary winding. The step-down transformer has more turns of wire in the primary winding than it has in the secondary winding. This difference will determine the stepped down secondary voltage.

stepped solvents – Solvents in a finish which have different rates of evaporation. Some evaporate almost instantly; others, more slowly. This provides the desired film.

stepped stud – A type of stud replacement used to replace a stud that has stripped the threads in a cast housing. The hole is drilled and tapped for the larger stepped stud.

step-up transformer — A device that steps up voltages and is made up of an iron core, a primary winding, and a secondary winding. The step-up transformer has less turns of wire in the primary winding than it has in the secondary winding. This difference, depending on the turns-ratio, will determine the step-up of the secondary voltage.

stiffener — A structural member attached to an aircraft skin for the purpose of making it more stiff. It is quite often an extruded angle or a formed hat section.

Stoddard solvent — A petroleum product similar to naphtha used as a solvent or cleaning agent.

stoichiometric — A chemical relationship in which all of the constituents are used in the reaction. In the case of a stoichiometric mixture, all of the oxygen and all of the hydrocarbon fuel are used. There is no oxygen or free carbon left.

STOL aircraft — Short Takeoff and Landing aircraft. *See also short takeoff and landing.*

stop — A device used to limit the throw or travel of a control.

stop countersink — A special countersink having a collar which will not allow the cutter to cut too deep into the metal skin.

stop drill — A hole drilled in the end of a crack in aircraft structural material to distribute the stresses and stop the crack from proceeding further.

stop nut — *See elastic stop nut.*

storage battery — A secondary cell. An electrical device in which electrical energy is converted into chemical energy and stored until needed, then converted back into electrical energy. It does not make electricity.

straight peen hammer — A metal beading hammer with one face flat and the other having a vertical edge.

straight roller bearings — Roller bearings which are used where the bearing is subjected to radial loads only.

straight shank drill — A form of twist drill with a straight shank. This is to distinguish it from a twist drill with a tapered shank.

straightedge — Wood, metal, or plastic having a perfectly straight edge used in drawing straight lines, or to check for the straightness of a piece of material.

straight-polarity arc welding – Electric arc welding where the electrode is connected to the negative terminal of the power supply.

straight-run gasoline – Gasoline that is refined from crude oil using the fractional distillation process to produce straight-run gasoline. In this method, the crude oil is heated at atmospheric pressure in a heating container. The various hydrocarbon liquids in the crude oil vaporizes first, followed by those of higher boiling points. Straight-run gasoline does not have a change in the chemical composition of any of the parts of the original crude petroleum.

strain –
[1]Deformation in a material that has been caused by stress.
[2]The process of exerting a force beyond the normal physical capacity of the material.

strain gage – An extremely tiny conductor that is bonded to the surface on which strain is to be measured. When the surface stretches, the cross section of the strain gage becomes smaller, and its resistance increases. This change in resistance is measured by an extremely sensitive instrument.

strain hardening – The increase in strength and hardness of a metal by work-hardening or cold-working. The strain-hardening is normally done after a piece of material has been heat treated, and is the only way a nonheat-treatment aluminum alloy may be hardened.

strained – A deformation, or a physical change, in a material that is caused by an excessive stress applied to it.

strainer – A very fine mesh screen located in the fuel system. Used to remove impurities from the fuel system.

stranded conductor – An electrical conductor made up of many strands of wire covered with an insulating material.

stranded wire – Electrical wire made up of many smaller wire strands.

strap pack – A tension-torsion system using sheet steel lamination to carry the loads of the rotor blades to the head. Used by Hughes Helicopters.

stratification – Formed in layers.

stratiform clouds – Clouds formed in layers.

stratocumulus – Low gray clouds formed in layers.

stratosphere – That portion of the earth's atmosphere beyond the tropopause.

stratus – Low gray uniform clouds.

streamline flow – A fluid flow in which the streamlines, except those very near a body and in a narrow wake, do not change with time.

streamlined – Having a shape or contour that presents a minimum resistance to the air with a minimum of turbulence.

streamlined body – *See streamlined.*

strength – The ability of a material to withstand forces which tend to deform the metal in any direction, or the ability of a material to resist stress without breaking.

strength-to-weight ratio – Ratio of the strength of a material to its weight.

stress – The internal force in a body that resists the tendency of an external force to change its shape.

stress analysis – A mathematical determination of the loads which will be applied to an aircraft structure.

stress corrosion – Corrosion of the intergranular type that forms within metals subject to tensile stresses which tend to separate the grain boundaries.

stress raiser – A scratch, groove, rivet hole, forging defect or other structural discontinuity causing a concentration of stress.

stress relieve – Heating to a suitable temperature, holding long enough to reduce residual stresses and then cooling to minimize the development of new residual stresses.

stress rupture – A condition seen as cracking and stretching or untwisting of both stationary and rotating airfoils. This condition occurs from pressure times area forces and G-forces.

stressed-skin structure – Aircraft skin designed to carry the tension and compression stresses of structural loads. Stressed-skin structured aircraft have few internal structural members.

stretching – A sheet metal forming operation in which the material is mechanically stretched over dies to form compound curves.

stringer – A thin metal or wood strip running the length of the fuselage to fill in the shape of the formers.

stripper bolt – Discarded term for a shoulder screw.

structural failure – When a structure fails to withstand the stresses imposed upon it and fails by bending or breaking.

structural machine screws – Machine screws which have an unthreaded portion of the shank and which are made of high-strength alloy steel. They may be used in place of an aircraft bolt to carry shear loads and some tensile loads.

structural member – Any part of an aircraft structure designed to carry loads or stress.

structural steel – An alloy steel used for parts of an aircraft which are subjected to high structural loads.

strut –
[1]A compression member in a truss.
[2]The external bracing on a noncantilever airplane.
[3]The stub wing assembly through which the thrust loads are transmitted from a pod-mounted turbine engine into the fuselage.

stub antenna – A short, UHF, quarter-wavelength antenna normally used for radar beacon transponders or distance measuring equipment.

stud – A headless bolt having threads on each end. One end normally has coarse threads for screwing into a casting, while the other end has fine threads to accept a nut.

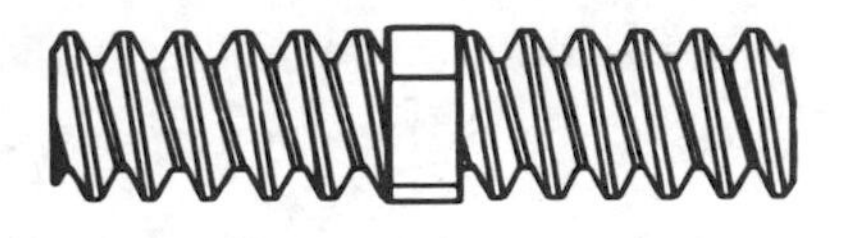

styrene – A liquid hydrocarbon used in the manufacture of certain synthetic resins to improve their workability.

Styrofoam – A rigid polymer of styrene plastic material.

subassembly – An assembly that is a component part of a larger assembly.

subfreezing temperature – Below freezing. The temperature that is needed to cause water to freeze.

sublimation – The process of purifying a solid material by heating it until it changes directly into a vapor without passing through the liquid stage.

submerged-arc welding – A method of electric arc welding in which a bare rod, covered with granulated flux, is used as the electrode. To form a uniform bead, the electrode must be moved along the plates to be welded at a constant speed in addition to the downward feed of the electrode. The granulated flux melts in the arc and flows ahead of the weld to prevent the formation of oxides getting into the bead.

subsonic flight — Aircraft flight in which the air flowing over the aircraft structure is moving slower than the speed of sound.

subsonic inlet — A divergent-shaped duct which acts as a subsonic diffuser.

subsonic speed — Speed below the speed of sound.

subsonic-diffuser — A divergent diffuser where the airstream spreads out to increase pressure as axial velocity decreases.

substandard — Unacceptable quality in a manufactured object.

substitute — The replacement of an object with the same or better quality material and which meets all of the specifications of the original.

substrate — The supporting material on which an integrated circuit chip is built.

subsystem — An operating unit or assembly that is a component part of a larger system.

subtrahend — A number that is subtracted from the minuend.

suction — The act of sucking or producing a negative pressure.

suction cup gun — A paint gun in which the material is held in a cup attached to the gun and drawn into the atomizing air by suction created by the atomizing airflow.

suction gage — An aircraft instrument used to measure the negative pressure or suction in an aircraft vacuum system.

suction relief valve — A control valve in the instrument pneumatic system that provides a constant negative pressure by opening the system to the outside air when the vacuum rises above the preset value.

sudden stoppage — A condition in which the aircraft engine has come to a complete stop in less than one revolution, usually caused by hitting an immovable object. Sudden stoppage requires a special inspection to determine internal engine damage.

sulfate radical — A combination of chemical elements which act as though they only were one atom. In this case, the SO_4 behaves in the chemical action of battery charging or discharging as though it were only one element.

sulfated — The condition of plates in a discharged lead-acid battery. The lead has turned to lead sulfate. If allowed to remain for a long period of time, this sulfate will become impossible to remove by normal charging action.

sulfur — A pale yellow, non-metallic chemical crystal element.

sum — The result when adding two or more numbers together; total.

sump —
[1]A low area in a fuel tank in which water will normally collect.
[2]That portion of an aircraft engine used to hold the lubricating oil.

sump jar — A small jar in the vent line of a battery box containing a pad. wet with a chemical such as bicarbonate of soda or boric acid. Fumes given off by the battery as it charges are neutralized by this material.

sun gear — The center gear in a planetary gear system around which the planetary gears rotate.

supercharger — An engine or exhaust driven air compressor used to provide additional pressure to the induction air so the engine can produce additional power.

supercharger control system — The system of controlling the supercharger to maintain a constant manifold pressure as the altitude changes.

supercharging — Increasing the cylinder pressure of a reciprocating aircraft engine by introducing compressed air into the cylinder on the intake stroke.

superconduction — The effect experienced at very low temperatures when atomic vibration ceases and conduction electrons are free to drift through conductors with no opposition and no loss of energy.

superconductivity — A reaction of certain chemical elements when they are cooled and held at or near absolute 0°. At absolute 0°, these elements lose almost all of their electrical resistance, and they become strongly diamagnetic

supercooled water — Water which exists in a liquid state in a cloud when the surrounding temperature is well below the freezing temperature. This will form solid ice as soon as it is disturbed.

superheat — Heat energy added to a gas after evaporation has been completed.

superheated vapor — Vapor which has been heated above its boiling point for a given pressure.

superheated water — The process of heating water to a temperature above 212°F when it normally changes its state from liquid to a vapor. Water may be superheated by heating it in a pressurized container, so that more heat energy will have to be added to it before its molecules move fast enough to become steam.

superheterodyne — A radio receiver circuit in which the received radio frequency signal is mixed with a frequency produced in a local oscillator to create an intermediate frequency.

superhigh radio frequency — Frequencies that are between 3.0 and 30.0 gigahertz and have wavelengths between 100 and 10 mm.

supersaturated — Any solution (such as sugar and water) in which a solid is dissolved in a liquid until the liquid will hold more than it normally would.

supersede — To set aside something in place of another. When a manufacturer issues new maintenance manuals, the new manuals supersede the old ones making the old manuals obsolete.

supersonic — A speed greater than the speed of sound.

supersonic aerodynamics — The branch of aerodynamics that deals with the theory of flight because the flight of an aircraft depends upon the laws of aerodynamics. Aero means pertaining to air; dynamics is that branch of physics which considers bodies in motion and the forces that produce changes of bodies in motion. Supersonic deals with aerodynamics at a speed that is greater than the speed of sound.

supersonic diffuser — A converging diffuser where the airstream pressure is raised as velocity decreases.

supersonic nozzle — A divergent shaped duct designed to allow gases to expand outward faster than they accelerate rearward.

supersonic speed — Mach 1.0 to Mach 5.0.

supersonic transport — The British-French Concorde and the Russian TU-144. The only supersonic commercial aircraft presently in use.

superstructure — Framework attached to an aircraft truss structure to provide the desired aerodynamic shape. It is usually covered with lightweight sheet metal or aircraft fabric.

Supplemental Type Certificate – Abbrev.: STC. A certificate authorizing an alteration to an airframe, engine, or component which has been granted an Approved Type Certificate.

support clamp – Clamp used to support various fluid lines or wire bundles to the aircraft structure.

suppressor grid – The electrode in a pentode vacuum tube used to suppress secondary emissions from the plate.

surface heat treatment – Surface heat treatment is a process that improves the hardness or other mechanical property of the fastener in any surface area.

surface tape – Strips of fabric made of the same material used to cover an aircraft structure. It is doped over all of the seams, rib-stitching, and edges to give the surface a smooth, finished appearance. Sometimes called finishing tape.

surface tension – A condition that exists on the surface of a liquid because of molecular attraction. It produces an effect similar to a film stretched over the surface.

surface treatment – Any treatment which changes the chemical, physical, or mechanical properties of a surface.

surficant – A partially soluble contaminant, a by-product of fuel processing or fuel additives which adheres to other contaminants, causing them to drop out of the fuel and settle to the bottom of the fuel tank as a sludge.

surge – *See compressor stall.*

surging – A change in engine RPM or engine power in an oscillatory manner. It is usually caused by a malfunction in the fuel control system.

sustaining speed – The speed of engine compressor and turbine at which the engine can keep itself running without having to depend on power from the starter to maintain suitable compressor pressure ratios.

swage – To squeeze together.

swaged terminals – Solderless terminals fastened to an electrical conductor by the swaging process.

sweat solder – A method of soldering two pieces of metal together in which both pieces are tinned with solder, and put together, and then the joint is heated. The bond is done with the solder on the surfaces without the use of additional solder.

Sweeny tool – Gear reduction-type torquing tools used in engine overhaul.

sweptback wing – A wing planform in which the tips of the wing are farther back than the wing root.

swirl frame – The inlet case on some turboshaft engines which act as an inlet particle separator.

swirl vanes – Air circulation vanes which surround fuel nozzles causing a vortex in which fuel vapor is made to recirculate and more completely ignite.

swiss pattern files – A set of precision files used for delicate metal work.

symbol – Graphic representation used to represent shape, size, or material on an aircraft drawing.

symmetrical – A condition in which both halves of an object are the same.

symmetrical airfoil – An airfoil in which the airfoil has the same shape on both sides of its center line. The center of pressure of a symmetrical airfoil has a very small change in the location of its center of pressure as its angle of attack changes.

symmetry check – A rigging check of an aircraft in which comparison measurements are made to determine that points on both sides of the airplane are the same distance from the center line.

synchro – A form of synchronous device in which a movable element is caused to follow a similar element in a master unit or transmitter.

synchro system – *See synchro.*

synchronize – To cause two events to occur at exactly the same time.

synchronous motor – A type of AC motor in which the rotor is an electromagnet, and the stator has a pulsating magnetic field from the AC flowing in it. The rotor must have a starting device, but once it is running, it will operate at a constant speed, determined by the frequency of the power source.

synchrophasing – A form of propeller synchronization in which not only the RPM of the engines are held constant, but also the position of the propellers with relation to each other.

synchroscope – An instrument showing the relationship between the speeds of the engines on a multi-engine aircraft.

synthetic fibers — Man-made products such as fiberglass, polyester, and polymid fibers used in the production of aircraft covering fabric.

synthetic oil — A lubricating oil with a synthetic rather than petroleum base. Has less tendency toward oxidation and sludge formation than petroleum oils. It is extensively used in turbine engines and is gaining popularity in reciprocating engines.

synthetic rubber — Any of several types of man-made products which have characteristics similar to natural rubber.

system — A group of parts or components working together to accomplish a purpose.

system discharge indicator — A yellow disc or blow-out plug on the side of an aircraft that blows out to indicate that the fire extinguishing system has been discharged by normal operation rather than by an overheat condition.

system pressure regulator — That hydraulic component which controls the hydraulic system pressure. It unloads the pump when a pre-selected pressure is reached and directs the pump back onto the line when the pressure drops to the desired kick-in pressure.

T

tab – A small auxiliary control surface which is hinged to an aircraft primary control surface. Tabs can be used to assist in the movement of a primary control surface or as a means of trimming an aircraft.

TACAN – An acronym for Tactical Air Navigation System. An electronic navigation system that provides a constant visual indication of the distance the aircraft is from a ground station. TACAN operates in the UHF range of the radio frequency spectrum.

tachometer – An instrument which measures the rotating speed of an engine in RPMs or in percent of the maximum RPM.

tachometer cable – The flexible cable used to drive a mechanical tachometer from the engine. It is made of two layers of steel wire, spiraled in opposite directions about a central core.

tachometer generator – A small electrical generator supplying current at amperages of frequencies proportional to the speed of the unit on which it is mounted.

tachometer generator or indicator – The generator has three-phase AC output of approximately 20V. It is a gearbox driven accessory producing power for a motor driven indicator. The indicator provides a RPM indication in the cockpit.

tack coat – A very light coat of material sprayed on a surface and allowed to stay until the solvents evaporate. It is then covered with the full wet coat of material.

tack rag – A rag, slightly damp with thinner, used to wipe a surface after it has been sanded to prepare it for the application of the next coat of finish.

tack weld – Small temporary welds along a welded seam made for the purpose of holding the pieces of metal in position until the weld is completed.

tacking – Hand-sewn temporary stitches that are removed prior to machine sewing.

tackle – A pulley in the form of a wheel mounted on a fixed axis and supported by a frame. The wheel, or disk, is normally grooved to accommodate a rope. The frame that supports the wheel is called a block. Block and tackle consist of a pair of blocks. Each block contains one or more pulleys and a rope connecting the pulley(s) of each block.

tacky — Gluey or sticky finish.

tag wire — Thin diameter wire used to tie identification tags to objects.

tail boom — A spar or outrigger connecting the tail surfaces to a pod-type fuselage.

tail cone —
[1]The conical-shaped portion of a turbine engine exhaust system which is used to produce the proper area increase for the gases as they leave the engine.
[2]The rearmost part of an aircraft fuselage.

tail load — A downward aerodynamic force produced by the tail of an airplane used to produce dynamic longitudinal stability.

tail pipe — That portion of the exhaust system of an aircraft engine through which the gasses leave the aircraft.

tail rotor — A rotor turning in a plane perpendicular to that of the main rotor of the helicopter and parallel to the longitudinal axis of the fuselage. This is used to control the torque of the main rotor and to provide movement on the yaw axis of the helicopter.

tail section — *See empennage.*

tail skid —
[1]On modern jet aircraft, a portion of the structure designed to absorb shock in case the tail strikes the runway during rotation for takeoff.
[2]A skid which supports the tail of an airplane on the ground. These were used before the advent of the tail wheel.

tail surface — A stabilizing or control surface in the tail of an aircraft.

tail wheel — A small wheel located at the rear of the fuselage of an airplane having a conventional landing gear. It is used as a support for the tail when the airplane is on the ground.

tail wind — An airplane that is flying in the same direction as the wind.

tail-heavy — A condition of balance in an aircraft in which the center of gravity is behind the aft limit.

tailpipe inserts — Small, sheet metal, wedge-shaped tabs that are inserted into the tailpipe of some older engines to reduce the nozzle opening and increase thrust. Adjustment of thrust is now done at the fuel control.

takeoff — The act of beginning flight in which an airplane is accelerated from a state of rest to that of normal flight. In a more restricted sense, the final breaking of contact with the land or water.

takeoff power –
[1]Takeoff power of a reciprocating engine is usually limited to a given amount of time such as one minute or five minutes.
[2]The brake horsepower approved by the engine manufacturer for takeoff. This may be limited in the time allowed.

takeoff thrust – The jet thrust approved for the normal takeoff and limited in the length of time it can be used.

takeoff weight – Weight of the aircraft at liftoff. Often used as maximum takeoff weight.

tandem – One behind the other.

tandem bearings – The placement of two ball bearings so that the thrust load is shared by both bearings.

tang – The projection point which is mounted onto the handle of a file or knife.

tangent –
[1]The position in which a straight line meets the circumference of a circle at a straight point.
[2]A trigonometric function which is the ratio of the lengths of the side opposite and the side adjacent to the angle in a right triangle.

tank – A container or reservoir for liquids.

tank circuit – A parallel resonant circuit including an inductance and a capacitance.

tank selector valve – A selector valve controlled by the pilot with which they can select the fuel tank from which they desire to operate the engine.

tantalum carbide – A rare, corrosion-resistant, metallic chemical element mixed with carbon used in cutting tools and instruments.

tap – A tool used to cut threads on the inside of a hole in metal, fiber, or other material.

tap drill – A type of twist drill used to drill a hole before it is tapped. Charts have been developed to determine the correct size tap drill to use to obtain the correct dimensions for a tap.

tap extractor – A tool equipped with projecting fingers which enter the flutes of the tap, which is backed out of the hole by turning the extractor with a wrench.

tape measure – A tape with graduations of inches, etc. used for measuring.

taper — A gradual decrease in width or thickness from one end of an object to the other.

taper in plan only — A gradual change (usually a decrease) in the chord length along the wing span from the root to the tip with the wing sections remaining geometrically similar.

taper in thickness ratio only — A gradual change in the thickness ratio along the wing span with the chord remaining constant.

taper pin — A device used for fastening concentric shafts together to prevent relative motion between them. It consists of a tapered pin pressed into a tapered hole.

taper reamer — A type of reamer used to smooth and "true" tapered holes and recesses.

taper tap — A form of hand-operated thread cutting tap used to start the tapping process in a drilled hole. The tap tapers for the first six or seven threads.

tapered crankshaft — A crankshaft to which the propeller is mounted by fitting it over a tapered end.

tapered propeller shaft — *See tapered crankshaft.*

tapered punch — A hand punch which is tapered in length and is used to start pins, bolts, or rivets from their holes.

tapered roller bearings — A form of anti-friction bearing made of hardened steel cylinders rolling between two cone-shaped, hardened steel races. These bearings are designed to carry both thrust and radial loads.

tapered-shank drill — A form of twist drill which has a tapered shank and is held in the chuck by friction.

tapped hole — A hole in a casting or other material which has threads cut on the inside.

tapped stud hole — A hole where threads have been cut for a stud to screw in.

tappet — The component in an aircraft reciprocating engine which rides on the face of the cam and transmits a reciprocating motion to the push rods to open the poppet valves in the engine cylinders. The hydraulic valve lifters normally fit inside the tappets.

tare weight – The weight of all items such as blocks or chocks used to hold an airplane on the scales when it is being weighed. Tare weight must be subtracted from scale reading to determine the weight of the aircraft.

target blade – The identification of one blade of a helicopter during electronic balancing. It is the blade with the double interrupter.

tarmac – A hard surfaced area of an airport used for aircraft parking, tie-down, and servicing.

tarnish – A stain, blemish, or dull surface.

tarpaulin – A canopy made of a large piece of heavy, waterproof canvas that is fitted along its edges with eyelets so ropes can be used for tying.

tautening dope – Aircraft dope consisting of nitrocellulose and a plasticizer, such as glycol sebacate, ethyl acetate, butyl acetate or butyl alcohol, or toluene. The nitrocellulose base is made by treating cotton in nitric acid. The plasticizer aids in producing a flexible film. Both the plasticizer and the solvents are responsible for the tautening action of the dope. The dope is applied to the fabric surface to produce tautness by shrinkage, to increase strength, to protect the fabric, to waterproof, and to make the fabric airtight.

taxi – To move an airplane on the ground under its own power.

taxi lights – Lights similar to landing lights on an aircraft which are aimed in such a way that will illuminate the runway or taxiway when the airplane is taxiing.

taxi weight – Maximum weight allowed for ground maneuvering. Sometimes referred to as ramp weight.

taxiway – An airport pavement that allows aircraft to taxi from the terminal or parking area to the runway.

technician – A person skilled in repairing aircraft who is authorized by the FAA .

tee fitting – A plumbing connector in the shape of a "T".

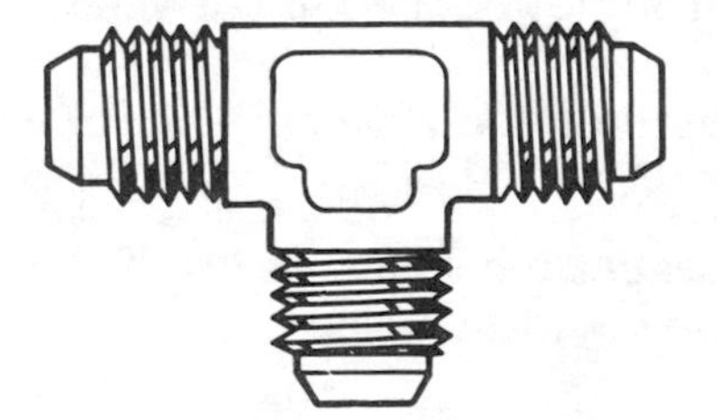

Teflon — A proprietary name for a fluorocarbon resin used to make hydraulic and pneumatic seals and backup rings.

teleflex cable — A type used to connect engine accessories to cockpit controls. Also a type of feedback cable.

telemetering — A system of sending measurements over great distances by radio.

telescoping gage — A form of precision measuring device which consists of a rod telescoping inside a tube and is spring-loaded outward. The gage is adjusted to the width of a boar or a hole and locked together. They are then removed from the hole and the length is measured with a micrometer caliper.

temper — The condition of hardness or softness of metal.

temperature — A measurement of heat intensity.

temperature amplifier — An electronic device used to amplify an exhaust temperature signal to the fuel control. Used to assist in scheduling fuel flow.

temperature bulb — The temperature sensor which is installed at the point temperature is to be measured and whose output is carried to the temperature measuring instrument in the aircraft cockpit.

temperature coefficient of resistance — Symbol: α. The rate of change in resistance per degree centigrade of temperature change.

temperature datum system — An electronic circuit in an electrohydromechanical fuel control. Used to assist in scheduling fuel flow.

temperature ratio — A ratio of two engine temperatures, used in certain performance calculations.

temperature scales —

[1]Celsius, formerly known as centigrade, is based on the freezing point of water as 0° and its boiling point as 100°. Absolute zero is –273°.

[2]Fahrenheit is based on the freezing point of water as 32° and its boiling point as 212°. Absolute zero is –460°.

[3]Kelvin is based on the freezing point of water as 273°K and its boiling point as 373°.

[4]Rankine is based on the freezing point of water as 492° and its boiling point as 672°.

tempering — A heat-treatment process in which some of the hardness is removed from the metal to increase its toughness and decrease its brittleness.

template – A pattern made of any suitable material to permit the layout of parts with a minimum expenditure of time and effort. It should be rigid and accurate and have pertinent data marked on it.

tensile load – An external force tending to lengthen or stretch a body.

tensile strength – The ability of an object to resist forces tending to stretch or lengthen it.

tensile stress – The stress forces that tries to pull an object apart.

tensiometer – A measuring instrument used to determine the installed tension of aircraft control cables.

tension – Stress produced in a body by forces acting along the same line but in opposite directions.

tension adjusters – *See tension regulators.*

tension regulators – Devices installed in an aircraft control system which maintain a constant cable tension, regardless of the dimensional changes in the airplane caused by expansion and contraction from temperature changes.

tension torsion bar – A strap made of layers of sheet used to absorb tension of centrifugal loads between the rotor blades and the hub, also the torque of blade pitch changes.

tension torsion strap – A strap made of wire used to serve the same purpose on a helicopter rotor head as the tension torsion bar.

terminal –
[1]A connecting fitting in the form of a ring which attaches to the end of a wire, battery, terminal strip, or other components.
[2]A keyboard and video monitor that allows a user to interface with the computer.

terminal strip – A strip of insulating material which contains terminal posts to which aircraft wiring is attached.

terminal velocity – The speed of an aircraft at which the drag has reached such an amount that the airplane will no longer accelerate.

terminating decimal – A decimal fraction that ends with a whole number.

terneplate – Lead-coated thin sheets of steel used in some of the older aircraft for the construction of fuel tanks and for the tipping on their wooden propellers.

tertiary – Something at a third level of importance.

test — To submit a unit, component, etc. to conditions which will show its quality. strength, etc.

test club — A special form of wide blade, short-diameter propeller used for applying a load to a freshly overhauled aircraft engine for its initial run-in. *See also club propeller.*

test stand — A stationary structure where engines, units, components, etc., may be mounted for testing.

test switch — A switch used to test a system to determine its operational condition.

tetraethyl lead — A heavy, oily, poisonous liquid, ($Pb(C_2H_5)_4$), mixed into aviation gasoline to increase its critical pressure and temperature.

tetrahedron — A large, triangular-shaped, kite-like object installed near the center of an airport near the runway. Tetrahedrons are mounted on a pivot and are free to swing with the wind to show the pilot the direction of the wind as an aid in takeoffs and landings.

tetrode — An electron tube having four active electrodes.

T-handle — A T-shaped handle used for turning sockets.

theodolite — An optical instrument used to measure vertical and horizontal angles. Theodolites are used for surveying and for weather observations.

theoretical pitch — An assumed pitch of the propeller blades. Same as geometric pitch.

therapeutic adapter — An adapter for a continuous-flow oxygen mask which flows approximately three times the normal rate. It is used for passengers having heart or respiratory problems.

thermal — Heat. Rising air as it is heated by the sun.

thermal anti-icing system — A heated leading edge of the wing and tail surfaces to prevent the formation of ice.

thermal barrier — Excessive heat produced by air friction over the surface of and aircraft. A thermal barrier limits the speed of an aircraft in flight.

thermal circuit breaker — A circuit breaker made up of a bimetallic set of contacts that opens a circuit when an excessive amount of current flows through it.

thermal coefficient of resistance — A material whose resistance changes with a change in temperature.

thermal conduction — The transfer of heat energy from one object in contact with another object.

thermal conductor — A material, such as metal, that can easily transfer heat energy.

thermal cutout switch — A type of circuit breaker, or switch that breaks the circuit at a predetermined temperature.

thermal decomposition — A chemical action in which a material is decomposed into simpler substances by the action of heat.

thermal efficiency — The ratio of the amount of heat energy converted into useful work, to the amount of heat energy in the fuel used.

thermal expansion — The expansion of any substance due to heat.

thermal fatigue — A condition in turbine metals caused by heating and cooling each time a power setting is changed.

thermal insulator — Materials such as paper, wood, etc. that are poor conductors of heat energy.

thermal output — Amount of heat being discharged.

thermal relief valve — A pressure relief valve installed in a static portion of a hydraulic system to relieve pressure built up by fluid expansion due to heat.

thermal runaway — A condition which exists in a nickel-cadmium battery when the cell resistances become unbalanced because of temperature. The resistance of some cells decreases and allows the cells to take more current; this lowers their resistance further, allowing more current. This action continues until the battery is seriously damaged, sometimes exploding.

thermal shock — A stress induced into a system or component due to a rapid temperature change.

thermal switch — A switch activated by heat.

thermionic — A term describing electron emission caused by an incandescent material.

thermionic current — Current flow in a conductor caused by heat.

thermistor — A semiconductor device with a core material, such as ceramic beads in and Inconel tube used in a continuous-loop fire detection system, whose electrical resistance changes with a change in temperature.

thermoammeter – An instrument for measuring RF alternating current in a circuit. The thermoammeter measures the RF alternating current by heating a wire of a known resistance. As the wire heats, it produces a current proportional to the amount of applied heat. This current is measured by the meter instrument.

thermocouple – A temperature measuring system consisting of two dissimilar metal wires joined at both ends. Current flows through the wires that is proportional to the difference in temperature between the two junctions. The circuit is opened and a current measuring device is inserted either to read the temperature or to actuate a sensitive relay.

thermocouple exhaust temperature probe – A bimetallic probe, generally of chromel and alumel alloy which is located in the exhaust stream. Heat causes a milliampere current to flow to a cockpit indicator which reads out in degrees Celsius.

thermocouple fire detector – A device which by thermal action produces an electrical current flow which ultimately illuminates a cockpit warning signal.

thermocouple oil temperature bulb – A thermocouple which is positioned in the oil flow to provide a cockpit indicator with an oil temperature indication. It is generally of bimetallic materials such as chromel and constantan.

thermodynamics – Branch of science that deals with mechanical actions caused by heat.

thermoelectricity – Electrical energy generated by the action of heat on the junction of two dissimilar metals.

thermoelectricity – Electricity caused by the addition of heat.

thermoelectromotive – Electron flow produced by a thermoelectric couple.

thermometer – An instrument for measuring temperatures. It consist of a graduated glass tube with a sealed bore with which mercury, colored alcohol, etc., rises and falls as it expand or contracts due to temperature changes.

thermopile – A collection of thermocouples connected in series used to measure minute changes in temperature or changes in the flow of electrical current.

thermoplastic material – A resin-based plastic material that can be softened by heat and cooled many times without losing its tensile strength.

thermoplastic resin – A resin material which will soften with the application of heat. Most aircraft windshields and side windows are made of this material.

thermosafety discharge indicator – A red blowout disc located on the outside of an aircraft fuselage or engine nacelle which will blow out to vent a cylinder of high-pressure gas in the event of an overpressure condition caused by heat.

thermosetting material – A plastic material that remains hard once it is hardened by chemical means or by heat and pressure.

thermosetting resin – A resin, widely used in today's plastics, that usually sets by chemical means and maintains its hardness even when heat is applied.

thermostat – A device that functions to establish or maintain a desired temperature produced by a heater or an air-conditioning system.

thermostatic bypass valve – A temperature-sensing valve in an engine oil cooler, used to direct the oil either through the core of the cooler or around the inside of the cooler shell to maintain the proper oil temperature.

thermoswitch fire detector – A device which by thermal expansion of metals closes an electrical contact in the presence of heat, illuminating a warning light in the cockpit.

thickness gages – A precision measuring tool consisting of a series of precision-ground steel blades of various thicknesses. It is used to determine the clearance or separation between parts.

thimble – That part of a micrometer caliper which is turned to rotate the spindle.

thinner – A solvent mixed with dope or paint to reduce its viscosity.

third-class lever – The third-class lever provides what is called a fractional disadvantage, i.e., one in which a greater force is required than the force of the load lifted. If a muscle pulls with a force of 1,800 lbs. in order to lift a 100 lb. projectile, a mechanical advantage of 100/1,800 is obtained. This is a fractional disadvantage, since it is less than 1.

thixotropic agent – A substance added to a resin to increase its resistance to flow.

thread –
[1]Projecting helical grooves that are cut around a bolt, fitting or a pipe by which matching threaded parts are cut into the inside, for purposes of screwing the parts together.
[2]A very fine strand of linen, cotton, or other material for making yarn or fabric.

thread chaser – A tool used to remove contamination from a threaded device.

thread gage – *See screw pitch gage.*

thread insert – A thread insert is an internally threaded bushing designed to be molded in or inserted into soft or brittle materials to provide greater strength and minimize wear of threaded assembly.

thread pitch – The distance from the peak of one thread to the peak of the next thread on a screw, bolt, or other thread fitting.

thread plug gage – A go/no-go type gage to be screwed into internal threads.

thread ring gage – A ring-type gage used for checking external threads.

thread snap gage – An adaptation of the plain snap gage having cone shaped or grooved anvils, threaded rolls, or wedge shaped pinions in place of the flat button anvils of the plain snap gage.

three-axis autopilot – An automatic flight control system which controls the airplane about all three axes.

three-d cam – Part of a linkage system, this multi-labeled cam can rotate and also move up and down to allow a cam follower riding on its surface to seek an infinite number of positions. Used to miniaturize linkage systems such as in fuel controls.

three-dimensional object – An object that has three dimensions namely, length, width, and depth.

three-phase system – An AC electrical system consisting of three conductors, each carrying a current 120° out of phase with each other. Three-phase systems are used extensively in modern electrical and electronic actuating systems.

three-point landing – The landing of an airplane in which all three main wheels of the landing system touch the ground at the same time. Three-point landings are not recommended.

three-pole, single-throw switch — An electrical switch with three contactors, or poles, each of which completes only one circuit, controlled by a single operating toggle.

three-view drawing — A type of orthographic projection drawing which uses three views to portray an object.

three-way light switch — An electrical switch wired in such a way that it allows the same light to be turned on or off from either of two separate locations.

throatless shear — A heavy-duty shear used for slitting large sheets of metal.

throttle — The valve in a carburetor or fuel control unit which determines the amount of fuel-air mixture that is fed to the engine.

throttle body — One of the units of a carburetor system in which all air entering the cylinders must flow through, and is the air control and measuring device. The airflow is measured by volume and by weight so that the proper amount of fuel can be added to meet the engine demands under all conditions. The throttle body contains the throttle valves, main venturi, boost venturi, and the impact tubes.

throttle ice — A form of carburetor ice which forms on the rear side of the throttle valve when the throttle is partially closed.

through bolts — Long bolts which pass completely through an object to hold its halves together.

throw — That part of a crankshaft to which the connecting rods are attached.

throwaway part — A part that is not economical to repair if it should fail, and must be thrown away and replaced with a new one.

thrust —
[1]A forward force that imparts momentum to a mass of air behind it.
[2]The forward force produced by a reaction to the exhaust gases escaping the nozzle of a jet engine, or produced by the aerodynamic force of a propeller.

thrust bearing — A bearing in a reciprocating aircraft engine which absorbs loads parallel to the length of the crankshaft.

thrust equivalent horsepower — Abbrev.: TEHP. Not in common use today. *See equivalent shaft horsepower.*

thrust, gross — *See gross thrust.*

thrust horsepower – The amount of horsepower the engine-propeller combination transforms into thrust.

thrust line – An imaginary line passing through the center of the propeller hub, perpendicular to the plane of the propeller rotation.

thrust loads – Loads imposed upon the engine crankshaft and bearings when the propeller is pulling or pushing the aircraft.

thrust, net – *See net thrust.*

thrust reverser – A mechanical device placed in the tail pipe of a turbojet engine to deflect the exhaust gases forward to increase the descent angle in flight, or to decrease the landing roll on the ground.

thrust specific fuel consumption – Abbrev.: TSFC.
1 An equation, $TSFC = W_f F_n$, where W_f is fuel flow in pounds per hour and F_n is the net thrust in pounds. Used to calculate the fuel consumption as a means of comparison between engines.
2 The amount of fuel that an engine must burn in one hour to produce one pound of thrust.

thruster – A miniature rocket engine that is fired to change or reposition of space vehicles and orbiting satellites.

thumb screw – A type of machine screw that has a round flat projection perpendicular to the screw shank. Thumb screws can be turned by hand, and are used on access panels where it is necessary to frequently open and close the panel.

thunder – A loud noise heard when lightening occurs between clouds, or between the cloud and the ground. The lightening generates instantaneous high heat which causes a violent expansion of the surrounding air. This expanding air causes the shock wave or noise that we hear.

thundercloud – A frequently used term for a cumulonimbus cloud.

thunderstorm – A cumulonimbus cloud charged with electricity and producing lightning, thunder, rain and hail.

thyratron – Gas-filled triode electron tube in which a continuous current is caused to flow by a momentary signal applied to the grid.

thyratron tube – A triode tube into which a gas has been introduced to change its operating characteristics.

tie rod – Tension rod used for internal and external bracing of various component parts. The ends are threaded for attachment and length adjustment.

tiedown – A special anchoring provision on an airport surface by which airplanes may be secured when they are parked.

tight-drive fit – An interference fit between parts which requires a sharp blow with a 12- to 14-ounce hammer to mate the parts.

time between overhaul – Abbrev.: TBO. A recommendation of the manufacturer of an aircraft engine as to the amount of time that the engine can operate under average conditions before it should be overhauled. Overhaul at this time will result in the most economical operation.

time change item – Any item, component, unit, etc. whose time in service is limited by hours, number of times the unit is operated on, or a calendar basis, and must be removed and replaced with a new or serviceable like item.

time in service – Time in service, with regard to maintenance records, is the time from the moment an aircraft leaves the surface of the earth until it touches it at the next point of landing.

time limited part – *See time change item.*

Time-Rite indicator – A piston position indicator, made by Time-Rite, used for locating the position of a piston in the cylinder of a reciprocating engine for the purpose of magneto or valve timing.

timing disc – A device or tool which may be mounted on an accessory drive or on the propeller to indicate the amount of crankshaft travel for ignition or valve timing.

tin – A soft, silver-white, metallic chemical element that is malleable, and used as an alloy in solders, utensils, and in making tin plate.

tin snips – A colloquial term for hand operated sheet metal shears.

tinned – A part which has been coated by soft solder.

tinned wire – Electrical wire that has been coated by a thin coating of soft solder.

tinner's rivet – A flat-headed solid rivet that is driven holding the flat head of the rivet on an anvil, and upsetting the rivet by peening the end over with a hammer.

tinplated – Sheet steel, coated with a thin layer of tin.

tip cap – A removable tip on the rotor blade tip. This cap is often used to hold spanwise balance weights.

tip path plane – The path followed by the tips of a propeller as it rotates.

tip pocket – An area provided at the tip of a helicopter rotor blade to place weight for spanwise balance.

tip speed – The speed of a rotating airfoil. Generally subsonic except for fan blade tips and the tip of centrifugal compression blades.

tip targets – *See tracking reflectors.*

tip weight – A weight placed in the tip of a helicopter rotor blade for spanwise balance.

tire – A ring or loop made of rubber compound for toughness and durability. Tires consist of tread, breakers the casing plies/cord body, and beads. Tires provide a cushion of air that helps absorb the shocks and roughness of landings and takeoffs. They support the weight of the aircraft while on the ground and provide the necessary traction for braking and stopping aircraft on landing.

tire bead – Bundles of steel wire embedded in the rubber around the inner circumference of an aircraft tire.

TIT – *See turbine inlet temperature.*

titanium – A dark-gray, silvery, lustrous, very hard, light, corrosion-resistant, metallic element.

title block – An information block in the lower right-hand corner of an aircraft drawing, in which the name of the part, the part number, and other pertinent information is displayed.

To-From indicator – An indicator on the course deviation indicator (CDI) During VOR operation the vertical needle of the CDI is used as the curse indicator. The vertical needle also indicates when the aircraft deviates from the course and the direction the aircraft must be turned to attain the desired course. The "TO-FROM" indicator presents the direction to or from the station along the omniradial.

toggle switch – An electrical switch in which a projecting knob or arm moving through a small arc causes the contacts to open or close rapidly.

toe-in – Aircraft wheels that tend to converge towards the front. Toe-in will cause the tires to wear excessively on the inside as they try to move closer together.

toe-out – Aircraft wheels that tend to diverge towards the rear of the wheels. Toe-out will cause the tires to wear excessively on the outside as they try to move apart.

tolerance – An allowable variation in dimensions of a part.

toluene – A colorless, water insoluble, flammable liquid ($C_6H_5CH_3$) used as a solvent, paint remover, and thinner.

toluol – A commercial grade of toluene which is a liquid aromatic hydrocarbon similar to benzene, but less volatile, flammable, or toxic.

ton of refrigeration – A measure of the cooling capacity of an air conditioning system. It is the same cooling effect as would be had by melting one ton of ice in 24 hours.

tone – A tint or shade of a color. A variation of a hue.

tool steel – Hard steel used in the making of tools.

top dead center – Abbrev.: TDC. The position of the piston within a cylinder when the piston has reached its furthest uppermost position.

top overhaul – The overhaul of the cylinders of an aircraft engine. It consists of grinding the valves, replacing the piston rings, and doing anything else necessary to restore the cylinders to their proper condition. The crankcase of the engine is not opened.

torching – Long plumes of flame extending from the exhaust stack. This is caused by an excessively rich mixture and consequent traces of unburned fuel remaining in the exhaust. This unburned fuel will not ignite until oxygen in the air of the exhaust system mixes with the charge.

toroidal wound coil – An electrical coil wound around a ring or dough-nut shaped core.

torque –
[1]A resistance to turning or twisting.
[2]Forces that produces a twisting or rotating motion.

torque limited – A limitation placed on the drive train of the helicopter in regards to power input.

torque links – The hinged linkage between the piston and cylinder of an oleo shock strut. The piston is allowed to move in an out, but is restrained from rotating. These are also called scissors or nutcrackers.

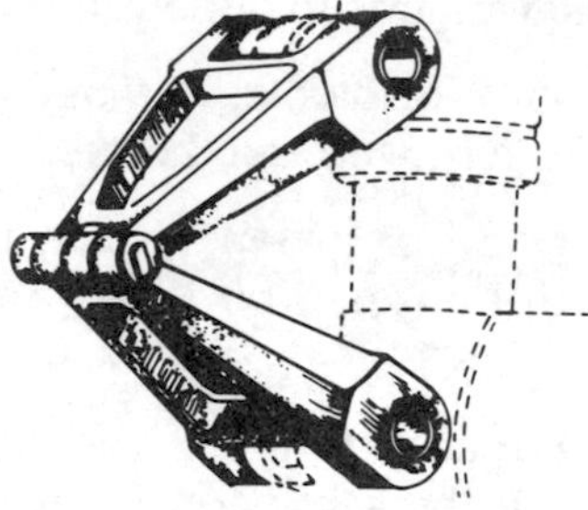

torque nose – A mechanism or apparatus at the nose section of the engine that senses the engine torque and activates a torquemeter.

torque tube – A tubular member of a control system used to transmit torsional movement to the control.

torque wrench – A precision hand tool used to measure the amount of torque applied to a bolt or a nut.

torquemeter – An indicator used on some large reciprocating engines or on turboprop engines to measure the reaction of the sun gear in the propeller reduction gear system, to indicate the amount of torque the engine is producing.

torsion – An external stress which produces twisting within a body.

torsional force – Twisting or being twisted. Torsional forces act on a rotating propeller in the form of aerodynamic twisting force and torsional forces from tensile forces that tries to pull it apart.

torsional strength – The strength of a material in such a direction that opposes a twisting force.

total pressure – The pressure a moving fluid would have if it were stopped. No losses are considered.

total temperature – A temperature measurement of air in motion. The total of static temperature plus temperature rise due to ram effect.

totalizer – A single fuel quantity gage that indicates the total of fuel in all of the fuel tanks.

touch and go – An aircraft training technique in which the pilot practices a series of takeoffs and landings without coming to a complete stop. As the airplane lands, the pilot advances the power for another takeoff and go-around.

toughness – The property of a metal which allows it to be deformed without breaking.

tow-axis autopilot – An automatic flight control system which controls the airplane about the roll and yaw axes.

towing eye – A ring or hook on an aircraft structure to which a tow bar may be attached for moving the airplane on the gound.

toxic – Poisonous.

track – The path followed by the tip of a propeller or rotor blade as it rotates.

tracking flag – Wooden pole supporting a white cotton flag, used to touch the operating rotor blades which have had their tips covered with colored chalk. The marks left on the nag will indicate track of the main rotor.

tracking reflectors – Reflectors placed on the blade tips to determine track with a spotlight or a strobe light.

tracking stick – A stick with a wick on one end used to touch the operating rotor blades and to determine blade track.

tracking targets – *See tracking reflectors.*

tractor propeller – A propeller that pulls the airplane through the air and is mounted to the front of the engine.

trade winds – Winds that blow toward the equator from the northeast on the north side of the equator and from the southeast on the south side. Trade winds are caused by the friction between the air and the earth, and by the rotation of the earth.

traffic pattern – The regulated movement of aircraft traffic that must fly along an established route when approaching or leaving an airport.

trailing edge – The aft edge of an airfoil or wing. Portion of wing that the air passes last.

trailing edge flap – Sections of the trailing edge of an airfoil which may be bent down or extended in flight to increase the camber of the airfoil, increasing both the lift and drag.

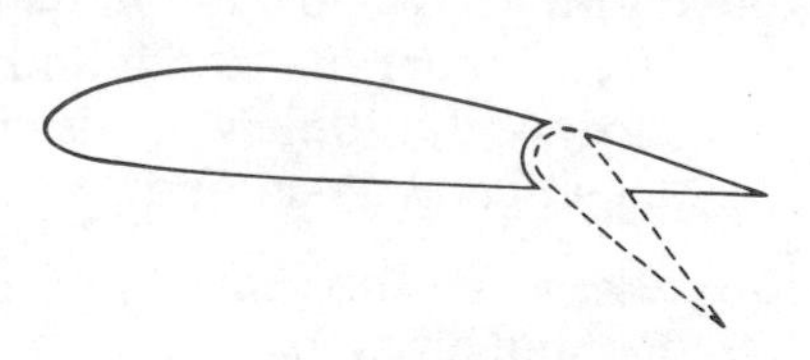

trailing finger – An electrode in the distributor of an ignition system which uses a booster magneto for starting. The high voltage from the booster magneto enters through a slip ring and is distributed through a finger which trails the normal ignition finger. In this way, the cylinder next in firing order receives the starting spark.

training manual – A technical publication used to explain the operation of a system or a component. It is general in nature, and is not considered FAA-approved data.

trammel points – Sharp points, usually mounted on a long bar, and used to transfer dimensions from one location to another.

tramming – A means of checking the alignment of an aircraft structure by making comparative measurements.

transceiver – A piece of electronic communications equipment in which the transmitter and receiver are housed in the same unit.

transconductance –Abbrev.: G_m. The ratio of a small change in plate current to a small change in grid voltage. The formula is $G_m = I_pE_g$ (plate voltage constant).

transducer – An electrical device that either takes electrical energy and changes it into mechanical movement or mechanical movement and changes it into electrical energy.

transfer – To carry, remove, or send from one place or position to another.

transfer gearbox – A gearbox which is driven from the main rotor shaft and which in turn drives the main (accessory) gearbox.

transfer punch – A special form of punch having an outside diameter the same as the rivet hole, and a sharp point at its exact center. Used to provide a punch mark for starting a drill in the exact center of the hole being transferred.

transformer – A special form of indicator in which a voltage is generated in one winding by mutual induction from another winding. There need be no electrical connection between the two windings.

transient conditions — Conditions which may occur briefly while accelerating or decelerating, or while passing through a specific range of engine operations.

transistor — A semiconductor device having three electrodes, similar in use to the vacuum tube.

transistor voltage regulator — Voltage regulator for DC alternators or generators which uses a transistor to control the field current. A zener diode senses the voltage to be controlled.

transistorized voltage regulator — A voltage regulator for DC generators or alternators which uses a transistor to control the flow of field current, but uses vibrating points to sense the voltage and control the transistor.

transit — An optical instrument used for measuring horizontal angles.

transition duct — A sheet metal rear support for can-annular combustion liners.

transition element — One of a series of elements which are consecutive in the periodic table and which possess similar chemical properties. All transition elements of a particular series posses the same number of valence electrons.

translating cowl — Portion of a turbine engine cowling which moves back to form an exhaust nozzle for thrust reverse air.

translational lift — The additional lift of a helicopter obtained when entering forward flight, due to the increased efficiency of the rotor system.

translucent — The condition of a material to be transparent, letting light pass through it but not diffusing it.

transmission line — A conductor, usually coaxial, used to join a receiver or a transmitter to the antenna.

transmissivity — Ratio of the amount of power transmitted through a radome to the amount of power that would be transmitted with the radome removed.

transmitter — Electric device whose function is to collect information from one point and sent it electrically to a remote indicator.

transom — The vertical bulkhead at the rear end of a seaplane float.

transonic flight — Aircraft in flight approximating the speed of sound in air (550-900MPH).

transonic speed – The speed of a body relative to the surrounding fluid which is in some places subsonic and in other places supersonic; usually from Mach 0.8 to 1.2.

transonic speed range – Generally stated at Mach 0.8 to 1.2 speed range, where some portions of the airfoil have subsonic flow and others supersonic flow.

transparent – The ability to view objects through a translucent material without being diffused.

transpiration cooling – Refers to internal cooling air which exits through porous walls of turbine blades and vanes.

transponder – Radar beacon transponder. A radar transmitter-receiver which transmits a coded signal every time it is interrogated by a ground radar facility.

transport category aircraft – An aircraft that is certificated under FAR Part 25.

transverse pitch – The perpendicular distance between two rows of rivets. This is also called gage.

transverse wave – A wave that moves in a perpendicular direction to the direction the wave is moving.

trapezoid – A plane four-sided geometric figure having only two sides parallel.

trapped fuel – *See undrainable fuel.*

traverse – Lying across a body.

TRF – *See tuned radio frequency receiver.*

triac – A semiconductor device similar to a silicon-controlled rectifier, that may be triggered by either a positive or a negative pulse applied to its gate.

triangle – Three-sided, enclosed plane figure.

triboelectric series – A list of materials that can produce static electricity by contact, friction, or induction. As an example of the friction method, a glass rod rubbed with fur becomes negatively charged, but if rubbed with silk, becomes positively charged. some materials that build up static electricity easily are flannel, silk, rayon, amber, hard rubber, and glass. In the following list, the material that is the higher in the list will become positive to the material that is lower in the list:

POSITIVE
glass
mica
nylon
fur
silk
paper
cotton
wood
acrylic
polystyrene
rubber
NEGATIVE

triboelectricity – The production of static electricity by contact or friction between different materials.

trickle charging – A constant current charging method that keeps cells on standby service at full charge by passing a small current through them until they are put into service.

tri-cresyl-phosphate –
[1]Plasticizer used in rejuvenator to restore resilience to the dope film.
[2]A toxic substance used in lubricants.

tricycle landing gear – The landing gear of an aircraft in which the main wheels are behind the center of gravity and the nose is supported with an auxiliary nose wheel.

trigger pulse – An electric pulse applied to certain electronic circuit elements to start, or trigger, an operation.

triggering transformer – High-voltage transformer connected in series with the igniter in a high-energy ignition system for a turbine engine. The transformer places a high voltage across the igniter, ionizing the gap and producing the triggering spark.

trigonometry – The branch of mathematics that deals with the ratios between sides of a right triangle, and the application of these facts in finding the unknown side of any triangle.

trijet – An aircraft that is propelled by three jet engines.

trim – The adjustment of the controls of an airplane to get a balanced or stable condition of flight.

trim tab — A small auxiliary hinged portion of a movable control surface that may be adjusted by the pilot in flight to a position which will result in a balance of control forces.

trimmer — A potentiometer, generally with a screwdriver adjusted slider used peaking or fine tuning a circuit.

trimming — Use of a trim curve to check the rated thrust output of an engine. The process involves operating the engine neat takeoff power and obtaining certain readings to compare to the trim chart target values.

trinomial — A mathematical term used to express the terms connected by plus or minus signs.

triode — A vacuum tube having three active electrodes.

trip-free circuit breaker — A circuit protection device which will open a circuit when a current overload exists regardless of the position of the control handle.

triphibian — An aircraft landing gear configuration that gives it the ability to operate from the ground, including snow and ice, and from water.

triplane — An airplane having three main supporting wing surfaces, usually located one above the other.

triple-slotted flap — A type of high lift device used to reduce the takeoff or landing speed by changing the lift characteristics of a wing during the landing or takeoff phases. When the flap is no longer needed, they are returned to a position within the wing to regain the normal characteristic of the wing. Triple-slotted flaps extend downward and rearward away from the position of the wing. The slots open from the trailing edge of a wing in three sections, and allows a flow of air over the upper surface of the flap. The effect is to streamline the airflow and to improve the efficiency of the flap.

triple-spool engine — Usually a turbofan engine design where the fan is the N_1 compressor, followed by the N_2 intermediate compressor and the N_3 high pressure compressor all of which rotate on separate shafts at different speeds.

tropopause — Boundary layer between the troposphere and the stratosphere.

troposphere — The layer of the earth's atmosphere immediately next to the surface and extending upward to the tropopause, approximately 36,000 ft.

troubleshooting – Systematic analysis of a malfunction in a system or component to determine the cause.

true airspeed – Calibrated airspeed corrected for non-standard temperature and altitude.

true airspeed indicator – An airspeed indicator which takes into consideration dynamic pressure, static pressure, and free air temperature to provide a display of the true airspeed.

true bearing – The direction measured in degrees clockwise from true north.

true course – A navigational direction or course of an aircraft in flight as measured from the geographic north pole.

true north – The direction on the earth's surface which points towards the exact geographic north pole.

true power – The power that actually exists in an AC circuit. It is the product of the voltage, current, and power factor; or is the product of the voltage and the current which is in phase with that voltage. It is expressed in watts.

truncated – Cut off as having the angles of an object cut off.

truss – A type of frame arranged together in such a manner that all members of the truss can carry both tension and compression loads with cross-bracing achieved by using solid rods or tubes.

truss fuselage – Fuselage usually constructed of welded steel tubing to carry the tensile and compressive loads. A superstructure or auxiliary framework is often attached to the truss to give the structure a desirable aerodynamic shape.

truss head – Low rounded top surface with a flat bearing surface.

T-square – An instrument used for making aircraft drawings. It consists of a head and a perpendicular blade and is shaped like the letter T.

tubing – A rigid, hollow piece of metal through which fluids or wiring is passed.

tubing cutter – A tool consisting of a sharp wheel and a set of rollers. The tube to be cut is clamped between the cutter cutting wheel, and the rollers and the tool are rotated around the tube. The cutting wheel is fed into the groove as the tube is cut.

tumble limit – The number of degrees of pitch or roll a gyro will tolerate before it reaches its gimbal stops. Beyond this point the gyro will tumble.

tumbling — Tumbling is the process of cleaning or abrading parts in a rotating container, either with or without cleaning or abrasive materials.

tuned radio frequency receiver — Abbrev.: TRF. A radio receiver in which tuning and amplification are accomplished in the RF section before the signal reaches the detector. After the detector, one or more stages of AF amplification are employed to increase the output sufficiently to operate a loudspeaker.

tungar rectifier — A high-capacity diode rectifier tube having a heated cathode and a graphite plate in an envelope filled with argon gas.

tungsten — A gray-white, heavy, high melting point, hard metallic element with steel to increase its hardness.

tungsten steel — Steel with which tungsten has been alloyed. It is used in the manufacturing of cutting tools because of its hardness.

tuning — The process of adjusting circuits to resonance at a particular frequency.

tunnel diode — A special form of semiconductor diode that exhibits a negative resistance characteristic. Under certain conditions, an increase in voltage across the tunnel diode results in a decrease in current through it.

turbine — A rotary wheel device fitted with vane like airfoils that is actuated by impulse or reaction of a fluid flowing through the vanes, or blades, arranged around a central shaft.

turbine bucket — A colloquial term for the blades on a turbine wheel.

turbine disc — The metal disc to which turbine blades are attached.

turbine discharge pressure — Symbol: PT7. The total pressure at the discharge of the low-pressure turbine in a dual turbine axial flow engine.

turbine efficiency — A ratio of actual work performed by the turbine wheel in ft.-lbs./Btu, and the laboratory standard of 778 ft.-lbs. of work in 1 Btu. Expressed as a percentage.

turbine engine — A type of aircraft engine which consists of an air compressor, a combustion section, and a turbine. Thrust is produced by increasing the velocity of the air flowing through the engine.

turbine inlet temperature — Abbrev.: TIT. Temperature taken in front of the first stage turbine nozzle vanes. The most critical temperature taken within the engine and one that is closely controlled by the fuel scheduling.

turbine nozzle — Orifice assembly through which exhaust gases are directed prior to passing into the turbine blades.

turbine nozzle vanes — Turbine stationary airfoils that precede each turbine blade set. They function to increase gas velocity and also to direct the gases into the turbine blade at the optimum angle.

turbine stage — A single turbine wheel in a turbine engine. Each turbine wheel constitutes a separate stage. *See also stage.*

turbine wheel — A rotating device actuated by either reaction, impulse or a combination of both. Used to transform some of the kinetic energy of the exhaust gases into shaft horsepower to drive the compressors and accessories.

turbocharger — An air compressor driven by exhaust gases, which increases the pressure of the air going into the engine through the carburetor or fuel injection system.

turbocompound engine — A high altitude flying reciprocating engine that supplements the internal supercharger by an external power recovery turbine (PRT) type turbosupercharger driven by a portion of the exhaust gas from the aircraft engine. The PRT is connected through a fluid coupling and gear arrangement to help drive the crankshaft.

turbojet engine — A turbine engine which produces its thrust entirely by accelerating the air through the engine.

turboprop engine — A turbine engine which drives a propeller through a reduction gearing arrangement. Most of the energy in the exhaust gases is converted into torque, rather than using its acceleration to drive the aircraft.

turboshaft — A gas turbine engine geared to an output shaft. Usually for rotorcraft installation but also for many marine and industrial uses.

turboshaft engines — A gas turbine engine that delivers power through a shaft to operate something other than a propeller.

turbulence — An occurrence in which a flow of fluid is unsteady.

turbulent flow — A flow of fluid in an unsteady state.

turn — To machine on a lathe.

turn and bank indicator — *See turn and slip indicator.*

turn and slip indicator — A flight instrument consisting of a rate gyro, to indicate the rate of yaw, and a curved glass clinometer to indicate the relationship between gravity and centrifugal force. It indicates the relationship between angle of bank and rate of yaw.

turn coordinator — A rate gyro which senses rotation about both the roll and yaw axes.

turnbuckle — A device used in a control system to adjust cable tension. It consists of a brass barrel with both left- and right-hand threaded terminals.

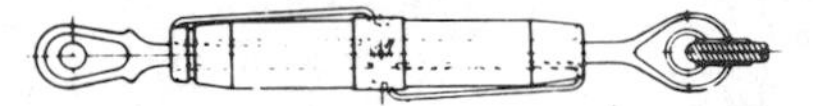

turning error — One of the errors inherent in a magnetic compass caused by the dip compensating weight. It shows up only on north or south headings and causes the compass to lead or lag the actual turn.

turpentine — A thinner and quick drying agent used in varnishes, enamels, and other oil-based paints.

turret — A tool mount on a special type of lathe. The turret holds several tools which may be turned into the work as needed.

turret lathe — A special type of metal-turning lathe in which the cutting tools are mounted in a turret.

tweek test — A test of the output of a wheel speed sensor made by tweaking or flipping the sensor blade with the fingers to rotate it enough for it to generate a voltage.

twelve point head — A standard head form for externally wrenched screws designed for use in counter-bored holes.

twelve-point socket — A type of socket wrench that has a 12-point double hex opening which makes it easy to position on a bolt or nut.

twenty-minute rating — The ampere hour rating of a battery indicating the amount of current that can be drawn from a battery in order to discharge it in twenty minutes.

twin-row radial engine — A radial engine having two rows of cylinders, one behind the other.

twist drill — A metal cutting tool which has a straight shank and deep spiral flutes in its side which provide a passage for the chips to be removed from the hole it is cutting.

two view drawing — A type of orthographic projection drawing which uses two views to portray an object.

two-cycle engine — A reciprocating engine in which a power impulse occurs on each stroke of the piston. As the piston moves outward, fuel-air mixture is drawn into the crankcase below the piston; while above the piston, the mixture compressed. Near the top of the stroke, ignition occurs. As the piston moves downward, power is produced by the crankshaft. Near the bottom of the stroke, exhaust action takes place on one side of the cylinder, and intake action occurs on the opposite side.

two-part adhesive — A special type of adhesive that consists of two parts. One part is the base and the other part is the accelerator. When the two are mixed together in the correct amounts the adhesive cures in a short time period.

two-state device — An electronic component which may be switched to a high or resistance state by a control signal.

two-stroke-cycle engine — *See two-cycle engine.*

two-way communications — The ability of both stations involved in communications to transmit and receive.

TYP — A mechanical drawing word that means "typical". It is used to show that the part symbolized with TYP is typical for more than one area or part of the drawing.

type — Specific make and basic model of aircraft which includes modifications which do not alter its handling or flight characteristics.

Type Certificate Data Sheets — The official specifications of an airplane, engine, or propeller. These are issued by the FAA and the device, in order to be airworthy, must conform to these specifications.

Ty-Rap — A patented nylon strap used to hold wire bundles together.

U

U-bolt — A rod threaded on both ends shaped like the letter U. U-bolts are used to fasten cables around a thimble.

UHF — *See ultrahigh frequency.*

ultimate load — The amount of load applied to a part, beyond which the part will fail.

ultrahigh frequency — Abbrev.: UHF. Radio frequencies between 300 and 3.000 MHz.

ultrasonic cleaner — Type of cleaning apparatus which transmits sound waves through a fluid. It is widely used for filter and bearing cleaning.

ultrasonic frequencies — Frequencies in the 20,000 Hz range which cannot be heard by the human ear.

ultrasonic inspection — A form of nondestructive inspection in which the condition of a material is determined by its ability to conduct vibrations normally at frequencies above those to which the human ear responds.

ultrasonic soldering — A method of soldering in which the tip of a special soldering iron is vibrated at an ultrasonic frequency. The vibration separates the oxide film on the surface of the metal. In addition, a flux is melted over the surface of the hot metal to prevent more oxide from forming.

ultraviolet lamp — A lamp that produces a light wavelength slightly shorter than the wavelength of visible light.

ultraviolet radiation — *See ultraviolet rays.*

ultraviolet rays — Radiation from the sun with wavelengths shorter than are visible. These rays damage organic materials.

umbilical cord — A cable used to carry power or life support to an astronaut for operating outside a space vehicle.

unbalanced cell — A condition in a nickel-cadmium battery in which one cell has discharged more than the other cells in the battery. This is the initial step in a thermal problem.

uncontrolled spin — A spin in an airplane in which the controls are of little or no use in effecting a recovery.

undercarriage — A term used to describe an airplane's entire landing gear.

undercurrent relay — An electrical circuit protection device that senses a specified low voltage condition and opens the circuit to protect it.

underpowered — An undesirable condition in which an engine does not have enough power to achieve the desired result.

undershoot — A term used to describe a condition of flight in which a landing aircraft touches the ground short of the runway or the landing strip.

underslinging — Placing the main rotor hub around and below the top of the mast as is done on semi-rigid rotor systems.

underspeed condition — Underspeeding results when the blades of a propeller have moved to a higher angle than that required for constant-speed operation. When the speed drops below the RPM for which the governor is set, the resulting decrease in centrifugal force exerted by the governor flyweights permits the speeder spring to lower the pilot valve, thereby opening the propeller-governor metering port. Oil is then directed to the prop in the proper direction to correct the underspeed condition.

undervoltage relay — An electrical circuit protection device that senses a specified low voltage condition and opens the circuit when voltage drops below a predetermined value.

underwing fueling — A method of fueling an aircraft from a single-point pressure fueling port located under the wing.

undrainable fuel — Amount of fuel that remains in the system after draining. This is considered a part of the empty weight of the aircraft.

undrainable oil — The oil remaining after draining the oil from an engine. This oil is considered a part of the empty weight.

unfeather — Unfeathering operation in which the prop blades are brought to low blade angle.

uniform acceleration — The uniform rate of increasing speed of an object.

uniform surface corrosion — A general covering of corrosion in which the action has been even. No pits or localized damage have formed.

unijunction transistor — A special form of transistor which allows flow between its two bases when an appropriate voltage is applied to its emitter.

union — Plumbing connectors, or fittings, that attach one piece of tubing to another or to a system units.

universal chuck — The three clamping jaws of a drill that move in and out simultaneously by turning it by hand or with a chuck key.

universal joint — A type of flexible joint that allows one shaft to drive another shaft at angles to each other.

universal motor — A series-wound motor that will operate on either alternating or direct current.

universal propeller protractor — A precision measuring device which is used to measure amount of blade angle.

unleaded gasoline — Reciprocating engine fuel that does not contain any tetraethyl lead.

unloading valve — A pressure control valve that is used in aircraft hydraulic systems to act as a pressure-limiting device to prevent excessive pressures from bursting lines and blowing out seals. When the pressure reaches a predetermined value, the valve reroutes the high output pressure back to the pump inlet. *See also system pressure regulator.*

unmetered fuel — Fuel which enters the fuel control from the fuel pump.

unscheduled maintenance — Maintenance performed as a result of discrepancies found by flight and ground personnel.

unstable air — Air with a temperature lapse rate that is greater than the surrounding air.

unstable oscillation — An oscillation whose amplitude increases from which there is no tendency to return toward the original attitude, the motion becoming a steady divergence.

unusable fuel — Fuel that cannot be consumed by the engine. This fuel is considered part of the empty weight of the aircraft.

unusable oil — Oil that cannot be used by the engine.

updraft carburetor — A carburetor mounted on the bottom of a reciprocating engine. All of the air entering the engine flows upward through the venturi.

upper deck pressure — It is the pressure of the air measured in inches of mercury between the compressor and the throttle plate. This air is used not only for the powerplant but also for pressurization of the cabin and the fuel injection system.

upset head – The end of the rivet that is not the manufactured head but is formed on the shank end of the rivet that protrudes through the hole. *See also upsetting.*

upsetting – The process of increasing the cross-sectional area by displacement of material longitudinally and radially, as in bucking a rivet.

upwind – Into the wind, or the direction from which the wind is blowing..

usable fuel – Portion of the total fuel load available for consumption by the aircraft in flight.

useful load –

[1]Weight of the occupants, baggage, usable fuel, and drainable oil. The difference between maximum and empty weight.

[2]The difference between the empty weight of the aircraft and the maximum weight of the aircraft. Sometimes referred to as the payload.

utility category airplane – An airplane certificated for flight which includes limited acrobatics. This includes spins, lazy eights, chandelles, and steep turns in which the bank angle exceeds 60°.

utility finish – A finish used on aircraft which provides the fabric with the necessary fill and tautness, but lacks the glossy appearance of a show-type finish.

V

vacuum — A negative pressure, or pressure below atmospheric. It is usually expressed in inches of mercury (in. Hg).

vacuum bottle — A container that has inner and outer walls with all of the air taken from the space between the container walls. A vacuum bottle is used for keeping liquids hot or cold.

vacuum distillation — A distillation process that boils a liquid at a lower temperature than would otherwise be possible to do without damaging the liquid.

vacuum forming — A type of thermoplastic forming that uses vacuum, and a heated die to form the required shape of the plastic.

vacuum pump — A pump mounted on an aircraft engine, used to move air. The negative pressure produced by this pump is used to drive some of the gyroscopic flight instruments.

vacuum tube — An evacuated glass or metal envelope containing a cathode. a heater, a plate, and often one or more grids. It serves as an electron control valve.

vacuum-tube voltmeter — Abbrev.: VTVM. An electronic voltage-measuring instrument used for electronic circuit testing. Its very high input impedance prevents it from drawing appreciable power from the circuit being tested.

valence electron — An outer electron which may be given up or gained in the process of ionic compound formation, or an outer electron which is shared with another atom in the process of covalent compound formation.

valve — A device that regulates a flow in one direction only by means of a flap, lid, plug, etc.

valve blow-by — *See blow-by, valve.*

valve clearance — The clearance between a valve head and rocker arm when the valve is seated.

valve core — A spring-loaded, resilient check valve inside the valve stem that allows air to flow into a tire to inflate it and then traps the air, preventing its leaking out. The pin of the valve core may be depressed to release the air from the tire.

valve duration – The length of time, measured in degrees of crankshaft rotation, a valve in an aircraft engine remains open.

valve face – The ground face of an intake or exhaust valve which forms a seal against the ground seat in the cylinder head when the valve is closed.

valve float – A condition in which the frequency of valve opening exactly corresponds to the resonant frequency of the valve spring. Under these conditions the valve spring will exert no closing force.

valve grinding – A process of removing part of the valve face with a precision grinding machine to restore the perfect seal between the valves and their seats.

valve guide – The component in an aircraft reciprocating engine cylinder head which guides the valve and holds the valve head concentric with the valve seat.

valve lag – The number of degrees of crankshaft rotation after top or bottom center at which the intake or exhaust valves open or close. For example, if the intake valve closes at 60° of crankshaft rotation after the piston passes over top center and starts down on its intake stroke, the exhaust valve lag is 60°.

valve lapping – A process in which the valve is fitted into the valve seat by using a fine abrasive compound between the two. Valve lapping removes any fine rough material to ensure an airtight seal is formed.

valve lead – The number of degrees of crankshaft rotation before top or bottom center at which the intake or exhaust valves open or close. For example, if an intake valve opens 15° before the piston reaches top center on the exhaust stroke, it is said to have a 15° valve lead.

valve lift – The distance that the valve is lifted off its seat when it is opened by the cam.

valve overlap – A method for increasing the efficiency of engine operation by allowing the low pressure caused by the exhaust gases leaving the cylinder to help the fresh charge of fuel/air that is in the induction system to start moving into the cylinder. Valve overlap is the angular distance of crankshaft rotation, of a four-stroke-cycle reciprocating engine, when the piston is passing top dead center on the exhaust stroke when both the intake and exhaust valves are open.

valve ports – The holes in the cylinder of an aircraft reciprocating engine, through which the intake gases enter the cylinder and the exhaust gases leave.

valve radius gage – A gage to determine if a valve has the proper radius between the stem and the head. An improper radius is an indication of a stretched valve.

valve seat – A hardened ring of steel or bronze, shrunk-fit into the cast aluminum cylinder head to provide a surface on which the poppet valve face can seat.

valve spring tester – A machine used in engine overhaul to test the condition of valve springs by measuring the force required to compress them to a specified height.

valve springs – Helical-wound steel wire springs which are used to close the poppet valves in an aircraft engine cylinder.

valve stem – That portion of a poppet valve which rides in the valve guide and maintains concentricity between the head of the valve and the valve seat.

valve stretch – Elongation of the valve stem due to an overheated condition.

valve stretch gage – A type of go/no-go tool that measures the radius between the valve head and the valve stem.

valve timing – The relationship between the crankshaft rotation and the opening and closing of the intake and exhaust valves.

valve-timing clearance – The clearance to which poppet valves using solid valve lifters are adjusted to set the cam for valve timing. After the timing is set, the valve clearance is adjusted to the cold, or running clearance.

vanadium – A malleable, ductile, silver-white metallic chemical element. Alloyed with steel, it is toughened and adds to the tensile strength to the steel.

vanadium steel – A steel alloyed with 0.10-0.15% vanadium to provide additional hardness and toughness to it. Vanadium steel is used in the manufacture of technicians tools.

vane – Term generally used for stationary airfoils within the engine.

vane-type pump — A form of constant displacement fluid-moving pump in which a rotor containing sliding vanes turns in an eccentric cavity to force the fluid through the pump.

vanishing point — Points in a drawing that converge to give the appearance of having depth.

vapor — Gaseous state of a material.

vapor degreasing — A method of degreasing in which the part to be cleaned is treated with hot vapors of some solvent such as trichlorethylene.

vapor lock — A condition in an aircraft fuel system in which liquid fuel has turned into a vapor in the fuel lines. This vapor prevents the flow of liquid fuel to the carburetor.

vapor pressure — The pressure exerted by the vapor above a liquid which prevents the release of additional vapor at any specific temperature.

vapor separator — A device in a pressure-type carburetor regulator unit that prevents air in the fuel from upsetting the metering of the carburetor. The vapor separator consist of a small float and needle valve positioned in the vapor separator chamber. When there are no vapors in the chamber, the float is raised and holds the needle valve closed. As vapors gather, the fuel level in the chamber drops, lowering the float until the needle valve opens releasing the vapors back to the fuel tank.

vapor trail — A moisture trail that high flying jet engines produce by the condensation of water vapor in the exhaust gases that leave the engine.

vapor-cycle air-conditioning system — A system for cooling the air in an aircraft cabin in which the cabin heat is absorbed into a liquid refrigerant, turning it into a vapor. This vaporized refrigerant is carried onto a condenser outside the airplane where the heat is given up to the outside air, causing the refrigerant to revert back to a liquid to begin the cycle over again.

vaporize — To change a liquid into a vapor.

vaporizing tube — A type of fuel nozzle which injects a fuel-air mixture into the combustor. This nozzle is in a system which operates at lower pressure than the atomizing type fuel nozzle system.

variable absolute pressure controller — Abbrev.: VAPC. The intelligence of the turbo system. It monitors compressor discharge pressure and limits the maximum pressure and also maintains the discharge pressure slightly higher than manifold pressure. The VAPC controls the discharge pressure by regulating the oil flow through the waste gate actuator. The variable portion of the pressure controller come about by being connected by mechanical linkage to the throttle. Opening the throttle to maintain a higher manifold pressure will also adjust the VAPC to maintain a higher compressor discharge pressure. The system's design prevents the turbocharger from working at maximum output when the excess pressure is not needed.

variable capacitor — A type of capacitor whose capacity can be changed by varying the distance between the plates. Variable capacitors are commonly used in radio or radar tuning devices.

variable cycle engine — Turbo-ram, a theoretical design capable of speeds in excess of Mach 5, probably to be fueled with hydrogen rather than hydrocarbon fuels. This engine functions as a turbojet up to perhaps Mach 3, then as doors close off the compressor inlet, it operates from a ramjet type duct surrounding the engine.

variable displacement pump — A pump whose output may be varied by the pressure on the system. For high pressure applications, this is usually done by varying the stroke, either actual or effective, of a piston-type pump.

variable geometry air inlet duct — The inlet duct on a supersonic turbojet aircraft whose area or shape may be varied in flight to provide the proper inlet pressure to the first stage of the compressor as the airspeed of the airplane changes.

variable geometry aircraft — A type of aircraft that has the ability to sweepback the wings after takeoff and for flying at supersonic flight speeds.

variable pitch propeller — A propeller whose pitch may be changed while the engine is operating.

variable pitch turbofan — An advanced technology designed engine which has controllable pitch blades. Only prototype models are in present use.

variable resistor — A resistor whose resistance may be varied by rotating an adjusting screw or shaft, or adjusting the position of a sliding contact. Rheostats and potentiometers fall into the category of variable resistors.

variable restrictor — A unit which may be adjusted to control the amount of fluid flow and thereby control the operating speed of a unit.

variable-angle stator vanes — Includes inlet guide vanes and compressor stator vanes which have a capability of changing their angle to the oncoming airstream. They operate to open as the power lever is advanced and to close as the power lever is retarded and control engine stall tendencies on acceleration and deceleration.

variable-MU tube — An electron tube having a control grid in which the grid wires are spaced less closely at the center than at the ends. This causes the amplification factor to change as grid bias is changed. Also called a remote-cutoff tube.

variation — A compass error caused by the compass magnet aligning with *magnetic* poles, while the aeronautical charts and maps are oriented to the *geographic* poles. Variation is determined by the geographic location of the airplane and is not affected by the airplane's heading.

vari-ramp — A movable ramp in a C-D inlet which controls supersonic diffusions and airflow velocity into the compressor.

varnish —

[1]Surface finish: A preparation of resin dissolved in oil or in alcohol. Used to give a glossy surface to wood, metal, etc.

[2]Engine deposit: A smooth oil baked deposit formed on reciprocating engine cylinder walls that operate at excessively high temperatures.

varsol — A petroleum product similar to naptha. Used as a solvent.

V-belt — A drive belt that has a cross section in the shape of a "V". V-belts are used to drive generators, pumps, and other accessories from a source of motorized power.

V-block — Hardwood block with a V-notch cut across it. It is used for shrinking or for stretching sheet metal angles or flanges.

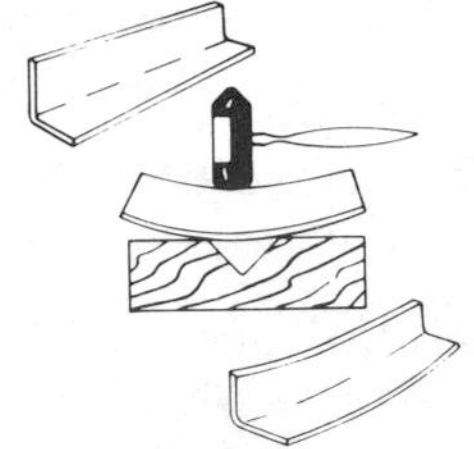

vector – A quantity with both magnitude and direction. The length of the line indicates magnitude; the arrowhead indicates direction of motion, or force.

velocity – The time-rate of linear motion in a given direction.

vendor – Someone who sells something such as goods or services.

veneer – Thin sheet of wood used in the manufacture of plywood.

V-engine – An engine with cylinders arranged in two rows, attached to the crankcase in the form of a "V", with an angle of between 45° and 60° between the banks.

vent – A small hole, or orifice, which allows the pressure on the inside of a compartment to be the same as that on the outside.

ventilate – To provide fresh air to the inside of a compartment.

ventral fin – A vertical, stabilizing fin on the lower rear portion of an airplane fuselage. Its purpose is to increase the directional stability of the airplane by increasing the area behind the vertical axis.

venturi – A specially shaped restrictor in a fluid flow passage, used to increase the velocity of the fluid and decrease its pressure.

venturi tube – A short tube with a large opening at both the front and rear end, and with a specially designed restrictor between them. Air flowing through this tube has its velocity increased and its pressure decreased as it passes the restriction.

vernier – A means of making extremely small divisions or measurements by measuring the difference in the reading of two scales occupying the same space, with one having one less graduation than the other.

vernier caliper – A precision measuring tool used to measure the inside or outside dimension of an object. An auxiliary or vernier scale is used to accurately divide the increments of the regular scale.

vernier micrometer caliper – A micrometer caliper with a special vernier scale which allows each one thousandth-inch increment to be broken down into ten equal parts so that 1/10,000″ may be accurately read.

vertical axis – The axis of an airplane extending vertically through the center of gravity. Also called the yaw axis.

vertical speed indicator — A sensitive differential pressure gage that indicates the rate at which an aircraft is climbing or descending. The vertical speed indicator is connected to the static system and senses the rate of change of static pressure. The vertical speed indicator dial is positive in climb and negative when descending in altitude. The dial pointer moves in either direction from the zero point, depending on whether the aircraft is going up or down. In level flight, the pointer remains at zero.

vertical stabilizer — The fixed vertical surface of an aircraft empennage to which the rudder is hinged. Also called vertical fin.

vertical takeoff and landing — Abbrev.: VTOL. A type of aircraft that has the ability to takeoff and land without any forward motion.

vertical vibration — A vibration in which the movement is in the vertical or up and down direction. One cause might be an out of track main rotor.

vertigo — The sensation of dizziness caused by the airplane in a spin, hard turn, or other vortex. Vertigo cause a person to experience spatial disorientation caused by the physical senses sending conflicting signals to the brain.

very-high frequency — Abbrev.: VHF. The frequency range between 30 and 300 MHz.

very-high frequency omniRange navigation equipment — Abbrev.: VHF VOR. The omnidirectional or all-directional range station provides the pilot with course from any point within its service range. It produces 360 usable radials or courses, any one of which is a radio path connected to the station. The radials can be considered as lines the extend from the transmitter antenna like spokes of a wheel. Operation is in the VHF portion of the radio spectrum (frequency range of 109.0-117.95 MHz).

very-low frequency — Abbrev.: VLF. The frequency band between 30 to 300 kHz.

vibrating-type voltage regulator — A voltage regulator for direct current generators or alternators which uses vibrating points to sense the voltage and provide a varying resistance for the generator field current.

vibration — Regular or irregular movement characterized by complete reversal of direction with time.

vibration insulator — A resilient support that tends to isolate a system from steady state excitation.

vibration isolator — A flexible shock mount type support installed between a component and the structure. Vibration isolators reduce damage to electronic units by keeping heavy vibrations in the structure from being transmitted into the unit.

vibration meter — A meter which reads out in MILS of inches or in./sec. of vibration occurring at the engine outer casings. Typical limit 3 to 4 MILS.

vibration pickup — Small electrical generator which transmits an engine vibration signal to a vibration meter either in the aircraft or in a test cell.

vibrator — A special form of relay which makes and breaks the flow of direct current to produce pulsating DC, which may be passed through an induction coil to change its voltage.

vibrator ignition system — An older type turbine engine ignition system which uses no storage capacitors.

vibratory torque control — Special patented coupling between the crankshaft and propeller shaft of a Continental Tiara engine. It incorporates a quill shaft to absorb torsional vibrations and a centrifugally actuated lock to lock out the quill shaft for operation when a solid shaft would be more advantageous.

video — A term describing electronic circuit components controlling or producing the visual signals displayed on a crt.

virga — Water that falls from clouds in the form of streaks but evaporates before it reaches the ground.

viscosimeter — An instrument used to measure the viscosity of a fluid.

viscosity — Property of a fluid that resists internal flow. The "stickiness" of a fluid.

viscosity index — The measure of change in viscosity of an oil with a change in temperature.

viscous — Thick, having a relatively high resistance to flow.

vise — An adjustable bench-mounted tool for holding material while it is being worked on.

visible light — Lightwave length between 4,000 and 7,700 angstroms that can be seen by the eye and can be perceived with the mind.

visible line — A line on an aircraft drawing that represents a portion of an object that may be seen.

visual flight rules — Abbrev.: VFR. FARs that govern the flight of aircraft when the visibility is less than the minimally accepted standards.

visual inspection — Inspection of a part or component by visual means only.

visual omnirange — *See very-high frequency omnirange navigation equipment.*

vitrify — To make a material into glass or a glass like substance by fusion due to high heat.

vivid color — One of the highly reflective colors used on aircraft for the maximum in visibility.

void — An unsatisfactory portion of a weld in which there are gaps in the weld area.

voids — Voids are internal fissures in ferrous materials. Sometimes called "fish eyes", "chrome checks", "shatter cracks", or "snowflakes".

volatile — Quickly evaporating.

volatile liquid — A fast evaporating fluid used for wiping surfaces prior to painting.

volatile mineral spirit — A fast evaporating fluid used for wiping surfaces prior to painting.

volatility — The ease with which the fluids change from a liquid to a vapor.

volt — Symbol: V. A unit of electromotive force or potential difference that is needed to force one ampere of electrical flow through a resistance of one ohm.

voltage — A term for electromotive force (emf). The force causing flow of electrons, measured in volts.

voltage amplifier — A circuit that is capable of increasing extremely small amounts of voltage.

voltage avalanche — The reverse voltage required to cause a zener diode to break down.

voltage divider — A series of resistors placed across the poles of a source to provide a number of different voltages.

voltage doubler — A circuit which produces an output voltage twice that of the input.

voltage drop –
[1]Decrease in voltage in an electrical circuit due to an increase in conductor length.
[2]The loss in potential energy in a circuit when a current is made to flow through a load and some of the energy is converted from electrical energy to some other form such as heat.

voltage dropping resistor – A resistor placed in series with some other component in order to reduce the terminal voltage across or limit the current through that component.

voltage regulation – The maintaining of a constant voltage level despite the variations in load current requirements.

voltage regulator – A circuit which maintains a constant-level voltage supply despite changes in input voltage or load.

voltage rise – The increase in potential energy caused by a source such as a chemical cell connected in series aiding with the general current direction in the circuit.

voltage spike – A burst of high voltage that last for only a short time.

voltage standing-wave ratio – Abbrev.: VSWR. The ratio of the maximum voltage to the minimum voltage along a coaxial cable.

voltage-fed antenna – An antenna fed at its end where the voltage is the highest.

voltaic cell – Sometimes called a galvanic cell. It consists of two dissimilar pole pieces separated by an electrolyte, i.e. ionic conducting solution.

volt-amperes – The product of the voltage and current in a circuit.

voltammeter – A D'Arsonval meter movement that can be used either as an ammeter or a voltmeter. An ammeter can be converted to a voltmeter by placing a resistance in series with the meter coil and measuring the current flowing through it. A voltmeter is a current-measuring instrument, designed to indicate voltage by measuring the current flow through a resistance of known value.

voltmeter sensitivity – A method used to determine the accuracy of a meter. The sensitivity of a voltmeter is given in ohms per volt and is determined by dividing the resistance of the meter plus the series resistance by the full scale reading in volts. Thus, the sensitivity of a 100 microampere movement is the reciprocal of 0.0001 ampere or 10,000 ohms per volt.

voltmeter – An electrical measuring instrument used to measure electrical pressure or voltage.

volt-ohm-milliammeter – A combination multirange electrical measuring instrument. The instrument can measure volts, amps, or resistance by selecting one of the instrument ranges which in turn changes the internal connections to measure a wide range of voltages.

volume – Space occupied. It is measured in cubic units, and it is found by multiplying the area of the base of the container by its height.

volume control – The circuit in a receiver or amplifier which varies loudness.

volumetric efficiency – The ratio of the volume of the charge taken into a cylinder, reduced to standard conditions, to the actual volume of the cylinder.

vortex – A whirling, circulatory fluid motion. *See also wing tip vortices.*

vortex compressor blade tip – A term to imply that this design provides a smooth airflow at its tip end. Same as profile-tip.

vortex generators – Small, low-aspect ratio airfoils mounted in pairs on the upper surface of high-speed aircraft wings. Their function is to bring high energy to the surface of the wing by keeping the air from separating from the surface of the wing, thereby delaying shock induced separation.

vortices – Turbulent air disturbances produced at the wing tips caused by the higher air pressure underneath the wing spilling over to the top of the lower pressure air on top of the wing. This spilling over produces tight wind vortexes that must be avoided by small airplanes which are flying near them.

V-tail surface – An empennage consisting of two fixed and two movable surfaces arranged in a V shape. These two surfaces have the same aerodynamic function as the more conventional three surfaces.

vulcanize – A process of treating crude rubber with sulfur and subjecting it to heat. This process increases the strength and elasticity of the rubber.

W

wafer – A thin, flat slice.

wafer-type selector switch – A multiple contact switch having the contacts arranged around the edge of a wafer and selected by a knob in the center.

waffle piston – A reciprocating engine piston that has fins cast on the bottom of the inside of the piston head, that appears like the surface of a waffle. This type of forging provides added strength and surface are for carrying heat away from the piston.

wake turbulence – The track of disturbed air left by the aircraft wing tip vortices. *See also vortices.*

walk-around bottle – A pressurized container of breathing oxygen that is small enough for a passenger or crew member to carry around the airplane. It consist of a strap for carrying, a mask, and regulator and volume indicator.

warm – Something that gives off a moderate amount of heat.

warm front – A warm air mass that flows over a colder air mass. Warm fronts bring low ceilings and rain.

warm-up time – The time used by a component for all of its parts to reach operating temperature.

warning area – A hazardous area.

warning lights – Annunciation lights in an aircraft cockpit to warn the flight crew of a dangerous situation or the failure of a system or component.

warp threads – Threads parallel to the length of the fabric.

Warren truss – A form of truss structure used for aircraft fuselages in which the diagonal members carry both tensile and compressive loads.

wash – The disturbance in the air produced by the passage of an airfoil. Also called the "wake" in the general case for any solid body.

wash primer – A self-etching primer used on aluminum or magnesium. It is often used to prepare the surface for zinc chromate primer.

washer — A flat metal disc with a central hole. It is used to provide a smooth surface for a nut or bolt to seat against, and to shim between a nut and the surface to compensate for an improper grip length of the bolt.

washer face — A washer face is a circular boss on the bearing surface of a bolt or nut.

wash-in — A condition of rigging in which a wing has an increase in its angle of incidence near the tip. The lift is increased on the side of the aircraft having wash-in.

wash-out — A condition of rigging in which a wing has decrease in its angle of incidence near the tip. Lift is decreased on the side of the airplane having wash-out.

Waspaloy — Trade name of Pratt and Whitney division of United Technologies Corp. A nickel-base alloy in the family of turbine super metals of high temperature strength.

waste gate — A controllable valve in the tailpipe of an aircraft reciprocating engine equipped with a turbocharger. The valve is controlled to vary the amount of exhaust gases forced through the turbocharger turbine. When the waste gate is fully closed, all of the gases must pass through the turbine.

waste gate actuator — This hydraulic actuator is used to position the waste gate. Engine oil pressure is used to move the internal piston against spring tension to close the waste gate. When oil pressure is released, the spring moves the piston toward the "open" waste gate position. Positioning of the waste gate actuator is accomplished by controlling the oil flow from the actuator. Oil that enters the actuator comes directly from the engine oil pump.

water collector sump — A low point in the tray below the evaporator in a vapor-cycle air conditioner, in which the water which has condensed on the evaporator coils may drain and be discharged overboard.

water injection — Water or an alcohol-water mixture that is injected either into the cylinders of a reciprocating engine or into the combustion section of a turbine engine to remove some of the heat that would cause damage under conditions of full power operation.

water soluble — Having the ability to dissolve in water.

waterline — A horizontal reference plane from which vertical measurements in an aircraft may be taken.

watt – The basic unit of electrical power. One ampere flowing under a pressure of one volt is equal to one watt. One watt equals 1/746 HP.

wattage rating – The maximum amount of power an electrical component needs to operate an appliance or device without damaging the device.

watt-hour – A unit of electrical power equal to one watt acting for one hour.

wattmeter – An instrument designed to measure electric power.

watt-second – A unit of electrical power equal to one watt acting for one second.

wave carrier – High-frequency AC which can be modulated to carry intelligence by propagation as a radio wave.

wave soldering – A mechanized, computerized method for soldering multiple components of an electronic printed circuit board.

waveform – The shape of an electrical signal as seen on an oscilloscope. Examples of wave forms are sine waves and square waves.

waveguide – A hollow metal tube designed to carry electromagnetic energy.

wavelength – The distance between the crests of a wave of energy. Wavelength is inversely proportional to the frequency of the wave.

way point – A designated point, position, or station in which a pilot an key in a radial from a VORTAC station on which the way point is to be located, and the distance in nautical miles from the station along a selected radial to locate a destination.

wear pads – Steel pads which are riveted to the surfaces of the stationary disks, the pressure plate, and the back plate to provide a wearing surface against the sintered material on the rotating disks. It is more economical to replace the wear pads than the disks and the plates themselves.

web – The portion of any beam or channel which lies between the flanges of a spar, rib, or channel selection which furnishes the strength necessary for longitudinal shear loads.

web of a beam – *See web.*

weber – Basic metric unit of magnetic flux. It is equal to that flux produced in a single turn of wire, when an emf of one volt is reduced to zero at a uniform rate of one ampere per second.

wedge — A tapered hard material (wood or metal) in which the thin edge can be driven into a narrow opening in order to separate the opening.

weighing points — Specific locations on an aircraft designated by the manufacturer at which the scales are placed for weighing the aircraft.

weight — The force by which a body is attracted toward the center of the earth by gravity. It is equal to the mass of the body times the acceleration due to gravity.

weight and balance records — Aircraft records that provide the information required on the weight of the empty aircraft and the location of its center of gravity (CG).

weld bead — The metal that is deposited in a welded joint to reinforce the joint.

weld fusion zone — The width of the bead made in welding.

weld procedures — Steps necessary in preparing to weld. These steps may include making certain the necessary equipment is available, that the welding connections are properly connected, the equipment is in good working order, and that the material to be welded is properly prepared.

welded patch — A patch of thin sheet steel welded over a dent in a steel tubular structure to reinforce the structure at the point of the damage.

welding — A method of joining materials in which a portion of each piece is melted and combined in its molten state. Filler material is usually added for extra mass at the joint.

welding flux — A material used in welding which melts and flows over the weld material to exclude oxygen from the surface of the molten metal, preventing oxides forming in the weld.

weldment — A welded together assembly.

Weston meter movement — A moving coil instrument movement.

wet cell — A chemical cell in which the electrolyte is in liquid form.

wet grinder — Form of precision grinding machine which uses a flow of liquid coolant over the stone to remove the heat caused by grinding. preventing damage to the material being ground.

wet head — A term used to describe a helicopter rotor head that uses oil as the lubricant.

wet sump engine — An engine in which all of the oil supply is carried within the engine itself.

wet takeoff — Engine operation during takeoff when water injection is used.

wet wing — An integral fuel tank in an aircraft wing made by sealing part of the structure to use as a fuel tank.

wet-bulb temperature — Temperature of the air modified by the evaporation of water from a wick surrounding the thermometer bulb.

Wheatstone bridge — An electrical measuring circuit in which the current through the indicator is determined by the ratio of the resistances of the four resistors which form the legs of the bridge.

wheel well — The part of the aircraft which receives or encloses the landing gear as it retracts.

whetstone — An abrasive stone used for sharpening knives.

whiffletree — A steering bell crank.

whip antenna — A quarter-wave antenna usually in the high- or very-high frequency range. It is normally vertically polarized.

Whitworth thread — A screw thread, also known as the British Standard Whitworth (B.S.W.), used principally in Great Britain.

wicking — Occurs when solder flows to the insulation of stranded electrical wire during the soldering process.

Wiggins coupling — A special fuel line connector, quite unlike a conventional AN fitting.

winch — A device with a crank handle and drum to which a cable or rope is wound. Used for transmitting motion by pulling or moving heavy loads.

wind chill factor — A way of measuring the velocity of the air and the temperature drop due to the cooling effect of the wind, and comparing this to the temperature if there were no wind.

wind shear — A dangerous condition in which two wind forces act at right angles to each other. Normally associated with weather fronts.

wind sock — A long hollow tube like device mounted on a free wheeling pivot that is turns in the direction from which the wind is blowing.

wind tunnel testing — A method of testing scale model airplanes for design aerodynamic characteristics. Wind tunnels are extremely large tubes with large high volume fans that blow air at controlled speeds to simulate the speed of an aircraft in flight. Airplane models are then tested at various speeds and angles of attack.

windmilling — The rotation of an aircraft propeller by air flowing over it with the engine not operating.

window de-mister — A system of keeping the windows of an aircraft free of condensed moisture by blowing warm air between the layers of transparent material.

windshield — A transparent screen device made of plastic or glass used for the protection of the occupants against the elements of wind, rain, and cold, and which allow for visual direction.

wing — Main aerodynamic supporting surface of an airplane.

wing area — The total wing area measured in square feet by multiplying the wing span by the wing chord. It is that area of the wing surface designed to obtain a desirable reaction from the air through which it moves.

wing chord — An imaginary line between the leading edge and the trailing edge of a wing airfoil section.

wing fillet — A streamlined fairing between a wing and the fuselage. Used to smooth out the airflow and minimize the interference drag caused by this junction.

wing flaps — Movable control surfaces on the trailing edge of a wing, inboard of the ailerons, which alter the airfoil section of the wing, increasing both the lift and the drag.

wing heavy — A condition of aircraft in flight in which one wing has a tendency to fly lower than the other wing about the longitudinal axis. This condition is corrected by properly adjusting the flight control rigging system.

wing loading — The ratio of the weight of the fully loaded aircraft to the total wing area.

wing nut — A nut with two protruding "wings" that allow it to be turned by hand.

wing panel — Removable access panels or wing sections attached with screw, bolts, or rivets.

wing profile — The outline of the wing section.

wing rib — A structural member which gives a wing its desired aerodynamic shape.

wing span — A distance measured from wing tip to wing tip.

wing tip vortices — An area of extreme turbulence below an aircraft in flight caused by the high-pressure air below the wing spilling over the wing tips into the low pressure above the wing.

wink Zyglo — A non-destructive inspection method that soaks the suspected part in a fluorescent penetrant liquid. The penetrant seeps into voids which extend to the surface of the part. The liquid is then washed from its surface and the part is placed in a fixture that vibrates while the part is observed under an ultraviolet light. If the vibration causes a crack that contains the penetrant to open up, the black light will cause the penetrant to glow giving the appearance of winking at the person inspecting the part each time the crack opens up to expose the penetrant.

wiper — A movable electrical contact used in an instrument.

wire braid — A woven, flexible metal covering over aircraft electrical wiring used to intercept and ground any radiated electrical energy from the wire to prevent radio interference.

wire bundle — A group of aircraft electrical wires tied together and secured to the structure.

wire cloth — A screen device used for filtering.

wire edge — A sharp burr on the edge of sheet metal that has been cut on a shear.

wire gage — A gage used to measure wire size.

wire group — Two or more wires going to the same location tied together to retain their identity.

wire stripper — A plier-like tool that has slots for wire size to be stripped.

wire-wound resistor — An electrical resistor whose base has a winding of high-resistance wire and covered with baked-on ceramic material.

wobble pump — A hand-operated fluid pressure pump. The name wobble comes from the movement of the pump handle back and forth as it pulls fluid into one side of the pump and forces it out the other side.

woodruff key — A hardened piece of metal shaped in a half circle on one side and flat on the other side. The key fits into a semi-circular furrow to prevent a wheel, disk or gear from turning on a shaft.

Wood's metal – An alloy of lead, tin, bismuth, and cadmium that melts at a temperature of 158°F.

work – The product of force and the distance through which the force acts.

workability – The ease with which wood, metal, or plastic may be formed or shaped.

work-hardening – *See strain-hardening.*

working voltage – The maximum amount of electrical voltage that can safely be applied to an appliance without damaging it.

worm gear – A gear mounted on a shaft that meshes with a spur gear. Rotating the worm gear on its shaft rotates the worm wheel.

wrench – A hand tool used for turning and holding any number of nuts or bolts.

wrinkle finish – A paint finish that wrinkles as it dries to give it a roughened appearance.

wrist pin – The hardened steel pin that attaches the small end of a connecting rod into a piston.

wrought iron – A type of iron that contains very little carbon. It is tough and hard, yet is pliable enough to be pounded tin to shape.

wrought metal – Metal which has been worked by rolling, drawing, or forging, and which has a different grain structure from that of cast metal.

wye connection – An electrical circuit connection that looks like the letter Y.

X-axis — The motion of the aircraft about the longitudinal axis for rolling and banking. The ailerons are used to control this movement.

X-band radar — Radar that operates in a frequency band of between 5.2 and 10.9 gigahertz.

xenon — A heavy, colorless, inert chemical gas element.

X-ray — An electromagnetic radiation of extremely short wave length, capable of penetrating solid objects and exposing photographic film.

X-ray inspection — A form of nondestructive inspection in which high-frequency, high-energy electromagnetic waves pass through the material and expose a piece of photographic film. Defects or discontinuities within the material show up as variations in the density of the image on the film.

xylol or xylene — A toxic, flammable, of extremely short wave length, aromatic hydrocarbon, similar to benzene, capable of penetrating solid objects. It is used as a solvent and for exposing photographic film.

Y

yard — A measure of length. One yard is equal to 3 ft., 36 in. or .914 m.

yardstick — A single-piece, graduated measuring device that is 3′ long and approximately 1½″ wide. Marked in inches and feet.

yaw — Rotation about the vertical axis of an aircraft.

yaw damper — An automatic control device used to keep the aircraft from yawing. When a wind is swept back, the effective dihedral increases rapidly with a change in the lift coefficient of the wing. Excessive dihedral effect can lead to Dutch Roll, difficult rudder coordination in rolling maneuvers, or place extreme demands for lateral control power during crosswind takeoff and landing. Yaw dampers overcomes the unwanted yawing condition.

Y-axis — The lateral, pitch axis, or pitching motion of the aircraft. The climb and dive control affected by elevator movement.

Y-connected circuit — A three-phase, or polyphase, alternating current circuit which has three single-phase windings spaced so that the voltage induced in each winding is 120° out of phase with the voltages in the other two windings.

yellow arc — A marking on an aircraft tank, and the lower arm is fitted with a instrument that indicates a caution valve, and from this point the oil may range of operation.

yield point — *See yield strength.*

yield strength — The load on a material, expressed in lbs./sq. in., which causes the initial indication of permanent distortion.

yoke — A cross member in a control system which links or joins something together. The control column in the airplane cockpit that connects and controls the movement of the elevators, and ailerons.

Y-valve — The oil drain valve for a dry sump engine. It derives its name from its shape. One arm of the "Y" goes to the pressure pump inlet, one arm to the oil tank, and the lower arm is fitted with a valve, and from this point the oil may be obtained from the tank. Fuel for oil dilution is introduced in the Y-valve.

Y-winding — A method of connecting the phase windings of a three-phase AC machine in which one end of each of the three phase windings is connected together to form a common point or a neutral terminal.

Z

Zahn cup – A special cup of definite size and shape, with a hole in its bottom. Used to measure the viscosity of a material by the number of seconds required for the cup to empty.

Z-axis – The vertical axis. Turning the nose of the aircraft causes the aircraft to rotate about its vertical axis. Rotation of the aircraft about the vertical axis is called yawing. This motion is controlled by using the rudder.

zener diode –
[1]A semiconductor diode having a specific peak inverse voltage. It acts as a check valve until this voltage is reached, and then it breaks down and allows flow.
[2]A diode rectifier designed to prevent the flow of current in one direction until the voltage in the reverse direction reaches a predetermined value. At this time, the diode permits a reverse current to flow.

zenith – A high point directly overhead.

zephyr – A soft breeze from the west.

Zeppelin – A dirigible airship.

Zerk fitting – A type of fitting with a ball head design feature used to fit the head of a grease gun. Grease, under pressure forces the internal ball check valve off its seat allowing grease to be pumped through the fitting into a bearing surface. Removal of the grease gun allows the check valve to reseat preventing grease from leaking out and dirt from entering the grease.

zero –
[1]Numerical: 0. Having no value. Used as a reference point.
[2]Temperature: A reference point on a Fahrenheit thermometer that is 32° below the freezing point of water.

zero adjustment – The adjustment on an instrument to a zero point, or to an arbitrary point from which all negative and positive measurements are to be adjusted.

zero fuel weight – The weight of the aircraft to include all useful load except fuel.

zero gravity – The effect of gravity when it has been nullified by parabolic flight.

zero lash — A condition in a valve train in a reciprocating engine in which all of the clearance is kept out of the valve train by the use of hydraulic valve lifters.

zero-lash valve lifter — A hydraulic device in a reciprocating engine that reduces the slack between a valve and the valve lifter due to changes in the engine operating temperatures.

zero-lift line — A line through an airfoil, along which a flow of relative wind will produce no lift.

zinc — A bluish-white, crystalline metal that is ductile in its pure state but quite brittle in its commercial form. It is used in the production of electrical batteries and for coating steel parts to protect them by sacrificial corrosion.

zinc chloride cell — A chemical cell using powdered manganese dioxide and zinc as its pole pieces and a solution of zinc chloride as its electrolyte.

zinc chromate primer — An alkyd resin, corrosion-inhibiting primer used on almost all metal surfaces. Moisture releases chromate ions to inhibit the formation of corrosion, and the alkyd resin forms a good bond for aircraft finishes.

zone numbers — Location marks on an aircraft drawing, both vertical and horizontal. They are used to locate detail parts on the drawing.

Zulu time — A universal time. Also called Greenwich mean time.

Zyglo inspection — A penetrant inspection system in which a fluorescent dye is drawn into surface defects in the material. It is pulled out of the defects by a powder-type developer.

Aviation Abbreviations

AAM ....air-to-air missile
AAS..........Airport Advisory Service
AC ....Advisory Circular
AC ....alternating current
AC ....Air Corps
A/C...........aircraft
A&P ....aircraft technician
ABC ....after bottom center
ACDO ......Air Carrier District Office
AD............Airworthiness Directive
AD............ashless dispersant
ADF..........automatic direction finder
ADIZ.........Air Defense Identification Zone
af..............audio frequency
AFC..........automatic frequency control
AFCS........automatic flight control system
AGL..........above ground level
AIM ....Airman's Information Manual
ALNOT ....Alert Notice
ALTRV......Altitude Reservation
AM ....amplitude modulation
AMC.........automatic mixture control
AN............Air Force-Navy Standard
APC..........absolute pressure controller
API ....American Petroleum Institute
APU ....auxiliary power unit
ARSR ....Air Route Surveillance Radar
ARTS........Automated Radar Terminal Systems
ARTTC .....Air Route Traffic Control Center
ASDE ....airport surface detection equipment
ASM ....air-to-surface missile
ASR..........Airport Surveillance Radar
ASTM ....American Society of Testing Materials
ATA ....Airline Transport Association
ATC ....after top center
ATCCC.....Air Traffic Control Command Center
ATIS ....Automatic Terminal Information Service
ATP ....Airline Transport Pilot
ATR ....Airline Transport Rating
AVC..........automatic volume control
AWG........American Wire Gage
BBC ....before bottom center
BCD ....binary coded decimal
BDC ....bottom dead center
BFO ....beat-frequency oscillator
BIM ....blade inspection method
BIS ....blade inspection system
BL ....bend tangent line
BMEP.......brake mean-effective pressure
BSFC........brake specific fuel consumption
BTC..........before top center
CCL..........Convective Condensation Level
CDI...........course deviation indicator
CAT ....clear air turbulence
CDP ....compressor discharge pressure
CERAP .....Combined Center-Rapcon
CFR..........Cooperative Fuel Research
CG............center of gravity
CIT ....compressor inlet temperature
CP ....center of pressure
CPU ....central processing unit
CRT..........cathode ray tube
CSD..........constant-speed drive
CTAF........Common Traffic Advisory Frequency
CVFP ....Charted Visual Flight Procedure
cw............continuous wave
cw............carrier wave
DC............direct current
DEWIZ.....Distant Early Warning Identification Zone
DH............Decision Height
DME.........distance measuring equipment
DOT ....Department of Transportation
DPDT.......double-pole, double-throw
DPST........double-pole, single-throw
DVFR ....Defense Visual Flight Rules
DVRCR ....differential voltage reverse-current relay
EARTS......En Route Automated Radar Tracking System
ECCM ......electronic counter-countermeasures
ECM ....electronic countermeasures
EFC ....Expect Further Clearance
EGT..........exhaust gas temperature
ELT...........emergency locator transmitter
emf ....electromotive force
EMSAW ...En Route Minimum Safe Altitude Warning
EPR..........engine pressure ratio
EPROM....enable programmable read-only memory
ESFC........equivalent specific fuel consumption
EWCG .....empty weight center of gravity
FAA..........Federal Aviation Administration

FAA-PMA. Federal Aviation Administration Parts Manufacturing Approval
FAD .... Fuel Advisory Departure
FARs .... Federal Aviation Regulations
FCC .... Federal Communications Commission
FET .... field effect transistor
FM .... frequency modulation
FMEP .... friction mean effective pressure
FOD .... foreign object damage
FSDO .... Flight Standards District Office
FSS .... Flight Service Station
GA .... general aviation
GADO ...... General Aviation District Office
GPA .... gas path analysis
GPU .... ground power unit
GSE .... ground support equipment
HIG .... hermetically sealed integrating gyro
HUD .... head-up display
IA .... Inspection Authorization
IC .... integrated circuit
ICAO .... International Civil Aviation Organization
IEPR .... integrated engine pressure ratio
if .... intermediate frequency
IFIM .... International Flight Information Manual
IFF .... Identification, Friend or Foe
IFR .... instrument flight rules
IGFET ...... insulated gate field effect transistor
IHP .... indicated horsepower
ILS .... instrument landing system
IMEP .... indicated mean effective pressure
IPB .... Illustrated Parts Breakdown
IPL .... Illustrated Parts List
IPM .... Illustrated Parts Manual
IR .... infra-red
IRAN .... Inspect and Repair As Necessary
IRBM .... intermediate-range ballistic missile
ITT .... intermediate turbine temperature
IVSI .... instantaneous rate of climb indicator
JATO .... Jet Assist Takeoff
JFC .... jet fuel control
JFET .... junction field effect transistor
KVAR .... kilovolt amperes reactive
LAAS .... Low Altitude Alert System
LASCR ..... light-activated silicon control rectifier
LCD .... liquid crystal display
LED .... light emitting diode
LEMAC .... leading edge of the mean aerodynamic chord
LH .... left-hand threads
LORAN .... long-range navigation
LOX .... liquid oxygen
LSB .... least significant bit
LSI .... large scale integration
MAA .... Maximum Authorized Altitude
MAC .... mean aerodynamic chord
MAP .... Missed Approach Point
MB .... marker beacon
MCA .... Minimum Crossing Altitude
MDA .... Minimum Descent Altitude
MEA .... Minimum En Route IFR Altitude
MEK .... methyl-ethyl-ketone
METO ...... maximum except takeoff power
MHA .... Minimum Holding Altitude
MIA .... Minimum IFR Altitudes
MIG .... metal inert-gas
MILSPEC . military specifications
MIS .... Meteorological Impact Statement
MLS .... Microwave Landing System
MM .... Middle Marker
mmf .... magnetomotive force
MOCA ...... Minimum Obstruction Clearance Altitude
MOS .... metal oxide semiconductor
MOSFET .. metal oxide semiconductor field effect transistor
MRA .... Minimum Reception Altitude
MS .... Military Standard
MSA .... Minimum Safe Altitude
MSB .... most significant bit
MSAW ...... Minimum Safe Altitude Warning
MSD .... most significant digit
MSI .... medium-scale integration
MSL .... mean sea level
MTI .... Moving Target Indicator
MTR .... Military Training Routes
MVA .... Minimum Vectoring Altitude
NAS .... National Airspace System
NAS .... National Aircraft Standards
NASA .... National Aeronautics and Space Administration
NAVAIDS . Navigational Aids
NFDC .... National Flight Data Center
NFDD .... National Flight Data Digest
OBS .... omni bearing selector
PCA .... Positive Control Area
PCB .... printed circuit board
pd .... potential difference
pH .... measure of ability
PIBAL ...... pilot balloon observation
PK .... Parker-Kalon

PLA..........power lever angle
PM ....phase modulation
PPI ....plan-position indicator
PROM......programmable read-only memory
PRT ....power recovery turbine
PVC..........polyvinyl chloride
QC ....quality control
QEC ....quick engine change
QECA.......quick engine change assembly
R-12..........refrigerant 12
RATO ....rocket-assisted takeoff
rf ....radio frequency
RFI ....radio-frequency interference
RHI...........range-height indicator scope
RMS ....root mean square
RNAV.......area navigation
RVR..........runway visual range
SAE ....Society of Automotive Engineers
SI..............international system of units
SSI............small-scale integration
SSU ....Saybolt Seconds Universal
STC ....Supplemental Type Certificate
STOL........short takeoff and landing
STOVL.....short takeoff and vertical landing
TACAN.....Tactical Air Navigation System
TBO ....time between overhaul
TCA..........terminal control area
TDC..........top dead center
TEMAC....trailing edge of the mean aerodynamic chord
TIG...........tungsten inert-gas
TIR ....Total Indicator Reading
TIT ....turbine inlet temperature
TRF ....tuned radio frequency receiver
TSFC........thrust specific fuel consumption
UHF ....ultrahigh frequency
UNC.........United National, coarse
UNF ....United National, fine
USS ....United States, standard
VAC..........volts of alternating current
VAPC ....variable absolute pressure controller
VAR..........volt-ampere reactive
VDC ....volts of direct current
VFO ....variable-frequency oscillator
VFR..........visual flight rules
VHF..........very-high frequency
VLF ....very-low frequency
VOR ....very high frequency omnirange navigation equipment
VORTAC..a VOR combined with UHF tactical air navigation
VSTOL.....vertical or short takeoff and landing
VSWR......voltage standing-wave rectifier
VTOL ....vertical takeoff and landing
VTVM.......vacuum-tube voltmeter
WHR........watthour

Aviation Symbols

α ....temperature coefficient of resistance
CO ....carbon monoxide
CO_2..........carbon dioxide
F_g ....static thrust or gross thrust
F_n.............net thrust
G_m ....transconductance
H_2.............hydrogen
HCl ....hydrochloric acid
k...............dielectric constant
N_f ....RPM of a free turbine
N_1.............RPM of a low-pressure compressor
N_2.............RPM of a high-pressure compressor
P...............power
Pb ....lead
P_b.............burner pressure
Pt_2............inlet pressure
Pt_4............compressor discharge pressure
PT_7...........turbine discharge pressure
R ....resistance
Tt_2............inlet temperature
V_1 ....takeoff decision speed
V_2 ....minimum takeoff safety speed
X_c.............capacitive reactance
Z...............impedance

Greek Alphabetical Symbols

Α	α	alpha
Β	β	beta
Γ	γ	gamma
Δ	δ	delta
Ε	ε	epsilon
Ζ	ζ	zeta
Η	η	eta
Θ	θ	theta
Ι	ι	iota
Κ	κ	kappa
Μ	μ	mu
Ν	ν	nu
Ξ	ξ	xi
Ο	ο	omicron
Π	π	pi
Ρ	ρ	rho
Σ	σ	sigma
Τ	τ	tau
Υ	υ	upsilon
Φ	φ	phi
Χ	χ	chi
Ψ	ψ	psi
Ω	ω	omega

Standard Symbols & Abbreviations

+ ....positive
– ....negative
Ω ....ohm
~ ....ampere
° ....degree
" ....inch(es)
' ....foot (feet)
amp ....ampere
Btu ....British thermal unit
C ....Celsius
cal. ....calorie(s)
Cal. ....large calorie(s)
cm ....centimeter
cos ....cosine
cu. cm .....cubis centimeter(s)
cu. in. ......cubis inch(es)
cu. ft. ....cubic foot (feet)
cu. m ....cubic meter(s)
dB ....decibel
deg...........degree
ESHP ....equivalent shaft horsepower
f................farad
F ....Fahrenheit
ft...............foot (feet)
ft.-lb. ....foot-pound
ft.-lbs........foot-pounds
g ....gram
gal. ....gallon
HP ....horsepower
hr..............hour
Hz ....hertz
in..............inch(es)
in. hg. ......inch(es) of mercury
IPS ....inches per second
k ....kilo
K ....Kelvin
kg ....kilogram
kHz ....kilohertz
km ....kilometer
kM............kilomega
kw............kilowatt
kw-hr. ......kilowatt hour
l ....liter
lb..............pound
lbs. ....pounds
m ....meter
mb ....milibar
mf or μf ...microfarad
mHz ....millihertz
MHz ....megahertz
mi. ....mile(s)
mm..........millimeter
MPH..........miles per hour
mv............millivolt
neg...........negative
oz. ....ounce(s)
pf or μμf ..picofarad
pos...........positive
PPH ....pounds per hour
PPM ....parts per million
PSI ....pounds per square inch
PSIA ....pounds per square inch absolute pressure
PSID ....pounds per square inch differential pressure
PSIG...........pounds per square inch gage
pt..............pint
qt..............quart
R...............Rankine
rev. ....revolution(s)
RPM ....revolutions per minute
sec. ....second
SHP............shaft horsepower
sin ....sine
sq. cm .....square centimeter(s)
sq. in........square inch(es)
sq. ft.........square foot (feet)
sq. m ....square meter
sq. mi. .....square mile(s)
sq. mil .....square mil
tan............tangent
TEHP..........thrust equivalent horsepower
THP ....thrust horsepower
U.S. ....United States
V...............volt
yd. ....yard

Chemical Elements

actinium	Ac
aluminum	Al
americum	Am
antimony	Sb
argon	Ar
arsenic	As
astatine	At
barium	Ba
berkelium	Bk
beryllium	Be
bismuth	Bi
boron	B
bromine	Br
cadmium	Cd
calcium	Ca
californium	Cf
carbon	C
cerium	Ce
cesium	Cs
chlorine	Cl
chromium	Cr
cobalt	Co
columbium	Cb
copper	Cu
curium	Cm
dysprosium	Dy
einsteinium	Es
erbium	Er
europium	Eu
fermium	Fm
fluorine	F
francium	Fr
gadolinium	Gd
gallium	Ga
germanium	Ge
gold	Au
hafnium	Hf
helium	He
holmium	Ho
hydrogen	H
indium	In
iodine	I
iridium	Ir
iron	Fe
krypton	Kr
lanthanum	La
lawrencium	Lr
lead	Pb
lithium	Li
Lutetium	Lu
magnesium	Mg
manganese	Mn
mendelevium	Md
mercury	Hg
molybdenum	Mo
neodymium	Nd
neon	Ne
neptunium	Np
nickel	Ni
niobium	Nb
nitrogen	N
nobelium	No
osmium	Os
oxygen	O
palladium	Pd
phosphorus	P
platinum	Pt
plutonium	Pu
polonium	Po
potassium	K
praseodymium	Pr
promethium	Pm
protactinium	Pa
radium	Ra
radon	Rn
rhenium	Re
rhodium	Rh
rubidium	Rb
ruthenium	Ru
samarium	Sm
scandium	Sc
selenium	Se
silicon	Si
silver	Ag
sodium	Na
strontium	Sr
sulfur	S
tantalum	Ta
technetium	Tc
telllurium	Te
terbium	Tb
thallium	Tl
thorium	Th
thulium	Tm
tin	Sn
titanium	Ti
tungsten	W
uranium	U
vanadium	V
wolfram	W
xenon	Xe
ytterbium	Yb
yttrium	Y
zinc	Zn
zirconium	Zr